Interdisciplinary Research in Engineering

Volume I

Interdisciplinary Research in Engineering
Volume I

Edited by **Michelle Vine**

CLANRYE
INTERNATIONAL

New Jersey

Published by Clanrye International,
55 Van Reypen Street,
Jersey City, NJ 07306, USA
www.clanryeinternational.com

Interdisciplinary Research in Engineering: Volume I
Edited by Michelle Vine

International Standard Book Number: 978-1-63240-316-2 (Hardback)

Printed in the United States of America.

Contents

Preface

The word 'engineering' is used almost continuously in our speech in today's world. What is engineering? It is the creative and imaginative application of scientific principles to design, conceive and develop machines, structures, manufacturing processes or apparatus. It also constructs or operates such instruments with full knowledge of their design and forecasts their performance under specific operating conditions. Engineering uses practical applications of science and mathematics to solve problems. It also uses mathematics and scientific fields such as physics to find appropriate solutions to specific problems or to make improvements to our daily lives. For all of us at every moment of time, engineering in its myriad forms improves the ways that we work, communicate, travel and entertain ourselves. Engineers can be termed as problem-solvers who work towards making things perform more efficiently, quickly and less expensively. Engineering is such a vast arena that it has innumerable interconnections with society and human behavior. Each and every product or construction used by man in modern society has been influenced by engineering. Engineering is a very powerful instrument to bring about changes in the environment, society and economies. Engineers thus have a wide range of study options and career opportunities that allow them to design, build and manage their ideas into reality.

I wish to personally thank all the contributing authors who not only shared their research work but also guided me from time to time in the editing process. I also wish to thank my publisher for giving me this unparalleled opportunity.

Editor

Latest Development on Membrane Fabrication for Natural Gas Purification: A Review

Dzeti Farhah Mohshim, Hilmi bin Mukhtar, Zakaria Man, and Rizwan Nasir

Chemical Engineering Department, Universiti Teknologi Petronas, Bandar Seri Iskandar, Perak Darul Ridzuan, 31750 Tronoh, Malaysia

Correspondence should be addressed to Dzeti Farhah Mohshim; dzetifarhah@gmail.com

Academic Editor: Hyun Seog Roh

In the last few decades, membrane technology has been a great attention for gas separation technology especially for natural gas sweetening. The intrinsic character of membranes makes them fit for process escalation, and this versatility could be the significant factor to induce membrane technology in most gas separation areas. Membranes were synthesized with various materials which depended on the applications. The fabrication of polymeric membrane was one of the fastest growing fields of membrane technology. However, polymeric membranes could not meet the separation performances required especially in high operating pressure due to deficiencies problem. The chemistry and structure of support materials like inorganic membranes were also one of the focus areas when inorganic membranes showed some positive results towards gas separation. However, the materials are somewhat lacking to meet the separation performance requirement. Mixed matrix membrane (MMM) which is comprising polymeric and inorganic membranes presents an interesting approach for enhancing the separation performance. Nevertheless, MMM is yet to be commercialized as the material combinations are still in the research stage. This paper highlights the potential promising areas of research in gas separation by taking into account the material selections and the addition of a third component for conventional MMM.

1. Introduction

Natural gas can be considered as the largest fuel source required after the oil and coal [1]. Nowadays, the consumption of natural gas is not only limited to the industry, but natural gas is also extensively consumed by the power generation and transportation sector [2]. These phenomena supported the idea of going towards sustainability and green technology as the natural gas is claimed to generate less-toxic gases like carbon dioxide (CO_2) and nitrogen oxides (NO_x) upon combustion as shown in Table 1 [3].

However, pure natural gas from the wellhead cannot directly be used as it contains undesirable impurities such as carbon dioxide (CO_2) and hydrogen sulphide (H_2S) [4]. All of these unwanted substances must be removed as these toxic gases could corrode the pipeline since CO_2 is highly acidic in the presence of water. Furthermore, the existence of CO_2 may waste the pipeline capacity and reduce the energy content of natural gas which eventually lowers the calorific value of natural gas [5].

Conventionally, natural gas treatment was predominated with some methods such as absorption, adsorption, and cryogenic distillation. But these methods require high treatment cost due to regeneration process, large equipments, and broad area for the big equipments [6]. With the advantages of lower capital cost, easy operation process, and high CO_2 removal percentage, membrane technology offers the best treatment for natural gas [6]. Natural gas is expected to contain less than 2 vol% or less than 2 ppm of CO_2 after the natural gas treatment in order to meet the pipeline and commercial specification [7]. This specification is made to secure the lifetime of the pipeline and to avoid an excessive budget for pipeline replacement.

Membrane technology has received significant attention from various sectors especially industries and academics in their research as it gives the most relevant impact in reducing the environmental problem and costs. Membrane is defined as a thin layer, which separates two phases and restricts transport of various chemicals in a selective manner

TABLE 1: Fossil fuel emission levels (pounds per billion Btu of energy input).

Fuel sources/pollutant (pound/BTU)	Natural gas	Oil	Coal
Carbon dioxide	117,000	164,000	208,000
Carbon monoxide	40	33	208
Nitrogen oxides	92	448	457
Sulphur dioxide	1	1,122	2,591
Particulates	7	84	2,744
Mercury	0.000	0.007	0.016

[8]. Membrane restricts the penetration of some molecules that have bigger kinetic diameter. The commercial value of membrane is determined by the membrane's transport properties which are permeability and selectivity. Major gap of the existing technologies is limited to low CO_2 loading (<15 mol%). Ideally, we required high permeability and high selectivity of membrane, but, however, most membranes exhibit high selectivity in low permeability and vice versa which make this is as a major tradeoff of membranes, and none of these technologies are yet to treat natural gas containing high CO_2 (>80 mol%) [9].

2. Membrane Technology Development

2.1. Early Membrane Development.
Membrane technology has been started as early as in 1850 when Graham introduced the Graham's Law of Diffusion. Then, gas separation utilization in membrane technology has been commercialized in late 1900's. Permea PRISM membrane was the first commercialized gas separation membrane produced in 1980 [2]. Summary of early development of membranes is shown in Figure 1. This innovation has led to the further membrane gas separation development. A lot of studies done by the researchers for various gas separation mostly focus on the natural gas purification.

Development of membrane for CO_2/CH_4 separation has been started since early 1990's. Numbers of membranes were fabricated using different kind of materials in the early stage of this membrane gas separation. The desirable material selected must be well suited to the separation performance by which mean separation of gases works contrarily in different materials. Excellent gas membranes separation should have the characteristic of high separation performance with reasonable high permeability, high robustness, chemically, thermally, and mechanically good and rational production cost [10, 11]. Two types of materials are practically used in gas separation: polymeric membrane and inorganic membrane and the comparison of both polymeric and inorganic membranes is showed in Table 2.

Gas separation using polymeric membranes has taken its first commercial scale in late 1970's after the demonstration of rubbery membranes back in 1830's [33]. Literally, the permeability of gas in a specific gas mixture varies inversely with its separation factor. The tighter of molecular spacing it has, the higher the separation characteristic of the polymer,

but, however, as the operating pressure increases, the permeability is decreasing due to experiencing lower diffusion coefficients [34]. Polymeric membranes that are commercially available for CO_2/CH_4 separation include polysulfone (PSU), polyetehrsulfone (PES), polyamide (PI) and many more. Generally, as the permeability of the gas increases, the permselectivity was attended to decrease in most cases of polymeric membranes [23].

Inorganic membrane like SAPO-34 could give higher separation performance compared to the polymeric membrane, but the separation performance is inversely proportional to the pressure loaded. This observation may create problem when we deal with high pressure natural gas well. The performance of both organic and inorganic membrane is summarized in Robeson's plot as in Figure 2 [35].

2.2. Conventional Mixed Matrix Membrane.
A lot of researches have been done to satisfy the needs of gas separation requirement through both polymeric and inorganic membranes. The deficiencies of these membranes have driven the researchers to develop an alternative material for membrane which is more mechanically stable and economic viable, and most important is having high separation performance. The combination of organic and inorganic material which is known as mixed matrix membrane (MMM) was then proposed in idea to get a better membrane gas separation performance at reasonable price [36]. The fabrication of MMM was a promising technology as this composite material has improved its mechanical and electrical properties [37], and it combines the exceptional separation ability and pleasant stability of molecular sieves with better processability of organic membrane [38]. The MMM is characterized by dispersing the inorganic material into the continuous phase of polymeric material which can be almost any polymeric material such as polysulfone, polyimide, and polyethersulfones [39, 40].

Various membrane materials can be selected based on the process requirement. Selected materials can be "tailored-made" in order to meet the specific separation purpose in a wide range of application [39]. There were many attempts of developing polymer-inorganic membrane that started few decades back then.

Based on Table 3, this was observed that the selection of materials is important, and it depends on the system requirement. Higher intrinsic diffusion selectivity characteristic of glassy polymer makes this material better than rubbery polymer [56]. Although MMM has proven an enhancement of selectivity, it was noticed that most MMMs were endured with poor adhesion between the organic matrix and inorganic particles [55]. Even MMM fabrication does have its disadvantages, but the research of MMM with different materials is worth to work on since it has proven its ability to have high separation performance.

2.3. Recent Development of Membrane Gas Separation

2.3.1. Ionic Liquid-Supported Membrane (ILSM).
In recent years, many researches have been evaluated on the ionic

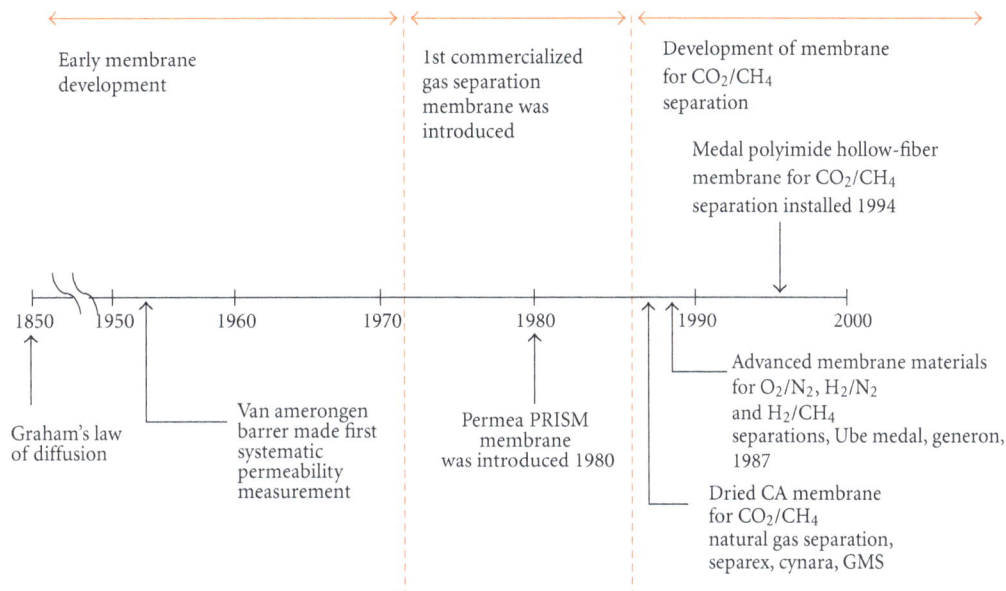

FIGURE 1: Membrane development timeline.

TABLE 2: Comparison between polymeric and inorganic membranes.

	Polymeric membranes	Inorganic membranes
Materials	Present in either rubbery or glassy type which depends on the operating temperature [12].	Made from inorganic-based material like glass, aluminium, and metal [13].
Characteristics	(i) Polymer is more rigid and hard in glassy state while in rubbery state it is more soft and flexible (ii) Glassy polymeric membranes exhibit higher glass transition temperature compared to rubbery membranes, and glassy types tend to have higher CO_2/CH_4 selectivity [14].	(i) Able to withstand with solvent and other chemicals and also susceptible to microbial attack. (ii) Comprise significantly higher permeability and selectivity, but they are also more resistant towards higher pressure and temperature, aggressive feeds, and fouling effects [15].
Disadvantages	(i) May have plasticization problem when handling high CO_2. (ii) Presence of CO_2 may result in membrane performance reduction at certain elevated pressure. (iii) As the membranes expose to CO_2, polymer network in the membrane will swell, and segmental mobility will also increase which consequently cause a rise in permeability for all gas components [16]. (iv) The components with low permeability characteristic will experience more permeability increment; thus, the selectivity of the membrane will definitely decrease [17–19].	(i) Inherent brittleness characteristic. (ii) Performed well under low pressure which does not suit the natural gas well which required high pressure for the exploration. (iii) High production cost which seems not practical for large industrial applications [20].
Examples	Polyethylene (PE), poly(dimethylsiloxane) (PDMS), polysulfone (PSU), polyethersulfone (PES), polyimide (PI) [21], polycarbonate [22], polyimide [23], polyethers [24], polypyrrolones [25, 26], polysulfones [27], and polyethersulfones [28].	Aminoslicate membrane [29], carbon-silicalite composite membrane [30], MFI membranes [31], and microporous silica membranes [32].

liquid supported membrane (ILSM) for gas separation membrane since ionic liquids are known materials that could dissolve CO_2 and stable at high temperature ranges [57]. To be specific, ionic liquids are molten salt that are liquid at room temperature [58]. Furthermore, ionic liquids are of particular interest for membrane gas separation application as they are inflammable, negligible vapour pressure, and nonvolatile

which make them also known as "green" solvents [58–60]. Extensive researches have been carried out to develop room temperature ionic liquid (RTIL)-based solvents for CO_2 separation with various types of ionic liquids such as pyridinium and imidazolium based. Among RTILs tested, imidazolium-based RTIL was chosen as the most feasible solvent for CO_2 separation as they are commercially viable

TABLE 3: Few researches of mixed matrix membranes.

| Year | Mixed matrix membrane (MMM) | | Observations | Ref. |
	Organic	Inorganic		
1973	Silicon rubber	Molecular sieves	Poor adhesion of organic and inorganic selected leads to poor separation performance. This poor interaction of both materials may result in nonselective voids present at the interface which consequently causes insufficient membrane performance [41–43].	[44]
1992	Polydimethylsiloxane (PDMS) Propylene diene rubber (EPDM)	Silicalite-1, 13X, KY, and zeolite-5A	Zeolite like silicalite-1, 13X, and KY have enhanced the separation performance of poorly selective rubbery membrane for the carbon dioxide (CO_2) and methane (CH_4) mixture. Zeolite-5A showed no change in gas selectivity with decrease permeability due to impermeable characteristic towards CO_2.	[45]
2000	Cellulose acetate (CA)	Silicalite, NaX, and AgX	Silicalite did in fact reverse the selectivity of CA membrane from H_2 to CO_2 for CO_2/H_2 separation.	[46]
2000	Polyvinyl acetate	4A	Formation of chemical bonds gave good adhesion, but there is still nonselective "leakage" from the existence of nanometric region.	[47]
2003	Matrimid	Carbon molecular sieves	Selectivity of CO_2/CH_4 mixture has increased up to 45%. Zeolites loading also affects both gas permeability and gas mixture selectivity. There were also a number of records where permeability increased with selectivity decreased as the zeolites loading was increased [48, 49] and vice versa [42].	[50]
2006	Polyethersulfone (PES)	Zeolite 4A	Due to low mobility of the polymer chain in glassy polymer such as to prevent them to completely cover the zeolites surface which resulted in void interface [51, 52].	[53]
2001	Polyimide (PI)	Zeolite 13X		[54]
2008	Polycarbonate	Zeolite 4A		[55]

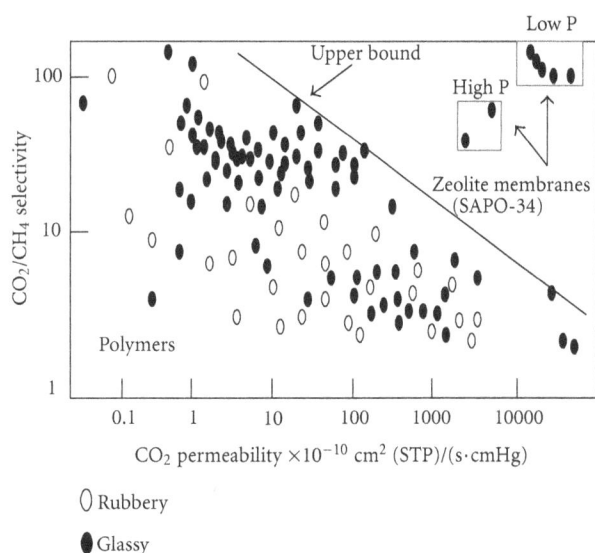

FIGURE 2: Zeolite (SAPO-34) membrane performance in Robeson's plot.

ILSMs have been proven that they offered an increase in permeability that outperforms many neat polymer membranes. ILSMs synthesized from poly(vinylidene fluoride) (PVDF) and 1-butyl-3-methylimidazolium tetrafluororate ($BMImBF_4$) showed high permeation performance of CO_2 and mechanically stable while operating at high pressure condition [63]. The consumption of RTILs showed an increment especially for 1-R-3-methylimidazolium (R-mim)-based RTILs as this type is preferable due to its properties of less viscous compared to other RTILs. In addition, gases like CO_2, nitrogen (N_2), and other hydrocarbons demonstrated high solubility in Rmim-based RTILs [64, 65]. Besides, the use of Rmim-based RTILs could calculate the latent permeability and selectivity of the mixture of given gases by using the molar volume of these RTILs [60]. RTIL can be functionalized and set up in according to the system requirement and application, and these researches could be good benchmark for designing the functionalized RTIL efficiently as showed in Table 4.

2.3.2. Polymerized Room Temperature Ionic Liquid Membrane (Poly(RTIL)). Comparatively, RTIL especially imidazolium based can be also polymerized into a solid, dense, and thin film membrane due to their modular nature [66–68]. It was a successful breakthrough when the researcher found

and easily tunable by tailoring the cation and anion to meet the system requirements [60].

TABLE 4: Effects of ionic liquid functionalization.

Functionalization	Effects
Nitrile and alkyne group	(i) Gas solubility and separation performance have been tailored. (ii) Functionalized RTIL solvents displayed a decreasing in CO_2, N_2, and CH_4 solubility, but, however, the selectivity of CO_2/N_2 and CO_2/CH_4 increased when compared to the nonfunctionalized RTIL [61].
Temperature	(i) As the temperature increases, the CO_2 solubility is decreasing while the CH_4 solubility remains unchanged. (ii) The ideal solubility selectivity of mix gases for CO_2/N_2, CO_2/CH_4, and CO_2/H_2 increased as the temperature decreased [62].

that polymer from ionic liquid monomer had higher CO_2 absorption capacity with faster absorption and desorption rate compared to the neat RTIL [69]. Moreover, poly(RTIL) is also attributed with higher mechanical strength [66]. These characters have proven that polymerized ionic liquid (poly(RTIL)) is also a promising material for membrane gas separation. Polymerization of RTIL monomer by varying the n-alkyl length also showed a pleasant result when increase of permeability of given gases like CO_2, N_2, and methane (CH_4) was observed as the n-alkyl group was lengthened [68]. Additionally, poly(RTIL) is also up to extend when it practically absorb about twice as much CO_2 as their liquid analogue which makes it much better than molten RTIL [68]. Apparently, performance of poly(RTIL) also depends on the substituent attached to it. In a research done on the inclusion of a polar oligo(ethylene glycol) on the cation side of imidazolium-based RTIL, the separation selectivity has seemed to increase [70].

As discussed earlier, mixed matrix membrane is a known membrane that composed of a compatible organic-inorganic pair which demonstrated having good separation properties subject to no interfacial adhesion problem. The improvement of separation performance is expected in an MMM comprising poly(RTIL) (polymer matrix) and zeolite (inorganic). In a very recent work, the benefit of MMM has become an idea to the researcher in ionic liquid membrane field. Hudiono and his coworkers have introduced a three-component mixed matrix membrane by utilizing the poly(RTIL), RTIL, and zeolite [71]. Their research was also based on a positive finding by Bara and his coworkers when they found that the addition of RTIL in poly(RTIL) has increased the gas permeability. This is due to that more rapid gas diffusion occurred as the free volume of membrane increased when RTIL was added [72].

On the other hand, Hudiono has used the RTIL to increase the membrane permeability and also to act as an aid for better interaction between the poly(RTIL) and zeolite (SAPO-34). The result was promising as the permeability of given gases like CO_2, N_2, and CH_4 increased accordingly. However, the selectivity was slightly decrease as they claimed

that the RTIL used which is emim[Tf_2N] was not selective towards CO_2/CH_4 separation [71]. Nonetheless, the result proved that the addition of RTIL could increase the polymer-zeolite adhesion in MMM as RTIL also acts as the wetting agent for the zeolite.

Hudiono again repeated the same experiment fabricating a three-component mixed matrix membrane but by varying the composition of RTIL and zeolite added in order to determine the optimum condition for the membrane. The CO_2 permeability seems to rise with the increasing amount of RTIL. The CO_2/CH_4 selectivity of the MMM also improved with the presence of SAPO-34 compared to neat poly(RTIL)-RTIL membrane as long as there is sufficient amount of RTIL as the wetting agent. Besides, the team also conducted an investigation of the separation performance by using the vinyl-based poly(RTIL). The addition of RTIL is not essential as they are structurally similar [73].

In contrast, a ternary MMM has been fabricated by Oral and his coworkers by using different materials. The project study on the effect of different RTIL loadings which are emim[Tf_2N] and emim[CF_3SO_3] towards MMM composed of polyimide-zeolite (SAPO-34). The addition of emim[Tf_2N] has performed as expected when the permeability of CO_2 increased while the incorporation of emim[CF_3SO_3] has increased the CO_2/CH_4 selectivity since emim[CF_3SO_3] is selective towards CO_2/CH_4 [74].

3. Conclusion

The escalating research in the membrane fabrication for gas separation applications signifies that membranes technology is currently growing and becoming the major focus for industrial gas separation processes. Latest research area using mixed matrix membranes combines the flexibility and low capital cost with improving selectivity, permeability, chemical, thermal, and mechanical strength. Material selection and method of preparation are the most important part in fabricating a membrane. So the next research must be very careful in determining the materials for gas separation and methods applied in the fabrication stage. Even the synthesized MMMs were only tested in a small scale, the research of MMMs is worth to be further explored since MMMs have shown better separation performance compared to polymeric and inorganic membranes.

References

[1] Soregraph, *Key World Energy Statistic*, The International Energy Agency, 2010.

[2] Longterm Outlook to 2030, *Natural Gas Demand and Supply*, The European Union of The Natural Gas Industry, 2010.

[3] "Natural Gas and Environment—Emission from the Combustion of Natural Gas," copyright 2004–2010, http://www.naturalgas.org/environment/naturalgas.asp#emission.

[4] A. Wan and A. Rusmidah, *Natural Gas*, Universiti Teknologi Malaysia, 2010.

[5] D. David and D. Kishore, *Recent Development in CO_2 Removal Membrane Technology*, UOP, 1999.

[6] M. I. Fauzi and A. Akkil, *Meeting Technical Challenge in Developing High CO$_2$ Gas Field Offshore*, Petronas Carigali Sdn. Bhd., 2008.

[7] Fuels Providers, *Natural Gas Specs Sheet*, The National Petroleum Agency, 2002.

[8] Separation Process, *Membrane Separation Process*, Membrane Properties, 1998.

[9] Separation Process, *Introduction to Membrane*, Chapter 1, 1998.

[10] K. Scott, *Membrane Separation Technology*, Scientific & Technical Information, Oxford, UK, 1990.

[11] H. Strathmann, "Membrane separation processes: current relevance and future opportunities," *AIChE Journal*, vol. 47, no. 5, pp. 1077–1087, 2001.

[12] S. Morooka and K. Kusakabe, "Microporous inorganic membranes for gas separation," *MRS Bulletin*, vol. 24, no. 3, pp. 25–29, 1999.

[13] A. F. Ismail and L. I. B. David, "A review on the latest development of carbon membranes for gas separation," *Journal of Membrane Science*, vol. 193, no. 1, pp. 1–18, 2001.

[14] W. A. W. Abdul Rahman, "Formation and characterization of mixed matrix composite materials for efficient energy gas separation," Project Report, Faculty of Chemical and Natural Resources Engineering, Universiti Teknologi Malaysia, 2006.

[15] J. A. Ritter and A. D. Ebner, "Carbon dioxide separation technology—R&D needs for the chemical and petrochemical industries," Chemical Industry Vision 2020, 2007.

[16] T. Visser and M. Wessling, "When do sorption-induced relaxations in glassy polymers set in?" *Macromolecules*, vol. 40, no. 14, pp. 4992–5000, 2007.

[17] A. Bos, I. G. M. Pünt, M. Wessling, and H. Strathmann, "CO$_2$-induced plasticization phenomena in glassy polymers," *Journal of Membrane Science*, vol. 155, no. 1, pp. 67–78, 1999.

[18] J. D. Wind, D. R. Paul, and W. J. Koros, "Natural gas permeation in polyimide membranes," *Journal of Membrane Science*, vol. 228, no. 2, pp. 227–236, 2004.

[19] J. D. Wind, S. M. Sirard, D. R. Paul, P. F. Green, K. P. Johnston, and W. J. Koros, "Relaxation dynamics of CO$_2$ diffusion, sorption, and polymer swelling for plasticized polyimide membranes," *Macromolecules*, vol. 36, no. 17, pp. 6442–6448, 2003.

[20] A. J. Bird and D. L. Trimm, "Carbon molecular sieves used in gas separation membranes," *Carbon*, vol. 21, no. 3, pp. 177–180, 1983.

[21] T. H. Kim, W. J. Koros, G. R. Husk, and K. C. O'Brien, "Relationship between gas separation properties and chemical structure in a series of aromatic polyimides," *Journal of Membrane Science*, vol. 37, no. 1, pp. 45–62, 1988.

[22] J. S. McHattie, W. J. Koros, and D. R. Paul, "Effect of isopropylidene replacement on gas transport properties of polycarbonates," *Journal of Polymer Science B*, vol. 29, no. 6, pp. 731–746, 1991.

[23] C. L. Aitken, W. J. Koros, and D. R. Paul, "Gas transport properties of biphenol polysulfones," *Macromolecules*, vol. 25, no. 14, pp. 3651–3658, 1992.

[24] L. A. Pessan and W. J. Koros, "Isomer effects on transport properties of polyesters based on bisphenol-A," *Journal of Polymer Science B*, vol. 31, no. 9, pp. 1245–1252, 1993.

[25] D. R. B. Walker and W. J. Koros, "Transport characterization of a polypyrrolone for gas separations," *Journal of Membrane Science*, vol. 55, no. 1-2, pp. 99–117, 1991.

[26] X. Gao, Z. Tan, and F. Lu, "Gas permeation properties of some polypyrrolones," *Journal of Membrane Science*, vol. 88, no. 1, pp. 37–45, 1994.

[27] J. S. McHattie, W. J. Koros, and D. R. Paul, "Gas transport properties of polysulphones: 2. Effect of bisphenol connector groups," *Polymer*, vol. 32, no. 14, pp. 2618–2625, 1991.

[28] Y. Liu, T. S. Chung, R. Wang, D. F. Li, and M. L. Chng, "Chemical cross-linking modification of polyimide/poly(ether sulfone) dual-layer hollow-fiber membranes for gas separation," *Industrial and Engineering Chemistry Research*, vol. 42, no. 6, pp. 1190–1195, 2003.

[29] G. Xomeritakis, C. Y. Tsai, and C. J. Brinker, "Microporous sol-gel derived aminosilicate membrane for enhanced carbon dioxide separation," *Separation and Purification Technology*, vol. 42, no. 3, pp. 249–257, 2005.

[30] L. Zhang, K. E. Gilbert, R. M. Baldwin, and J. Douglas Way, "Preparation and testing of carbon/silicalite-1 composite membranes," *Chemical Engineering Communications*, vol. 191, no. 5, pp. 665–681, 2005.

[31] M. P. Bernal, J. Coronas, M. Menéndez, and J. Santamaría, "On the effect of morphological features on the properties of MFI zeolite membranes," *Microporous and Mesoporous Materials*, vol. 60, no. 1-3, pp. 99–110, 2003.

[32] C. Y. Tsai, S. Y. Tam, Y. Lu, and C. J. Brinker, "Dual-layer asymmetric microporous silica membranes," *Journal of Membrane Science*, vol. 169, no. 2, pp. 255–268, 2000.

[33] R. W. Baker, E. L. Cussler, W. Eykamp, W. J. Koros, R. L. Riley, and H. Strathmann, *Membrane Separation Systems—Recent Developments and Future Directions*, Noyes Data Corporation, 1991.

[34] D. E. W. Vaughan, "The synthesis and manufacture of zeolites," *Chemical Engineering Progress*, vol. 84, no. 2, pp. 25–31, 1988.

[35] M. A. Carreon, *Novel Membranes for Efficient CO$_2$ Separation*, University of Lousville, 2011.

[36] S. Kulprathipanja, R. W. Neuzil, and N. N. Li, "Separation of fluids by means of mixed matrix membranes in gas permeation," US Patent 4,740,219, 1988.

[37] T. M. Gür, "Permselectivity of zeolite filled polysulfone gas separation membranes," *Journal of Membrane Science*, vol. 93, no. 3, pp. 283–289, 1994.

[38] L. Yi, *Development of Mixed Matrix Membrane for Gas Separation Application*, Tsinghua University, 2006.

[39] C. M. Zimmerman, A. Singh, and W. J. Koros, "Tailoring mixed matrix composite membranes for gas separations," *Journal of Membrane Science*, vol. 137, no. 1-2, pp. 145–154, 1997.

[40] R. Mahajan, C. Zimmerman, and W. Koros, *Fundamental, Practical Aspects of Mixed Matrix Gas Separation Membranes*, ACS Symposium Series, 1999.

[41] V. Bhardwaj, A. MacIntosh, I. D. Sharpe, S. A. Gordeyev, and S. J. Shilton, "Polysulfone hollow fiber gas separation membranes filled with submicron particles," *Annals of the New York Academy of Sciences*, vol. 984, pp. 318–328, 2003.

[42] R. Mahajan, R. Burns, M. Schaeffer, and W. J. Koros, "Challenges in forming successful mixed matrix membranes with rigid polymeric materials," *Journal of Applied Polymer Science*, vol. 86, no. 4, pp. 881–890, 2002.

[43] M. G. Süer, N. Baç, and L. Yilmaz, "Gas permeation characteristics of polymer-zeolite mixed matrix membranes," *Journal of Membrane Science*, vol. 91, no. 1-2, pp. 77–86, 1994.

[44] D. R. Paul and D. R. Kemp, "The diffusion time lag in polymer membrane containing adsorptive fillers," *Journal of Polymer Science C*, no. 41, pp. 79–93, 1973.

[45] J. M. Duval, B. Folkers, M. H. V. Mulder, G. Desgrandchampsb, and C. A. Smolders, "Adsorbent filled membranes for gas separation. Part 1. Improvement of the gas separation properties of polymeric membranes by incorporation of microporous adsorbents," *Journal of Membrane Science*, vol. 80, no. 1, pp. 189–198, 1992.

[46] S. Kulprathipanja, "Review of recent progress in mixed matrix membranes," *Membrane Technology*, vol. 105, pp. 6–8, 2000.

[47] R. Mahajan and W. J. Koros, "Factors controlling successful formation of mixed-matrix gas separation materials," *Industrial and Engineering Chemistry Research*, vol. 39, no. 8, pp. 2692–2696, 2000.

[48] J. M. Duval, *Adsorbent filled polymeric membranes [Ph.D. thesis]*, The University of Twente, 1995.

[49] Z. Huang, J. F. Su, X. Q. Su, Y. H. Guo, L. J. Teng, and C. M. Yang, "Preparation and permeation characterization of β-zeolite-incorporated composite membranes," *Journal of Applied Polymer Science*, vol. 112, no. 1, pp. 9–18, 2009.

[50] D. Q. Vu, W. J. Koros, and S. J. Miller, "Mixed matrix membranes using carbon molecular sieves: I. Preparation and experimental results," *Journal of Membrane Science*, vol. 211, no. 2, pp. 311–334, 2003.

[51] M. D. Jia, K. V. Peinemann, and R. D. Behling, "Preparation and characterization of thin-film zeolite-PDMS composite membranes," *Journal of Membrane Science*, vol. 73, no. 2-3, pp. 119–128, 1992.

[52] T. W. Pechar, S. Kim, B. Vaughan et al., "Preparation and characterization of a poly(imide siloxane) and zeolite L mixed matrix membrane," *Journal of Membrane Science*, vol. 277, no. 1-2, pp. 210–218, 2006.

[53] Z. Huang, Y. Li, R. Wen, M. M. Teoh, and S. Kulprathipanja, "Enhanced gas separation properties by using nanostructured PES-zeolite 4A mixed matrix membranes," *Journal of Applied Polymer Science*, vol. 101, no. 6, pp. 3800–3805, 2006.

[54] H. H. Yong, H. C. Park, Y. S. Kang, J. Won, and W. N. Kim, "Zeolite-filled polyimide membrane containing 2,4,6-triaminopyrimidine," *Journal of Membrane Science*, vol. 188, no. 2, pp. 151–163, 2001.

[55] D. Sen, *Polycarbonate based zeolite 4A filled mixed matrix membranes: preparation, characterization and gas separation performances [Ph.D. thesis]*, Middle East Technical University, 2008.

[56] D. R. Paul and D. R. Kemp, "Diffusion time lag in polymer membranes containing adsorptive fillers," *Journal of Polymer Science C*, no. 41, pp. 79–93, 1973.

[57] J. D. Figueroa, T. Fout, S. Plasynski, H. McIlvried, and R. D. Srivastava, "Advances in CO_2 capture technology-The U.S. Department of Energy's Carbon Sequestration Program," *International Journal of Greenhouse Gas Control*, vol. 2, no. 1, pp. 9–20, 2008.

[58] M. Smiglak, W. M. Reichert, J. D. Holbrey et al., "Combustible ionic liquids by design: is laboratory safety another ionic liquid myth?" *Chemical Communications*, no. 24, pp. 2554–2556, 2006.

[59] M. J. Earle, J. M. S. S. Esperança, M. A. Gilea et al., "The distillation and volatility of ionic liquids," *Nature*, vol. 439, no. 7078, pp. 831–834, 2006.

[60] D. Camper, J. Bara, C. Koval, and R. Noble, "Bulk-fluid solubility and membrane feasibility of Rmim-based room-temperature ionic liquids," *Industrial and Engineering Chemistry Research*, vol. 45, no. 18, pp. 6279–6283, 2006.

[61] T. K. Carlisle, J. E. Bara, C. J. Gabriel, R. D. Noble, and D. L. Gin, "Interpretation of CO_2 solubility and selectivity in nitrile-functionalized room-temperature ionic liquids using a group contribution approach," *Industrial and Engineering Chemistry Research*, vol. 47, no. 18, pp. 7005–7012, 2008.

[62] A. Finotello, J. E. Bara, D. Camper, and R. D. Noble, "Room-temperature ionic liquids: temperature dependence of gas solubility selectivity," *Industrial and Engineering Chemistry Research*, vol. 47, no. 10, pp. 3453–3459, 2008.

[63] Y. I. Park, B. S. Kim, Y. H. Byun, S. H. Lee, E. W. Lee, and J. M. Lee, "Preparation of supported ionic liquid membranes (SILMs) for the removal of acidic gases from crude natural gas," *Desalination*, vol. 236, no. 1-3, pp. 342–348, 2009.

[64] D. Camper, C. Becker, C. Koval, and R. Noble, "Low pressure hydrocarbon solubility in room temperature ionic liquids containing imidazolium rings interpreted using regular solution theory," *Industrial and Engineering Chemistry Research*, vol. 44, no. 6, pp. 1928–1933, 2005.

[65] P. Scovazzo, J. Kieft, D. A. Finan, C. Koval, D. DuBois, and R. Noble, "Gas separations using non-hexafluorophosphate [PF6]-anion supported ionic liquid membranes," *Journal of Membrane Science*, vol. 238, no. 1-2, pp. 57–63, 2004.

[66] H. Ohno, M. Yoshizawa, and W. Ogihara, "Development of new class of ion conductive polymers based on ionic liquids," *Electrochimica Acta*, vol. 50, no. 2-3, pp. 255–261, 2004.

[67] X. Hu, J. Tang, A. Blasig, Y. Shen, and M. Radosz, "CO_2 permeability, diffusivity and solubility in polyethylene glycol-grafted polyionic membranes and their CO_2 selectivity relative to methane and nitrogen," *Journal of Membrane Science*, vol. 281, no. 1-2, pp. 130–138, 2006.

[68] J. E. Bara, S. Lessmann, C. J. Gabriel, E. S. Hatakeyama, R. D. Noble, and D. L. Gin, "Synthesis and performance of polymerizable room-temperature ionic liquids as gas separation membranes," *Industrial and Engineering Chemistry Research*, vol. 46, no. 16, pp. 5397–5404, 2007.

[69] J. Tang, W. Sun, H. Tang, M. Radosz, and Y. Shen, "Enhanced CO_2 absorption of poly(ionic liquid)s," *Macromolecules*, vol. 38, no. 6, pp. 2037–2039, 2005.

[70] J. E. Bara, C. J. Gabriel, S. Lessmann et al., "Enhanced CO_2 separation selectivity in oligo(ethylene glycol) functionalized room-temperature ionic liquids," *Industrial and Engineering Chemistry Research*, vol. 46, no. 16, pp. 5380–5386, 2007.

[71] Y. C. Hudiono, T. K. Carlisle, J. E. Bara, Y. Zhang, D. L. Gin, and R. D. Noble, "A three-component mixed-matrix membrane with enhanced CO_2 separation properties based on zeolites and ionic liquid materials," *Journal of Membrane Science*, vol. 350, no. 1-2, pp. 117–123, 2010.

[72] J. E. Bara, D. L. Gin, and R. D. Noble, "Effect of anion on gas separation performance of polymer-room-temperature ionic liquid composite membranes," *Industrial and Engineering Chemistry Research*, vol. 47, no. 24, pp. 9919–9924, 2008.

[73] Y. C. Hudiono, T. K. Carlisle, A. L. LaFrate, D. L. Gin, and R. D. Noble, "Novel mixed matrix membranes based on polymerizable room-temperature ionic liquids and SAPO-34 particles to improve CO_2 separation," *Journal of Membrane Science*, vol. 370, no. 1-2, pp. 141–148, 2011.

[74] C. A. Oral, R. D. Noble, and S. B. Tantekin-Ersolmaz, "Ternary mixed-matrix membranes containing room temperature ionic liquids," in *Proceedings of the North American Membrane Society Conference (NAMS '11)*, 2011.

2

Fast Far Field Computation of Single and Dual Reflector Antennas

Marcos Arias-Acuña, Antonio García-Pino, and Oscar Rubiños-López

Departamento de Teoría de la Señal y Comunicaciones, Universidade de Vigo, 36310 Vigo, Spain

Correspondence should be addressed to Marcos Arias-Acuña; marcos@com.uvigo.es

Academic Editor: Karim Kabalan

The physical optics (PO) method has been widely used for the analysis of the electromagnetic behavior of single and dual reflector antennas. An extensive work has been done by the authors of this paper in order to increase the speed for obtaining far field patterns from single and dual geometries and also in order to increase the accuracy of the method. This paper reviews these contributions and improves the existing published work with the physical interpretation of the radiation from a single patch and the computer implications when using acceleration techniques such as OpenMP.

1. Introduction

In engineering electromagnetic reflector antennas are commonly used for long-range radio communications, in applications such as satellite communications, radiolinks, and space exploration. In these applications it is necessary to accurately estimate the radiation pattern of the antenna to fulfill requirements such as gain, sidelobe levels, and cross-polar radiation.

The electromagnetic scattering by metallic surfaces has been extensively treated in the literature with special emphasis on the reflector antenna application case. Usually, the radiated integral of the current on the metallic surface is numerically solved with the aid of some mathematical and physical approximations.

The earliest method was based on series representation [1] in order to develop the radiated field. This mathematical approximation was then adapted to the case of an offset paraboloid using Jacobi polynomial series method [2].

The second type of mathematical approximation, based on numerical integration, was presented in [3], where it is demonstrated that the convergence depends on the choice of the integration grid coordinate system. In [4] popular numerical integration methods applied to the reflector antenna problem are compared. It was shown that the most accurate

one to develop the surface integral consisted of using Gauss-Zernike polynomial integration for the radial coordinate and the trapezoidal rule for the angular coordinate along the circumference.

Other very popular physical approximations are based on geometrical optics with aperture integration (GO + AI) and physical optics (PO). These techniques are compared in [5] for offset parabolic antenna. An alternative consists of combining GO with the geometrical theory of diffraction (GTD) to represent the reflector rim. In [6] it is demonstrated that the PO solution agrees well with the GO + GTD for the copolar component.

In [7] the integration across the main reflector of a Cassegrain antenna is transferred to the aperture of the feed. When using several reflector surfaces, the interaction among surfaces must be taken into account. For example, in [8] a ray tracing technique is described in order to model the propagated field through a multireflector system that is numerically specified.

In [9] the authors presented a comparison between using direct numerical integration and using triangular or rectangular patches. In that book chapter, the methods that used patches demonstrated to be faster than those using numerical integration. In Section 2, a brief summary of this book chapter is included. This method can be described as

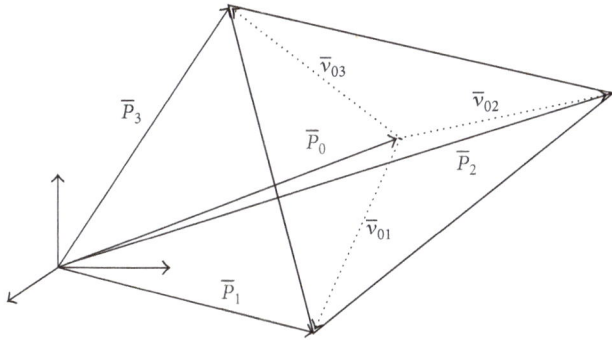

FIGURE 1: Triangular points and vectors.

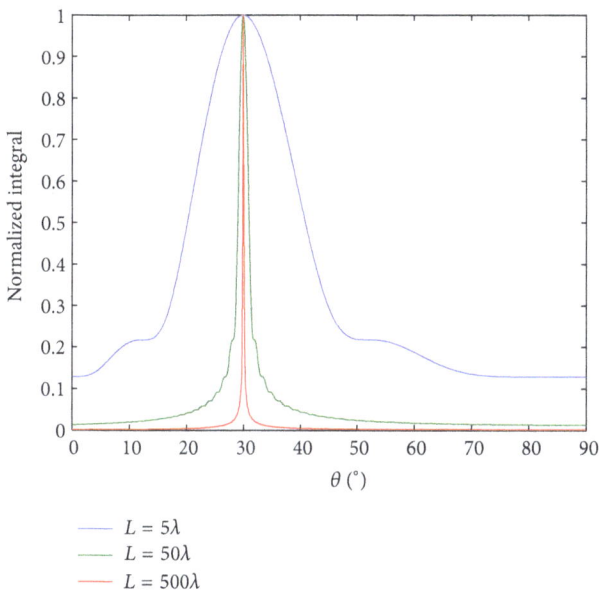

FIGURE 2: Reflection versus length of one triangle.

an approximation of the magnetic field integral equation (MFIE). This approximation consists of neglecting the interactions between the patches on the same surface. This is a good approximation when the surfaces are flat enough as in reflector antennas.

In Section 3, a new physical interpretation of the triangular patch current integration is provided.

In Section 4, the case of dual reflector geometries is addressed. Computing the PO currents induced by the feed on the subreflector and those induced by the subreflector on the main reflector surface, in addition to the feed fields, provides an estimation of the radiated fields including the spillover effect. The blockage introduced by the subreflector can be simulated by a second iteration consisting of including the radiation of a new set of blocking currents across the subreflector. The PO blocking currents are derived from the fields produced by the main reflector into the subreflector. Using the same approximation of the MFIE as with single reflector antennas and assuming that the incident field only exists on the subreflector, the process can be reduced to

invert a small matrix, which is possible to perform in modern computers. This method was introduced in [10] and improved in [11–13].

In Section 5 some techniques to improve the speed of the computation are presented. The first [14] is based on precomputing the complex exponentials and on using OpenMP [15]. And the second [16] is based on using memory hierarchy techniques.

Finally, in Section 6 some comparative results are shown.

Two computer programs were developed for educational purposes as described in [17, 18]. The final version is downloadable from the web for free since 2006 (http://www.com.uvigo.es).

2. Physical Optics for Reflector Antennas

In this section a method based on magnetic field integral equation (MFIE) [19, 20] will be used to estimate the fields radiated by a reflector antenna.

MFIE model is obtained from equivalent currents on the surface [21], by a similar way as classical high frequency methods such as physical optics (PO) but is more accurate than PO because it calculates the coupling among surfaces. In this paper, MFIE model is modified in order to adapt it to the particular problem with smooth (almost flat) and electrically big surfaces [12].

2.1. MFIE Field Formulation. The current $\bar{J}(\bar{r}_S)$ on a PEC with smooth and closed surface can be obtained using

$$\bar{J}(\bar{r}_S) = 2\hat{n}(\bar{r}_S) \times \left(\overline{H}^i(\bar{r}_S) + \oint_{S'} \bar{J}(\bar{r}_S') \times \nabla' g(\bar{r}_S, \bar{r}_S') dS' \right), \tag{1}$$

where $\bar{r}_S$ is the observation point, $\bar{r}_S'$ is the source point, $\overline{H}^i(\bar{r}_S)$ is the incident magnetic field at point $\bar{r}_S$, symbol $\oint$ means the principal value of the integral, and $g(\bar{r}_S, \bar{r}_S')$ is free space Green function, defined as

$$g(\bar{r}_S, \bar{r}_S') = \frac{e^{-jk|\bar{r}_S - \bar{r}_S'|}}{4\pi |\bar{r}_S - \bar{r}_S'|}. \tag{2}$$

Taking into account (2), (1) can be rewritten as

$$\bar{J}(\bar{r}_S)$$
$$= 2\hat{n}(\bar{r}_S)$$
$$\times \left(\overline{H}^i(\bar{r}_S) + \oint_{S'} \bar{J}(\bar{r}_S') \times \overline{R}_S \frac{1 + jkR_S}{R_S} \frac{e^{-jkR_S}}{4\pi R_S^2} dS' \right), \tag{3}$$

where $\overline{R}_S = \bar{r}_S - \bar{r}_S'$ and $R_S = |\bar{r}_S - \bar{r}_S'|$.

After obtaining equivalent currents, the radiated magnetic field at points $\bar{r}$ not belonging to the surface can be obtained by

$$\overline{H}^s(\bar{r}) = \frac{1}{4\pi} \int_{S'} \bar{J}(\bar{r}_S') \times \hat{R} \frac{1 + jkR}{R^2} e^{-jkR} dS', \tag{4}$$

where $\overline{R} = \bar{r} - \bar{r}_S'$, $R = |\bar{r} - \bar{r}_S'|$ and $\hat{R} = \overline{R}/R$.

2.2. PO Field Formulation. Physical optics (PO) is a simplification of (3) when the surface is flat enough to consider that the currents $\overline{J}(\overline{r}'_S)$ are on the same plane as the vector position $\overline{R}_S$ pointing to another point on the surface. In this case, their cross-product $\overline{J}(\overline{r}'_S) \times \overline{R}_S$ is colinear with the normal vector $\hat{n}(\overline{r}_S)$ and their cross-product tends to zero. With this approximation, (3) is simplified as

$$\overline{J}^{\text{PO}}\left(\overline{r}_S\right) = 2\hat{n}\left(\overline{r}_S\right) \times \overline{H}^i\left(\overline{r}_S\right) \tag{5}$$

and the radiated magnetic field at points not belonging to the surface can be obtained with physical optics by considering

$$\overline{H}^s\left(\overline{r}\right) = \frac{1}{4\pi} \int_{S'} \overline{J}^{\text{PO}}\left(\overline{r}'_S\right) \times \hat{R} \frac{1 + jkR}{R^2} e^{-jkR} dS'. \tag{6}$$

When the observation point is at a distance $r = |\overline{r}| \gg |\overline{r}'_S|$ for every source point, then the distance can be simplified as follows:

$$R = \left|\overline{r} - \overline{r}'_S\right| = \sqrt{r^2 + \overline{r}'^2_S - 2\overline{r} \cdot \overline{r}'_S} \approx r - \hat{r} \cdot \overline{r}'_S, \tag{7}$$

where the term $\hat{r} \cdot \overline{r}'_S$ is important only for the phase calculation but it can be neglected in the amplitude terms. With this approximation, (6) for points far from the surface can be obtained as

$$\overline{H}^s\left(\overline{r}\right) = \frac{1}{4\pi} \frac{(1 + jkr) \cdot e^{-jkr}}{r^2} \left[\int_{S'} \overline{J}^{\text{PO}}\left(\overline{r}'_S\right) e^{jk\hat{r} \cdot \overline{r}'_S} dS' \right] \times \hat{r}. \tag{8}$$

If the distance $r = |\overline{r}| \gg \lambda$, then (8) can be simplified by

$$\overline{H}^s\left(\hat{r}\right) = \frac{j}{2\lambda} \frac{e^{-jkr}}{r} \left[\int_{S'} \overline{J}^{\text{PO}}\left(\overline{r}'_S\right) e^{jk\hat{r} \cdot \overline{r}'_S} dS' \right] \times \hat{r}. \tag{9}$$

In the next subsections, (8) is used to obtain the radiated fields from a small patch at any point on other surfaces.

2.3. Patch Discretization. The radiated field at an observation point $\overline{r}_k$ can be obtained from (8) by dividing the surface in subdomains (patches) small enough that we can consider that the point $\overline{r}_k$ is far from each patch. The magnetic field $\overline{H}_k$ at the point $\overline{r}_k$ can be represented as

$$\overline{H}_k = \sum_i \overline{H}_{ik}, \tag{10}$$

where $\overline{H}_{ik}$ is the magnetic field radiated by the patch i at $\overline{r}_k$. The field radiated by each patch will be

$$\overline{H}_{ik} = \frac{1}{4\pi} \frac{(1 + jkr_{ik}) \cdot e^{-jkr_{ik}}}{r^2_{ik}} \left[\int_{S_i} \overline{J}^{\text{PO}}\left(\overline{r}'_S\right) e^{jk\hat{r}_{ik} \cdot \overline{r}'_S} dS_i \right] \times \hat{r}_{ik}, \tag{11}$$

where S_i represents the surface of the considered patch i.

PO approximation can be used when the patch surface is flat enough to consider that the whole patch has the same

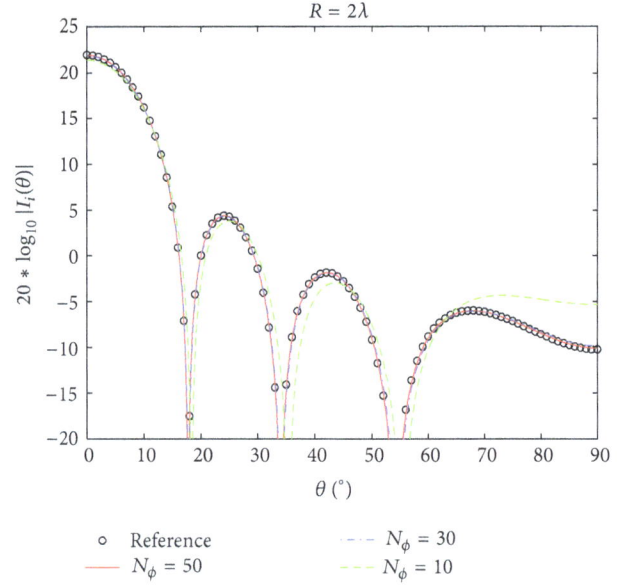

FIGURE 3: Convergence when $R = 2\lambda$.

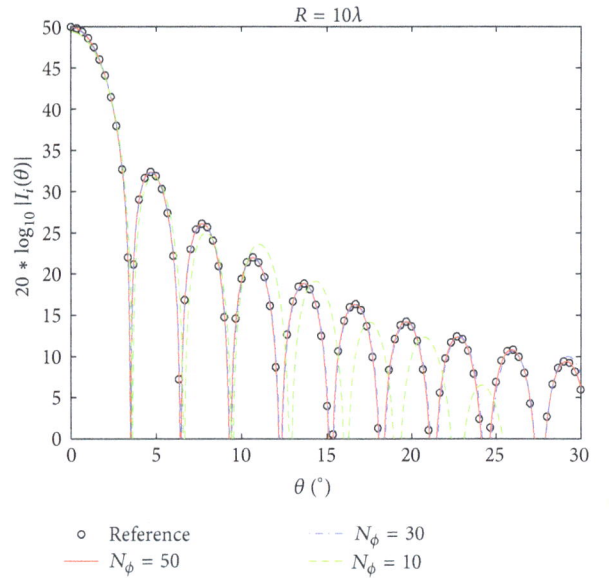

FIGURE 4: Convergence when $R = 10\lambda$.

normal vector $\hat{n}_i$. Inserting this vector into (11) produces the following expression:

$$\overline{H}_{ik} = \frac{1}{2\pi} \frac{(1 + jkr_{ik}) \cdot e^{-jkr_{ik}}}{r^2_{ik}} \left[\hat{n}_i \times \int_{S_i} \overline{H}^i\left(\overline{r}'_S\right) e^{jk\hat{r}_{ik} \cdot \overline{r}'_S} dS_i \right]$$
$$\times \hat{r}_{ik}. \tag{12}$$

2.4. Physical Approximation. Equation (11) is difficult to compute due to the existence of the incident field inside the integral. But if the patch is flat and small, the incident field on this patch can be considered locally as a plane wave. With

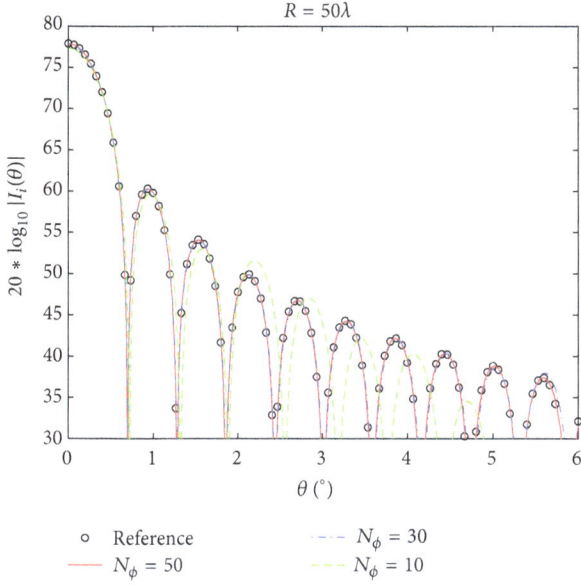

FIGURE 5: Convergence when $R = 50\lambda$.

this approximation, the amplitude and the direction of the field are assumed to be constant on the patch, and the phase will vary according to the direction of the propagation of the incident plane wave $\hat{p}_i$. Consider that

$$\overline{r}'_S = \overline{r}_i + \overline{r}''. \tag{13}$$

Then,

$$\overline{H}^i\left(\overline{r}'_S\right) \approx \overline{H}_{i0} \cdot e^{-jk\overline{r}''\cdot\hat{p}_i}. \tag{14}$$

Including (13) and (14) into (12),

$$\overline{H}_{ik} \approx \frac{1}{2\pi} \frac{(1 + jkr_{ik}) \cdot e^{-jkr_{ik}}}{r_{ik}^2} \left[\hat{n}_i \times \overline{H}_{i0} \cdot I_i\right] \times \hat{r}_{ik}, \tag{15}$$

where

$$I_i = \int_{S_i} e^{jk(\hat{r}_{ik} - \hat{p}_i)\cdot\overline{r}''} dS_i. \tag{16}$$

Then, the scattered far field will now be obtained as

$$\overline{H}\left(\hat{r}\right) = \frac{j}{\lambda}\frac{e^{-jkr}}{r} \left[\sum_i \hat{n}_i \times \overline{H}_{i0} \cdot I_i \cdot e^{jk\hat{r}\cdot\overline{r}_i}\right] \times \hat{r}. \tag{17}$$

2.5. Triangular Patches. Let us consider a triangular flat patch as seen in Figure 1. The triangle is defined by three points $\overline{P}_1$, $\overline{P}_2$, and $\overline{P}_3$. $\overline{P}_0$ is the point where $\overline{H}_{i0}$ is calculated which is assumed to be at the barycenter $P_0 = (P_1 + P_2 + P_3)/3$. If we define $\overline{v}_{mn} = \overline{P}_n - \overline{P}_m$, the local coordinates of the triangle can be described with two variables u and v as

$$\overline{r}'' = \overline{v}_{01} + u \cdot \overline{v}_{12} + v \cdot \overline{v}_{13}. \tag{18}$$

The normal vector and the area of the triangle are related to the cross-product of $\overline{v}_{12}$ and $\overline{v}_{13}$ as follows:

$$\overline{v}_{12} \times \overline{v}_{13} = 2A_i \cdot \hat{n}_i. \tag{19}$$

Using this coordinate system, (16) is then given by

$$I_i = 2Ae^{-j((\alpha+\beta)/3)} \int_0^1 \int_0^{1-u} e^{j(\alpha u + \beta v)} dv\, du, \tag{20}$$

whose solution is

$$I_i\left(\hat{r}_k\right) = 2Ae^{-j((\alpha+\beta)/3)} \left[\frac{\alpha e^{j\beta} - \beta e^{j\alpha} + \beta - \alpha}{(\alpha - \beta)\alpha\beta}\right], \tag{21}$$

where

$$\alpha = k\overline{v}_{12} \cdot \left(\hat{r}_{ik} - \hat{p}_i\right), \tag{22}$$

$$\beta = k\overline{v}_{13} \cdot \left(\hat{r}_{ik} - \hat{p}_i\right). \tag{23}$$

Equation (21) has the following singular values:

$$\alpha = 0 \implies I_i\left(\hat{r}_k\right) = 2Ae^{-j(\beta/3)}\left[\frac{1 + j\beta - e^{j\beta}}{\beta^2}\right],$$

$$\beta = 0 \implies I_i\left(\hat{r}_k\right) = 2Ae^{-j(\alpha/3)}\left[\frac{1 + j\alpha - e^{j\alpha}}{\alpha^2}\right],$$

$$\alpha = \beta \implies I_i\left(\hat{r}_k\right) = 2Ae^{-j(2\alpha/3)}\left[\frac{e^{j\alpha}\left(1 - j\alpha\right) - 1}{\alpha^2}\right], \tag{24}$$

$$\alpha = \beta = 0 \implies I_i\left(\hat{r}_k\right) = A.$$

Form small values of α and β, a second order series approximation can be used:

$$I_i\left(\hat{r}_k\right) \approx A\left(1 - \frac{\alpha^2 + \beta^2 - \alpha\beta}{36}\right). \tag{25}$$

Finally, for directions close to the main lobe or for very small patches, the expression (25) could be approximated as

$$I_i\left(\hat{r}_k\right) \approx A. \tag{26}$$

3. Physical Interpretation

In this section the relationship between this integral representation and other methods such as geometrical optics is explained.

3.1. Vertex Contributions. Equation (21) can be reformulated in order to clarify its physical meaning:

$$I_i\left(\hat{r}_k\right)$$

$$= -2A$$

$$\times \left[\frac{e^{(-\alpha-\beta)/3}}{\alpha \cdot \beta} + \frac{e^{(2\alpha-\beta)/3}}{(\beta-\alpha)\cdot(-\alpha)} + \frac{e^{(-\alpha+2\beta)/3}}{(\alpha-\beta)\cdot(-\beta)}\right]. \tag{27}$$

(a)

Reference
$N_r = 24$

(b)

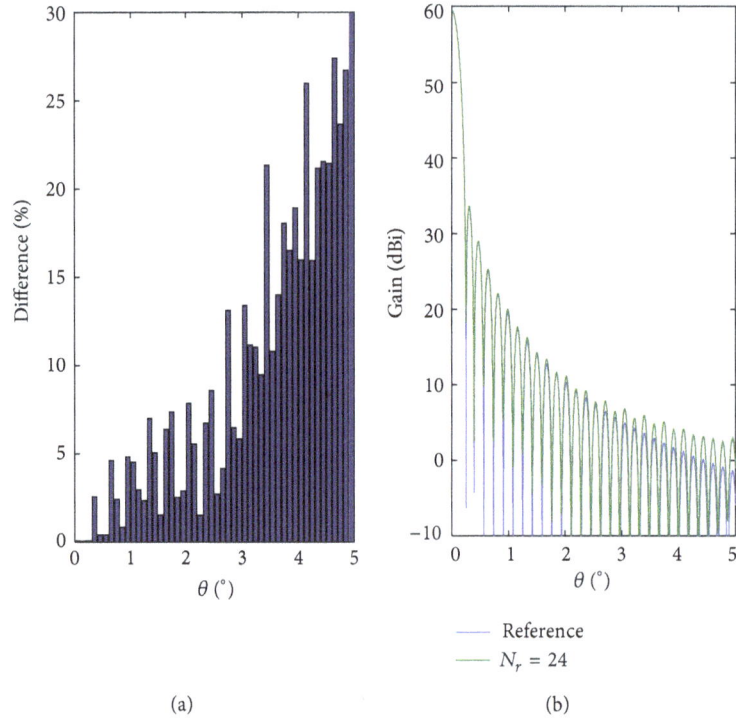

FIGURE 6: Difference using (26): at (a) the difference and at (b) the pattern.

Each of the three terms of (27) can be interpreted as the contribution of a vertex to the radiation integral I_i. Taking into account the definitions of $\overline{v}_{0n}$ it can be stated that

$$\overline{v}_{01} = \overline{P}_1 - \frac{\overline{P}_1 + \overline{P}_2 + \overline{P}_3}{3} = \frac{-\overline{v}_{12} - \overline{v}_{13}}{3},$$

$$\overline{v}_{02} = \overline{P}_2 - \frac{\overline{P}_1 + \overline{P}_2 + \overline{P}_3}{3} = \frac{2\overline{v}_{12} - \overline{v}_{13}}{3}, \qquad (28)$$

$$\overline{v}_{03} = \overline{P}_3 - \frac{\overline{P}_1 + \overline{P}_2 + \overline{P}_3}{3} = \frac{-\overline{v}_{12} + 2\overline{v}_{13}}{3}.$$

So, expression (27) can be transformed into

$$I_i\left(\widehat{r}_k\right) = -2A \left[\frac{e^{\overline{v}_{01} \cdot (\widehat{r}_{ik} - \widehat{p}_i)}}{a_{12} \cdot a_{13}} + \frac{e^{\overline{v}_{02} \cdot (\widehat{r}_{ik} - \widehat{p}_i)}}{a_{23} \cdot a_{21}} + \frac{e^{\overline{v}_{03} \cdot (\widehat{r}_{ik} - \widehat{p}_i)}}{a_{32} \cdot a_{31}} \right], \quad (29)$$

where each phase term $e^{\overline{v}_{0n} \cdot (\widehat{r}_{ik} - \widehat{p}_i)}$ is due to the displacement from the barycenter to the vertex n and the amplitude terms depend on the vectors from the vertex n to the other two vertices, according to the values of α and β given by (22) and (23):

$$a_{nm} = k\overline{v}_{nm} \cdot \left(\widehat{r}_{ik} - \widehat{p}_i\right). \qquad (30)$$

3.2. Reflection Point. The theory of geometrical optics states that the contributors for the radiated field are the reflection and the diffraction points. The diffraction comes from the border of the object, and if the surface is divided in patches, then each patch creates also diffraction fields that can be

compensated with the diffraction generated by the surrounding patches. But (27) and (29) do not represent separately diffraction and reflection terms, as it has been done in [22]. However, it is possible to study the reflection effect, that appears in the direction $\widehat{r} = \widehat{p}_i$ or $\widehat{r} = \widehat{p}_i - 2\widehat{n} \cdot (\widehat{p}_i \cdot \widehat{n})$.

If the triangle is big enough compared with the wavelength, then the main effect is the reflection, as it can be seen in Figure 2 where the incident field forms 30 degrees with the plane of the triangle. It can be observed that when the size of the triangle increases, then the radiated field tends to the geometrical optics effect, represented by a narrow beam at $30°$.

3.3. Discretization Effect as a Function of the Size of the Object. Methods like MoM require discretization less than $\lambda/10$ in order to converge to the real solution. With PO the size of the triangles depends on the required accuracy in the angular domain, as it can be seen in the next example.

Consider a circle of radius R on the plane XY with a plane wave normal to the circle as incident field. The scattered field for $\phi = 0$ will be

$$I_i(\theta) = \int_0^R \int_0^{2\pi} e^{jk\rho \cos\phi \cdot \sin\theta} \rho \, d\phi \, d\theta = 2\pi R \frac{J_1(kR\sin\theta)}{k\sin\theta} \quad (31)$$

with a maximum value $I_i(\theta)|_{\max} = I_i(0) = \pi R^2$.

Adapting (27) to this example, where the phase reference point will be at the center of the circle and $\overline{v}_{12} = R \cdot \widehat{\rho}_\phi$ and

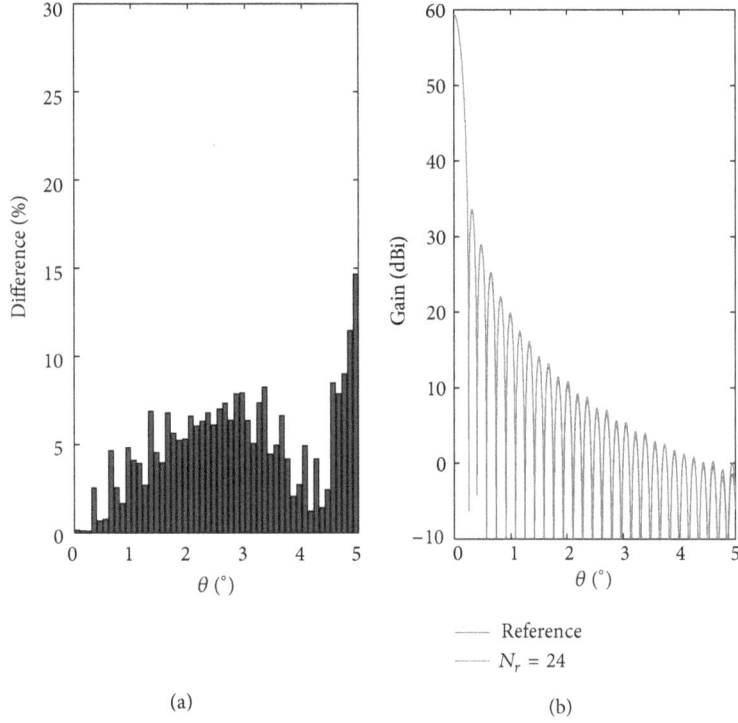

FIGURE 7: Difference using (25): at (a) the difference and at (b) the pattern.

$\bar{v}_{13} = R \cdot \hat{\rho}_{\phi+\Delta\phi}$, this leads to $\alpha_n = R \cos\phi_n \sin\theta$ and $\beta_n = R\cos(\phi_n + \Delta\phi)\sin\theta$. Then, the integral will be

$$I_i(\theta)$$

$$= \frac{\sin\Delta\phi}{k^2\sin^2\theta}$$

$$\times \sum_{n=0}^{N-1} \left[\frac{e^{jkR\sin\theta\cos((n+1)\Delta\phi)} - 1}{\cos((n+1)\Delta\phi)\cdot(\cos(n\Delta\phi) - \cos((n+1)\Delta\phi))} \right.$$

$$\left. - \frac{e^{jkR\sin\theta\cos(n\Delta\phi)} - 1}{\cos(n\Delta\phi)\cdot(\cos(n\Delta\phi) - \cos((n+1)\Delta\phi))} \right], \tag{32}$$

where $\Delta\phi = 2\pi/N$.

The limit of this expression tends to the integral:

$$I_i(\theta) = \frac{1}{k^2\sin^2\theta} \int_0^{2\pi} \frac{e^{jkR\sin\theta\cos\phi}(1 - jkR\sin\theta) - 1}{\cos^2\phi} d\phi. \tag{33}$$

Now, we are going to determine the number of angular sectors N_ϕ that are necessary to obtain an accurate solution when the radius R varies. In Figures 3, 4, and 5 the convergence can be seen when the radius is 2λ, 10λ, and 50λ.

The convergence behaviour shown in these figures is similar because in each case the radiation pattern is calculated in a margin according to the size of the circle, where the radiation is in a margin of 40 dB less than the maximum. Greater areas imply lower beam widths, and the number of triangles required for the same accuracy is almost the same.

But if the same accuracy is required in the whole θ angular view, then the number of triangles must be increased in the same rate as the radius is increased.

4. PO for Dual Reflector Antennas

When there is more than one surface, it is not possible to eliminate the radiation of a patch over the patches on the other surfaces. Applying this situation to a dual reflector antenna, the radiation between both surfaces must be calculated.

4.1. MFIE Approximation for Dual Reflector Antennas. If the patches are small enough to use (26), then the amplitude and also the phase at each patch can be considered constant. So the current at patch k will be a vector $\bar{J}_k$. This simplifies the integral assuming that $\bar{R}_S \approx (\bar{r}_k - \bar{r}_n) = \bar{r}_{nk}$ and $|\bar{r}_{nk}| = r_{nk}$, so (3) can be obtained as

$$\bar{J}_k = 2\hat{n}_k \times \overline{H}_k^i + \frac{1}{2\pi}\hat{n}_k \times \sum_n \bar{J}_n \times \frac{\bar{r}_{nk}}{r_{nk}^2}\left(jk + \frac{1}{r_{nk}}\right)e^{-jkr_{nk}}A_n, \tag{34}$$

where A_n is the area of nth-triangle.

The equivalent current in (34) is due to incident (PO) currents and to the coupling among surfaces. We can define

$$\bar{J}_k^{PO} = 2\hat{n}_k \times \overline{H}_k^i, \tag{35}$$

$$\bar{c}_{nk} = \bar{r}_{nk}\frac{e^{-jkr_{nk}}}{2\pi r_{nk}^2}\left(jk + \frac{1}{r_{nk}}\right) \tag{36}$$

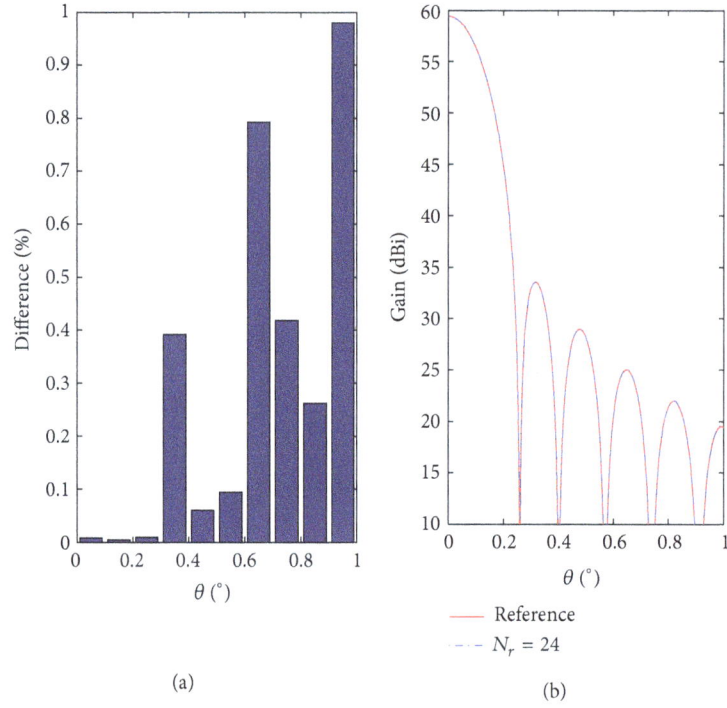

FIGURE 8: Difference when $N_r = 42$: at (a) the difference and at (b) the pattern.

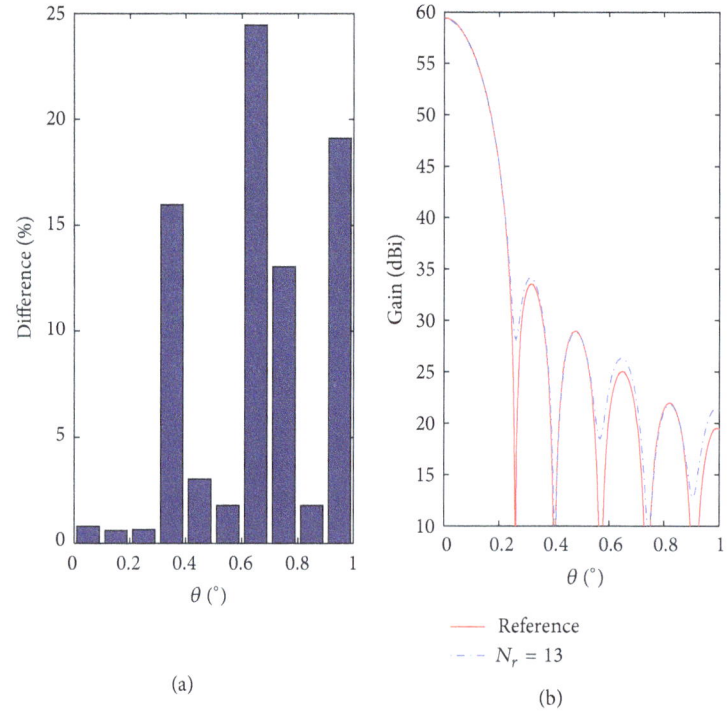

FIGURE 9: Difference when $N_r = 13$: at (a) the difference and at (b) the pattern.

and now the equivalent current can be expressed as

$$\bar{J}_k = \bar{J}_k^{\text{PO}} + \hat{n}_k \times \sum_n A_n \bar{J}_n \times \bar{c}_{nk}, \tag{37}$$

where k and n are for points on different surfaces.

When the currents on both surfaces are obtained, the scattered far fields can be calculated at a direction $\hat{r}$ using

$$\overline{H}^s(\hat{r}) = \frac{j}{2\lambda} \frac{e^{-jkr}}{r} \sum_i A_i \bar{J}_i e^{jk\bar{r}_i \cdot \hat{r}}, \tag{38}$$

where i is intended for all the main reflector and subreflector patches and

$$\overline{E}^{s}\left(\hat{r}\right) = \eta\overline{H}^{s}\left(\hat{r}\right) \times \hat{r}. \tag{39}$$

4.2. Iterative Solution for MFIE Currents. The solution to (37) can be obtained with an iterative method or with a matrix method. From this point, k will be for reflector patches and n for main reflector patches.

With the iterative method, the currents can be obtained using an iterative process in which the first step is using PO currents:

$$\overline{J}_{k}^{0} = \overline{J}_{k}^{\mathrm{PO}} = 2\hat{n}_{k} \times \overline{H}_{k}^{i},$$

$$\overline{J}_{n}^{1} = \overline{J}_{n}^{\mathrm{PO}} + \hat{n}_{n} \times \sum_{k} A_{k}\overline{J}_{k}^{0} \times \overline{c}_{kn}$$

$$= 2\hat{n}_{n} \times \overline{H}_{n}^{i} + \hat{n}_{n} \times \sum_{k} A_{k}\overline{J}_{k}^{0} \times \overline{c}_{kn}, \tag{40}$$

$$\overline{J}_{k}^{N} = \overline{J}_{k}^{\mathrm{PO}} + \hat{n}_{k} \times \sum_{n} A_{n}\overline{J}_{k}^{N} \times \overline{c}_{nk},$$

$$\overline{J}_{n}^{N+1} = \overline{J}_{n}^{\mathrm{PO}} + \hat{n}_{n} \times \sum_{k} A_{k}\overline{J}_{k}^{N} \times \overline{c}_{kn}$$

where $\overline{c}_{kn} = -\overline{c}_{nk}$.

This iterative process converges in few steps, usually less than six.

4.3. Matrix Formulation. In order to obtain a matrix formulation, we need to define the local coordinates on the patches. A flat patch can be defined with two unitary and orthogonal vectors $\hat{u}_{1k}$ and $\hat{u}_{2k}$ for kth-patch. Using triangular patches, these vectors can be obtained as

$$\overline{v}_{12} = \overline{P}_{2} - \overline{P}_{1}; \qquad \overline{v}_{13} = \overline{P}_{3} - \overline{P}_{1},$$

$$\hat{n}_{k} = \frac{\overline{v}_{12} \times \overline{v}_{13}}{\left|\overline{v}_{12} \times \overline{v}_{13}\right|}, \tag{41}$$

$$\hat{u}_{1k} = \frac{\overline{v}_{12}}{\left|\overline{v}_{12}\right|}; \qquad \hat{u}_{2k} = \hat{n}_{k} \times \hat{u}_{1k}.$$

The currents defined in (37) can be expressed in terms of $\hat{u}_{1k}$ and $\hat{u}_{2k}$ as

$$\overline{J}_{k} = J_{1k}\hat{u}_{1k} + J_{2k}\hat{u}_{2k} \tag{42}$$

and now (37) can be divided into two equations:

$$J_{1k} = \overline{J}_{k} \cdot \hat{u}_{1k} = \overline{J}_{k}^{\mathrm{PO}} \cdot \hat{u}_{1k} - \sum_{n} \hat{u}_{2k} \cdot \left(A_{n}\overline{J}_{n} \times \overline{c}_{nk}\right),$$

$$J_{2k} = \overline{J}_{k} \cdot \hat{u}_{2k} = \overline{J}_{k}^{\mathrm{PO}} \cdot \hat{u}_{2k} + \sum_{n} \hat{u}_{1k} \cdot \left(A_{n}\overline{J}_{n} \times \overline{c}_{nk}\right). \tag{43}$$

$\overline{J}_{n}$ can also be divided into two components. After some manipulations, (43) can be expressed as

$$J_{1k}^{\mathrm{PO}} = J_{1k} + \sum_{n} J_{1n}\left(\hat{u}_{2k} \times \hat{u}_{1n}\right) \cdot \overline{c}_{nk}$$

$$+ \sum_{n} A_{n}J_{2n}\left(\hat{u}_{2k} \times \hat{u}_{2n}\right) \cdot \overline{c}_{nk},$$

$$J_{2k}^{\mathrm{PO}} = J_{2k} - \sum_{n} J_{1n}\left(\hat{u}_{1k} \times \hat{u}_{1n}\right) \cdot \overline{c}_{nk}$$

$$- \sum_{n} A_{n}J_{2n}\left(\hat{u}_{1k} \times \hat{u}_{2n}\right) \cdot \overline{c}_{nk}. \tag{44}$$

The mutual coupling between the surfaces can be expressed as

$$J_{1k}^{\mathrm{PO}} = J_{1k} + \sum_{n} A_{n}J_{1n}\left(\hat{u}_{2k} \times \hat{u}_{1n}\right) \cdot \overline{c}_{nk}$$

$$+ \sum_{n} A_{n}J_{2n}\left(\hat{u}_{2k} \times \hat{u}_{2n}\right) \cdot \overline{c}_{nk},$$

$$J_{1n}^{\mathrm{PO}} = J_{1n} + \sum_{k} A_{k}J_{1k}\left(\hat{u}_{2n} \times \hat{u}_{1k}\right) \cdot \overline{c}_{kn}$$

$$+ \sum_{k} A_{k}J_{2k}\left(\hat{u}_{2n} \times \hat{u}_{2k}\right) \cdot \overline{c}_{kn},$$

$$J_{2k}^{\mathrm{PO}} = J_{2k} - \sum_{n} A_{n}J_{1n}\left(\hat{u}_{1k} \times \hat{u}_{1n}\right) \cdot \overline{c}_{nk}$$

$$- \sum_{n} A_{n}J_{2n}\left(\hat{u}_{1k} \times \hat{u}_{2n}\right) \cdot \overline{c}_{nk},$$

$$J_{2n}^{\mathrm{PO}} = J_{2n} - \sum_{k} A_{k}J_{1k}\left(\hat{u}_{1n} \times \hat{u}_{1k}\right) \cdot \overline{c}_{kn}$$

$$- \sum_{k} A_{k}J_{2k}\left(\hat{u}_{1n} \times \hat{u}_{2k}\right) \cdot \overline{c}_{kn}. \tag{45}$$

This equation system can be expressed with matrix formulation:

$$\begin{bmatrix} \overline{\overline{0}}_{k \times k} & \overline{\overline{M}}_{2k2n} & \frac{1}{A_{k}}\overline{\overline{I}}_{k \times k} & \overline{\overline{M}}_{2k1n} \\ \overline{\overline{M}}_{2n2k} & \overline{\overline{0}}_{n \times n} & \overline{\overline{M}}_{2n1k} & \frac{1}{A_{n}}\overline{\overline{I}}_{n \times n} \\ \frac{1}{A_{k}}\overline{\overline{I}}_{k \times k} & \overline{\overline{M}}_{1k2n} & \overline{\overline{0}}_{k \times k} & \overline{\overline{M}}_{1k1n} \\ \overline{\overline{M}}_{1n2k} & \frac{1}{A_{n}}\overline{\overline{I}}_{n \times n} & \overline{\overline{M}}_{1n1k} & \overline{\overline{0}}_{n \times n} \end{bmatrix} \cdot \begin{bmatrix} \overline{A_{k}J_{2k}} \\ \overline{A_{n}J_{2n}} \\ \overline{A_{k}J_{1k}} \\ \overline{A_{n}J_{1n}} \end{bmatrix} = \begin{bmatrix} \overline{J}_{1k}^{\mathrm{PO}} \\ \overline{J}_{1n}^{\mathrm{PO}} \\ \overline{J}_{2k}^{\mathrm{PO}} \\ \overline{J}_{2n}^{\mathrm{PO}} \end{bmatrix}, \tag{46}$$

where

$$\overline{\overline{\overline{M_{1k1n}}}} = -(\widehat{u}_{1k} \times \widehat{u}_{1n}) \cdot \overline{c}_{nk}$$

$$\overline{\overline{\overline{M_{1n1k}}}} = -(\widehat{u}_{1n} \times \widehat{u}_{1k}) \cdot \overline{c}_{kn} = \overline{\overline{\overline{M_{1k1n}}}}^{T}$$

$$\overline{\overline{\overline{M_{1k2n}}}} = -(\widehat{u}_{1k} \times \widehat{u}_{2n}) \cdot \overline{c}_{nk}$$

$$\overline{\overline{\overline{M_{2n1k}}}} = (\widehat{u}_{2n} \times \widehat{u}_{1k}) \cdot \overline{c}_{kn} = -\overline{\overline{\overline{M_{1k2n}}}}^{T}$$

$$\overline{\overline{\overline{M_{2k1n}}}} = (\widehat{u}_{2k} \times \widehat{u}_{1n}) \cdot \overline{c}_{nk} \qquad (47)$$

$$\overline{\overline{\overline{M_{1n2k}}}} = -(\widehat{u}_{1n} \times \widehat{u}_{2k}) \cdot \overline{c}_{kn} = -\overline{\overline{\overline{M_{2k1n}}}}^{T}$$

$$\overline{\overline{\overline{M_{2k2n}}}} = (\widehat{u}_{2k} \times \widehat{u}_{2n}) \cdot \overline{c}_{nk}$$

$$\overline{\overline{\overline{M_{2n2k}}}} = (\widehat{u}_{2n} \times \widehat{u}_{2k}) \cdot \overline{c}_{kn} = \overline{\overline{\overline{M_{2k2n}}}}^{T}.$$

In (47), the first two subscripts of M are for rows and the third and fourth subscripts are for the columns.

With dual reflector antennas, we can assume that the incident field on main reflector is negligible. Then, in order to solve (46) the only matrix to be inverted is the one with subreflector currents:

$$\begin{bmatrix} \overline{J_{2k}} \\ \overline{J_{1k}} \end{bmatrix} = \begin{bmatrix} \overline{\overline{M1}} & \overline{\overline{M2}} \\ \overline{\overline{M2}} & \overline{\overline{M3}} \end{bmatrix}^{-1} \cdot \begin{bmatrix} \overline{J_{1k}^{PO}} \\ \overline{J_{2k}^{PO}} \end{bmatrix}, \qquad (48)$$

where

$$\overline{\overline{M1}} = -\overline{\overline{M_{2k2n}A_n}} \cdot \overline{\overline{M_{1n2k}A_k}} - \overline{\overline{M_{2k1n}A_n}} \cdot \overline{\overline{M_{2n2k}A_k}},$$

$$\overline{\overline{M2}} = \overline{\overline{I_{k\times k}}} - \overline{\overline{M_{2k2n}A_n}} \cdot \overline{\overline{M_{1n1k}A_k}} - \overline{\overline{M_{2k1n}A_n}} \cdot \overline{\overline{M_{2n1k}A_k}},$$

$$\overline{\overline{M3}} = -\overline{\overline{M_{1k2n}A_n}} \cdot \overline{\overline{M_{1n1k}A_k}} - \overline{\overline{M_{1k1n}A_n}} \cdot \overline{\overline{M_{2n1k}A_k}} \qquad (49)$$

and the components of the main reflector currents are calculated directly with (50).

$$\overline{J_{1n}} = -\overline{\overline{M_{2n2k}}} \cdot \overline{A_k J_{2k}} - \overline{\overline{M_{2n1k}}} \cdot \overline{A_k J_{1k}},$$

$$\overline{J_{2n}} = -\overline{\overline{M_{1n2k}}} \cdot \overline{A_k J_{2k}} - \overline{\overline{M_{2n1k}}} \cdot \overline{A_k J_{1k}}. \qquad (50)$$

Finally, all the currents must be transformed to the general coordinate system, using (42).

5. Computer Acceleration

The methods described here for obtaining the radiation pattern of single and dual reflector antennas are really fast, even for electrically big surfaces. The results can be observed in the next section.

But there are some new computer techniques that can reduce even more the computer time. The first is related to the size of the memory. The second consists of precalculating the exponential functions. And the third is related to working with several computer cores at the same time.

5.1. Memory Size. With single reflector antennas, almost the whole computer time is spent for obtaining the integral (from (21) to (23)), mainly with the three exponential terms. These exponentials must be computed for all the triangles and all the directions, so it can be implemented with a matrix expression or in a loop expression. With matrices, the programs can use acceleration techniques such as BLAS [23] but require more size in the computer memory.

Using loops offers two possibilities. The first is to obtain for each direction the contribution of the scattering of all the patches and then adding their contributions. The second one consists of obtaining for each patch the scattering in all the directions and then adding these vectors to obtain the radiation in each direction.

In real examples, the number of triangles is usually larger than the number of directions, so the first loop solution needs more memory than the second loop solutions. But in any case, memory requirements are much lower than the matrix implementation.

5.2. Complex Exponential. Most of the execution time is used in the calculation of complex exponentials.

For single reflector antennas, the exponential term $e^{jk\widehat{r}\cdot\overline{r}_i}$ and those that appear in the calculation of the integral (21) must be computed for each direction and each triangle. In order to reduce the computation time, the exponentials can be precomputed previously using the property that the complex exponential is periodical with period 2π. Using the function Int to extract the integer part of the real numbers, the exponential must be computed in three steps:

$$\cos_{pre}(n) = \cos\left(\frac{n}{N} \cdot 4\pi\right); \qquad \sin_{pre}(n) = \sin\left(\frac{n}{N} \cdot 4\pi\right), \qquad (51)$$

$$\text{arg} = \left[\frac{\alpha}{\lambda} - \text{Int}\left(\frac{\alpha}{\lambda}\right)\right] \cdot \frac{N}{2} + \frac{N}{2}, \qquad (52)$$

$$e^{jk\alpha} = \cos_{pre}(\text{arg}) + \sin_{pre}(\text{arg}). \qquad (53)$$

When using a precomputed table of 3000 values, the execution time is reduced by a factor of 2.5 appearing as a slight quantization error.

For dual reflector antennas, the same problem appears for calculating the scattering fields. But obtaining $\overline{c}_{nk}$ in (36) spends much more time because this exponential must be computed for each patch on the subreflector radiating at each patch on the main reflector. The proposed method to reduce the computer time of the factor $(1 - j/kr) \cdot e^{-jkr}/r^2$ has the following steps:

$$a = \frac{N}{r_{max}^2 - r_{min}^2}; \qquad b = a \cdot r_{min}^2;$$

$$r_2 = r_{min}^2 + \frac{n}{a}; \qquad r_n = k \cdot \sqrt{r_2}, \qquad (54)$$

$$\text{real}_{\text{pre}}(n) = \frac{\cos r_n - \sin r_n / r_n}{r_2};$$

$$\text{imag}_{\text{pre}}(n) = -\frac{\sin r_n + \cos r_n / r_n}{r_2},$$

(55)

$$\text{arg} = \text{Int}\left(r^2 \cdot a - b\right), \quad (56)$$

$$\frac{e^{-jkr}}{r^2}\left(1 - \frac{j}{kr}\right) = \text{real}_{\text{pre}}(\text{arg}) + j \cdot \text{imag}_{\text{pre}}(\text{arg}), \quad (57)$$

where $r_{\min}$ is the minimum value of r and $r_{\max}$ the maximum value. The term r^2 is calculated directly from the components $r_x^2 + r_y^2 + r_z^2$, so we avoid to use the square root function.

When using a precomputed table of 12000 values, the execution time is reduced by a factor of 3.4 appearing as a slight quantization error.

5.3. OpenMP. In [13] OpenMP has been used to parallelize the code and memory-hierarchy-based optimization techniques to reduce the computer time of the code. Using these techniques, the computer time can be reduced in a factor very close to the number of cores of the CPU (usually 4 or 8).

The main problem that appears using OpenMP is the order of the loops. If the directions of the observation points are used as outer loop, then each core can compute the scattered field created by all the triangles in one direction, and at the end, it should store the result in a position of the output vector. But if the index of the triangles is used as the outer loop, then each core must compute the radiation in all directions and then use the reduction method to add all the results. Unfortunately, the reduction method is not well implemented for vectors in OpenMP, and each core must wait for the others to write their results.

6. Results

In this section, some results are shown to present the last improvements in the physical optics technique.

6.1. Single Reflector Antenna. As an example, an antenna of 0.5 m of radius, 0.25 m of offset height, and 0.5 m of focal length was studied. The frequency is 100 GHz and it has a cos-q type feed with taper 12 dB. The scattered field is obtained for $\phi = \pi/2$ and θ from 0 to 1° or 5°.

The convergence result is obtained raising the number of sectors N_r until the difference will be negligible. Then, the difference in directivity is accounted for in natural units, using

$$\text{Difference (\%)} = \frac{\text{mean}\left(\left|D_{N_r} - D_{\text{ref}}\right|\right)}{\text{max}\left(\left|D_{N_r} - D_{\text{ref}}\right|\right)} \cdot 100. \quad (58)$$

The integral is approximated by the area as stated in (26), because with this small angular margin, it provides less error than the other approximations, and it is the fastest (about 30% when using (25)). These two differences can be observed in Figure 6 for the area and in Figure 7 for the approximation (25) of the integral.

This difference with the convergence is measured in sectors with width 0.1°. With $N_r = 42$ (area of the triangles of $8\lambda^2$), the difference is below 1%. With $N_r = 24$ (area of the triangles of $25\lambda^2$), the difference is below 5%. And with $N_r = 13$ (area of the triangles of $86\lambda^2$), the difference is below 25%. In Figure 8 the first case is represented, and Figure 9 represents the third case.

The fastest version of the antenna analysis has been obtained using C language and OpenMP directives, for using all the cores of the computer, and (53). With these improvements, the case with $N_r = 42$ is obtained in 0.023 seconds in a quad-core 2.6 GHz computer.

6.2. Dual Reflector Antenna. The antenna used for this example was a Cassegrain reflector at 10 GHz. The main reflector had a radius of 0.5 m and the subreflector had a radius of 0.1162 m. The number of rings used was 25 for the main reflector and 10 for the subreflector. In order to maximize the coupling between the surfaces, the antenna had no offset height.

The results show that the iterative method converges in 6 steps, so 12 near field calculations were required. Using the matrix formulation (48), the Matlab code spent 4.61 seconds, because of the matrix inversion. Using (37) 12 times, the result was obtained in 3.73 seconds with Matlab. If the matrix $\bar{c}_{nk}$ is first computed and then used 12 times, in Matlab the spent time was reduced to 1.27 seconds. Finally, if the code is developed in C with OpenMP and the technique (57), the spent time was reduced up to 0.18 seconds.

7. Conclusion

A review of previous achievements regarding the analysis of single and dual reflector antennas by physical optics is presented. It includes the PO method as a simplification of the more general MFIE formulation. The discretization of the surfaces in triangles is addressed including a new physical interpretation of the meaning of the triangle integral. The method has been applied to single and dual reflector antennas. In the dual reflector case a new contribution has been reported to improve the method accuracy by using mutual coupling between reflectors from the MoM formulation.

Finally, some improvements in the calculation speed of the radiation pattern have been presented. The use of OpenMP has been shown to be of special interest to reduce the computation time.

Acknowledgment

This work was supported by the Spanish Government Grants TEC2011-28683-C02-02 and CONSOLIDER-INGENIO CSD2008-00068.

References

[1] V. Galindo-Israel and R. Mittra, "A new series representation for the radiation integral with application to reflector antennas,"

IEEE Transactions on Antennas and Propagation, vol. 25, no. 5, pp. 631–641, 1977.

[2] R. Mittra, Y. Rahmat-Samii, V. Galindo-Israel, and R. Norman, "An efficient technique for the computation of vector secondary patterns of offset reflectors," *IEEE Transactions on Antennas and Propagation*, vol. 27, no. 3, pp. 294–304, 1979.

[3] J. R. Parkinson and M. J. Mehler, "Convergence of physical optics integral by Ludwig's technique," *Electronics Letters*, vol. 22, no. 22, pp. 1161–1162, 1986.

[4] W. L. Stutzman, S. W. Gilmore, and S. H. Stewart, "Numerical evaluation of radiation integrals for reflector antenna analysis including a new measure of accuracy," *IEEE Transactions on Antennas and Propagation*, vol. 36, no. 7, pp. 1018–1023, 1988.

[5] Y. Rahmat-Samii, "A comparison between GO/aperture-field and physical optics methods for offset reflector," *IEEE Transactions on Antennas and Propagation*, vol. 32, no. 3, pp. 301–306, 1984.

[6] N. C. Albertsen and K. Pontoppidan, "Analysis of subreflectors for dual reflector antennas," *IEE Proceedings H*, vol. 131, no. 3, pp. 205–213, 1984.

[7] J. Chen and Y. Xu, "Analysis and calculation of radiation patterns of Cassegrain antennas," *IEEE Transactions on Antennas and Propagation*, vol. 38, no. 6, pp. 823–830, 1990.

[8] P. S. Kildal, "Analysis of numerically specified multireflector antennas by kinematic and dynamic ray tracing," *IEEE Transactions on Antennas and Propagation*, vol. 38, no. 10, pp. 1600–1606, 1990.

[9] A. M. Arias-Acuña, J. O. Rubiños-López, I. Cuiñas-Gómez, and A. García-Pino, "Electromagnetic scattering of reflector antennas by fast physical optics algorithms," in *Recent Res. Devel. Magnetics*, vol. 1, pp. 43–63, 2000.

[10] F. Obelleiro, J. M. Taboada, J. L. Rodríguez, J. O. Rubiños, and A. M. Arias, "Hybrid moment-method physical-optics formulation for modeling the electromagnetic behavior of on-board antennas," *Microwave and Optical Technology Letters*, vol. 27, no. 2, pp. 88–93, 2000.

[11] M. Graña, M. Arias, O. Rubiños, A. García-Pino, and J. A. Martínez-Lorenzo, "Iterative physical optics solution for MFIE on dual reflector antennas," in *Electromagnetic Theory Symposium (EMTS '07)*, pp. 1–3, Ottawa, Canada, 2007.

[12] M. Graña, M. Arias, O. Rubiños, and A. García-Pino, "Asymptotic MFIE for optimizing dual reflectors pattern," in *Proceedings of the 2nd European Conference on Antennas and Propagation (EuCAP '07)*, pp. 1–5, Edinburgh, UK, 2007.

[13] M. Graña-Varela, M. Arias, O. Rubiños, and A. García-Pino, "Rapid dual reflector shaping using ant colony optimization, fast iterated PO and asymptotic MFIE," in *Proceedings of the 3rd European Conference on Antennas and Propagation (EuCAP '09)*, pp. 2731–2735, Berlin, Germany, March 2009.

[14] M. Graña-Varela, M. Arias-Acuña, O. Rubiños, and A. García-Pino, "Shaped solid contour beam reflector antennas that adjust to varying operating conditions with just a few mechanical actuators," *IEEE Antennas and Wireless Propagation Letters*, vol. 8, pp. 839–842, 2009.

[15] http://openmp.org/wp/.

[16] H. Gómez-Sousa, J. A. Martínez-Lorenzo, O. Rubiños-López et al., "Strategies for improving the use of the memory hierarchy in an implementation of the modified current approximation (MECA) method," *ACES*, vol. 25, no. 10, pp. 841–852, 2010.

[17] I. Cuiñas, A. M. Arias, A. G. Pino, and A. Ramos, "Educational software for single- and dual-reflector antennas," *Computer Applications in Engineering Education*, vol. 7, no. 1, pp. 23–29, 1999.

[18] J. A. Martínez Lorenzo, A. G. Pino, I. Vega, M. Arias, and O. Rubiños, "ICARA: induced-current analysis of reflector antennas," *IEEE Antennas and Propagation Magazine*, vol. 47, no. 2, pp. 92–100, 2005.

[19] N. Morita, N. Kumagai, and J. R. Mautz, *Integral Equation Methods for Electromagnetics*, Artech House, Norwood, Mass, USA, 1987.

[20] R. Harrington, "Boundary integral formulations for homogeneous material bodies," *Journal of Electromagnetic Waves and Applications*, vol. 3, pp. 1–15, 1989.

[21] C. A. Balanis, *Engineering Electromagnetics*, John Wiley & Sons, New York, NY, USA, 1st edition, 1989.

[22] Y. M. Wu, L. J. Jiang, and W. C. Chew, "An efficient method for computing highly oscillatory physical optics integral," *Progress in Electromagnetic Research*, vol. 27, pp. 211–257, 2012.

[23] http://www.netlib.org/blas/.

3

Thermomechanical Impact of Polyurethane Potting on Gun Launched Electronics

A. S. Haynes, J. A. Cordes, and J. Krug

US Army RDECOM ARDEC, Picatinny Arsenal Building 94, Morris County, NJ 07806, USA

Correspondence should be addressed to A. S. Haynes, aisha.s.haynes.civ@mail.mil

Academic Editor: Sadhan C. Jana

Electronics packages in precision guided munitions are used in guidance and control units, mission computers, and fuze-safe-and-arm devices. They are subjected to high g-loads during gun launch, pyrotechnic shocks during flight, and high g-loads upon impact with hard targets. To enhance survivability, many electronics packages are potted after assembly. The purpose of the potting is to provide additional structural support and shock damping. Researchers at the US Army recently completed a series of dynamic mechanical tests on a urethane-based potting material to assess its behavior in an electronics assembly during gun launch and under varying thermal launch conditions. This paper will discuss the thermomechanical properties of the potting material as well as simulation efforts to determine the suitability of this potting compound for gun launched electronics. Simulation results will compare stresses and displacements for a simplified electronics package with and without full potting. An evaluation of the advantages and consequences of potting electronics in munitions systems will also be discussed.

1. Introduction

The weapons community currently has a strong need to design and field reliable electronics for precision munitions. These electronic systems and subsystems are used in guidance and control units, mission computers, sensors, and fuze-safe-and-arm devices. While modern technology has evolved to produce high-precision and highly reliable electronic systems, they are not constructed to survive the harsh environments of gun launch. The current issues faced are high vibratory loads, high-g-accelerations, pyrotechnic shocks during flight, and high-impact loads through hard targets. During each phase, these electronics are expected to retain full functionality. In an effort to mitigate the severity of these loads, electronic components are often potted to enhance structural support, dampen large dynamic vibrations due to gun launch [1]. Electronic components are also potted in order to protect sensitive equipment from environmental conditions (such as moisture), as well as to insulate electrical leads in the event that other components fail [2].

Potting of electronics has become one of the most viable and cost-effective solutions to enhance electronic package survivability. It is important that engineers understand the severity of gun launch and the harsh thermal loads to which the potting material and the potted assemblies are subjected to [3]. To improve tolerance to high-shock and vibratory loads, viscoelastic materials are traditionally employed as potting compounds. This is because they possess both the ability to store and dissipate energy. The relative contribution of each component dictates the material's effectiveness as a shock dampener [4].

Projectile live-fire testing is customarily used to build reliability and optimize design. However, this testing is costly and, in many cases, impractical for defining gun-launch-induced stresses on many organic materials used in subassemblies. For this reason, structural analysts employ finite element analyses (FEAs) to predict the impact of gun launch on these subassemblies and components. Dynamic FEA helps to build reliability and provides significant insight into systems survivability for munitions designers. Moreover, it saves programs millions of dollars in testing, as multiple designs and materials can then be simulated in a matter of days and for a fraction of the cost of live-fire testing. By coupling this modeling effort with testing throughout the life of a program, both time and money can be saved in the long run.

The fidelity of a simulation is related to the fidelity of the material models that are employed. The constitutive models are based on mechanical test data, which reflect the material behavior under varying load profiles. In addition to mechanical testing, thermal characterizations of materials are also required, since projectile flight time contains thermal loading on internal and external components.

Most studies on the use of FEA to model-potted electronics focus on the thermomechanical impact of potting materials on the integrated circuits and electronic components [3, 5, 6]. It has been well documented that under high thermal loads, the potting material imparts large stresses on the electronic components. These stresses derive from coefficient of thermal expansion (CTE) mismatches between the potting material and the electronic components, as well as from variations in the potting material stiffness as operating temperatures traverse the glass transition temperature (T_g).

This paper will detail the development of finite element constitutive material models, the dynamic structural behavior of the potted materials, as well as the impact of gun launch on a potted electronic assembly. A previous study compared potted to unpotted electronics at ambient temperatures assuming that the potting did not stick to key components [7]. In the previous study, the stresses in the chips were higher for a ringed model (unpotted-ring represent elastomeric type o-ring components) compared to a fully potted model. For both models, stresses were within acceptable bounds for ambient temperature. In this study, similar comparisons are made at temperature extremes to assess differences with the potting in the glassy and rubbery phase. This study also assumes that the potting adheres to the chips, the board, and the can.

2. Potting Testing and Modelling

2.1. Material Testing. A polyurethane-based potting with a shore D hardness of 45 was acquired from Alchemie Ltd (trade name Alchemix). Due to the nature of the application, the formulation is proprietary. Dynamic Mechanical Analyses (DMAs) operated in 3-point-bend mode. Uniaxial tension and uniaxial compression testing were performed by the US Army ARDEC. The DMA samples were subjected to multiple frequency oscillations under varying temperatures from $-101°C$ to $71°C$. The uniaxial compression and tension tests were performed on ASTM D638 type 1 tensile bars and ASTM D695 compression prisms. These samples were conditioned and tested at $-51°C$, ambient, and $71°C$.

2.2. Mechanical Behavior of Filled Potting. The uniaxial data collected suggests that within the strain range analyzed there is significant change in mechanical behavior as temperature varies from $-51°C$ to $71°C$. This is a consequence of testing through the materials glass transition ($-10°C$). As with most polymers the material exhibits glassy behavior at low temperatures below its glass transition and rubbery behavior at high temperatures above its glass transition. The strength data acquired demonstrates this: at high temperatures the material strength drops to 30% of its strength at ambient

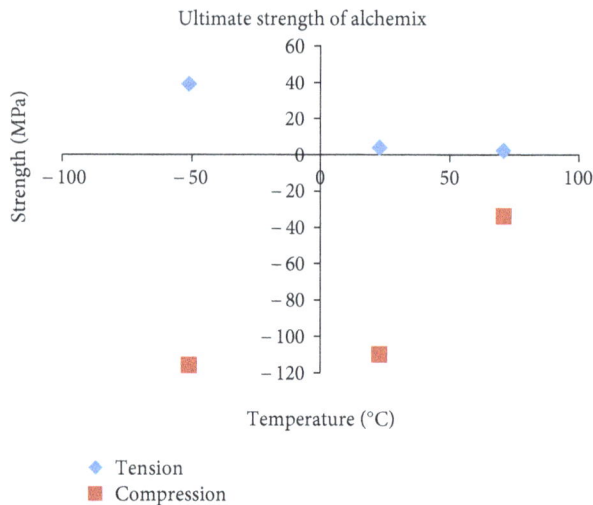

Figure 1: Ultimate strength of Alchemix in tension and compression.

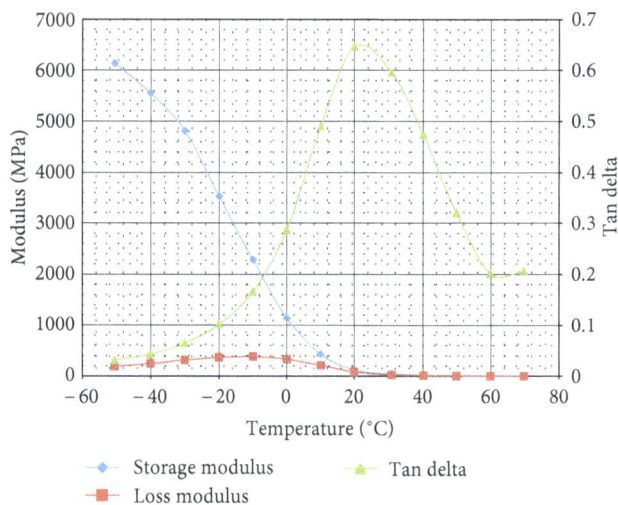

Figure 2: 3-point-bend DMA.

under compressive loading and down to 60% of its ambient strength under tensile loading. At low temperatures the strength of the material increases by 5% of its ambient strength under compressive loading and 90% of its ambient strength under tensile loading. Figure 1 displays the strength data for Alchemix in tension and compression at varying temperatures.

DMA of the potting material also reveals a significant change in mechanical behavior with temperature. Figure 2 shows that the storage modulus decreases significantly as temperature increases from $-101°C$ to $71°C$. Again, this is due to the material's phase transition from glassy to rubbery. Below the glass transition temperature molecular mobility is limited yielding an increase in material stiffness and stored energy. Close to and above the glass transition temperature polymer chains are flexible and mobile [4].

Also observed from DMA testing is that the peak of the tan delta curve occurs at $20°C$. Tan delta is a measure of how the material performs as a shock dampener. The higher the tan delta, the more effective the material is as a dampener. From this data the material will perform best at 20°C. The loss modulus is a depiction of the material's viscous behavior and ability to dissipate energy. The amplitude of the curve provides information on molecular mobility. The peak of the curve represents the temperature at which there is increased chain movement. This typically signals a phase transition in the material. The peak of the loss modulus is chosen to represent the glass transition.

2.3. Potting Constitutive Models. Linear elastic and viscoelastic models were parameterized using the mechanical test data acquired. The linear viscoelasticity model is chosen because the potting is considered confined by the electronics housing and thus will be subjected to finite strains. If the assembly had not been confined, a nonlinear elastic/viscoelastic material model would be more appropriate. Although the material is confined, the gun launch load applied in the finite element analysis (FEA) occurs within 15 ms imparting high-frequency forces on the potting material. During this dynamic event, the change in frequency will impact the molecular mobility of polymer chains and overall viscosity of material. For most materials, this essentially leads to a stiffer material at high frequencies. The viscoelastic constitutive model is expected to capture the dynamic changes in overall structural behavior. The input for time domain viscoelasticity is the normalized shear modulus data as a function of frequency which is directly inputted in the property module in tabular form. The shear modulus is determined by (1):

$$G = \frac{E}{2 \cdot (1 + \mu)},\qquad(1)$$

where μ is Poisson's ratio and E is the elastic modulus (measured using DMA). Tables 1, 2, and 3 display the data inputted into ABAQUS for calibration at cold (−40°C), ambient, and hot (60°C) temperatures, respectively.

ωg^*real, ωg^*imaginary, ωk^*real, and ωk^*imaginary represent the frequency-dependant real and imaginary components for the shear and bulk moduli. "ω" (omega) represents the circular frequency.

For this analysis, the potting material is assumed to be isotropic and incompressible ($\nu = 0.5$). Under this assumption, bulk modulus approaches infinite and ABAQUS considers it a negligible quantity. Another assumption is that the calculated shear modulus at 1 Hz represents the long-term shear modulus (G_∞). With this assumption the parameters ωk^*real, ωg^*imaginary are inputted into ABAQUS using ωg^*real $= Gl/G\infty$ and ωg^*imaginary $= 1 − G_s/G_\infty$ [8]. G_l and G_s represent the loss and storage shear moduli, respectively. The calculated parameters are used by ABAQUS to relate the shear and loss moduli to Prony series functions employed to evaluate the time-dependant volumetric and shear behavior of the potting material [8].

3. Finite Element Modeling and Simulation

3.1. General Finite Element Method. The purpose of this modeling and simulation effort is to (1) identify the impact of potting on an electronics assembly during gun launch, (2) assess differences in stresses and displacements at temperature extremes, and (3) determine if a ringed support system is adequate thereby limiting some of the deleterious effects resulting from temperature changes in fully potted electronics.

The modeling and simulation were completed using the general purpose finite element package ABAQUS, version 6.10.1 [8]. The following summarizes the analysis:

(i) general Purpose Finite Element Software: Abaqus Explicit 6.10.1;

(ii) analysis: dynamic, nonlinear materials, nonlinear geometry;

(iii) parts: All parts are representative, not actual;

(iv) elements: 8-node brick elements, reduced integration, hourglass control. The fully potted model also used 10-node tetrahedral elements;

(v) materials: elastic and elastic/plastic and visco/elastic;

(vi) friction: frictionless, all contact surfaces;

(vii) damping: no applied damping currently in the model except for what's imparted by the viscoelastic constitutive model;

(viii) initial conditions: no initial displacement or velocity; initial temperature: 60°C and −40°C, the extremes of the operating range. (The storage temperature range is larger.)

3.2. Internal Constraints, Assumptions. The following items were tied: board to solder, solder to chips, can bottom to can top, and potting parts to one another. In the potted model, the potting was tied to the chips, the board, and the can. In the ringed model, the potting was tied to the boards and can. All other contacts were applied as frictionless contact.

3.3. Applied Load and Output. Measured acceleration data was applied to the bottom of the can as a three-dimensional acceleration, Figure 3. The acceleration sequence was recorded on a live-fire shot of a precision projectile near where a critical electronics package is located. The on-board recorder assembly was design to be a rough match of the electronics package mass and stiffness. Axial G-forces in the range of 15-kgs and transverse force in the range of 3-kgs are expected, applied with different frequency content [9]. Components are typically designed for the maximum G-force which occurs in the gun tube at around 0.0038 sec. Carlucci et al. observed, however, that many electronics components fail during the muzzle exit event, around 0.011 sec in Figure 3 [10]. Contour plots of stress and strain were recorded at 98 increments of the load from Figure 3. A Python script and Fortran program were used to determine the maximum stress and the time at which it occurred for each of the chips in the model.

3.4. Potted versus Ringed Geometries. Two geometries were compared. First, a fully potted assembly was evaluated,

TABLE 1: Input for viscoelastic prony parameter calibration, cold.

ωg^* real	ωg^* imaginary	ωk^* real	ωk^* imaginary	Frequency (Hz)
0.0303	0.00000	0.00000	0.00000	1.0
0.03271	−0.007392	0.00000	0.00000	1.5
0.02989	−0.01671	0.00000	0.00000	2.5
0.02855	−0.02411	0.00000	0.00000	3.9
0.02761	−0.03325	0.00000	0.00000	6.3
0.02535	−0.04129	0.00000	0.00000	10.0
0.0259	−0.05033	0.00000	0.00000	15.8
0.01834	−0.05893	0.00000	0.00000	25.1
0.02134	−0.07476	0.00000	0.00000	39.8
0.009863	−0.08714	0.00000	0.00000	63.0
0.01946	−0.1178	0.0000	0.00000	100

TABLE 2: Input for viscoelastic prony parameter calibration, ambient.

ωg^* real	ωg^* imaginary	ωk^* real	ωk^* imaginary	Frequency (Hz)
0.64605	0.00000	0.00000	0.00000	1.0
0.74276	−0.15751	0.00000	0.00000	1.5
0.86017	−0.35639	0.00000	0.00000	2.5
0.98753	−0.58518	0.00000	0.00000	3.9
1.12523	−0.84795	0.00000	0.00000	6.3
1.26766	−1.14182	0.00000	0.00000	10.0
1.41668	−1.47368	0.00000	0.00000	15.8
1.56600	−1.84187	0.00000	0.00000	25.1
1.72882	−2.30529	0.00000	0.00000	39.8
1.89617	−2.82774	0.00000	0.00000	63.0
2.06754	−3.41998	0.00000	0.00000	100.0

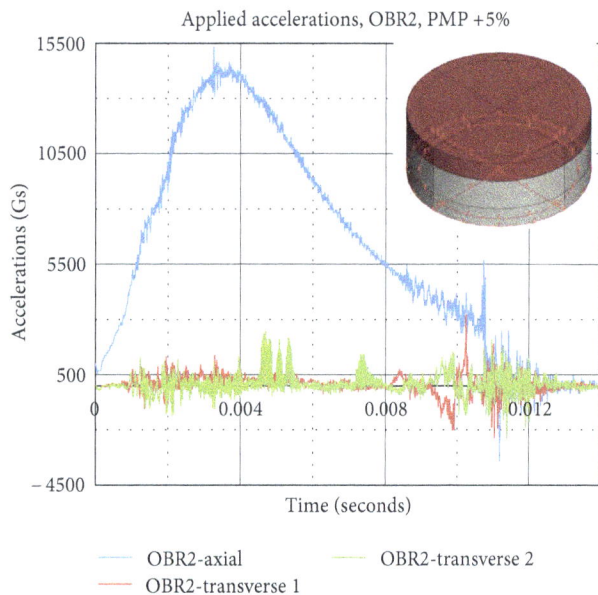

FIGURE 3: Example gun launch G-forces, 3-dimensional.

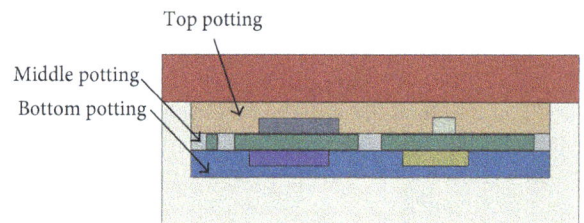

FIGURE 4: Geometry of potted model.

(0.005 inch) thick solder loads under the chips. The can, chips, board, and solder were the same for both geometries. Second, the same potting material was modeled as rings that gripped the outside of the circuit board. In the second analysis, the chips and most of the board were free from contact with potting. Central grommets were employed to reduce the relative deflections across the chips. Figure 5 shows the geometry with ring and grommet supports. Figure 6 shows the solder geometry for the top and bottom of the board.

Figure 4. The model includes an aluminum supporting can, an aluminum lid, a circuit board, 8 chips with solder, and 3 tied-together sections of potting. The model has 0.127 mm

3.5. *Material Assumptions.* Figure 7 shows the material assumptions. Table 4 displays the material properties. The electronic chips on the bottom were modeled as ceramics.

TABLE 3: Input for viscoelastic prony parameter calibration, hot.

ωg^* real	ωg^* imaginary	ωk^* real	ωk^* imaginary	Frequency (Hz)
0.2061	0.00000	0.00000	0.00000	1.0
0.1274	−0.02513	0.00000	0.00000	1.5
0.1464	−0.04945	0.00000	0.00000	2.5
0.1723	−0.07912	0.00000	0.00000	3.9
0.2074	−0.1149	0.00000	0.00000	6.3
0.252	−0.1598	0.00000	0.00000	10.0
0.3111	−0.2163	0.00000	0.00000	15.8
0.3957	−0.2878	0.00000	0.00000	25.1
0.4926	−0.3817	0.00000	0.00000	39.8
0.6204	−0.4866	0.00000	0.00000	63.0
0.8108	−0.5726	0.00000	0.00000	100.0

TABLE 4: Material assumptions.

Part	Material	Young's modulus MPa	Poisson's ratio	Mass density Kg/m^3	Ref. temperature °C	Material model
Solder	Solder-sn60-pb40	$3E + 05$	0.4	8598	All	Linear elastic plastic
Chips	Ceramic_X7R	$1.05E + 05$	0.3	5624	All	Linear elastic plastic
Can	Aluminum	$7E + -4$	0.3	2716	All	Linear elastic plastic
Potting	Alchemix	6131	0.49	1604	−51	Linear elastic/viscoelastic
Potting	Alchemix	120	0.49	1604	23	Linear elastic/viscoelastic
Potting	Alchemix	7.2	0.49	1604	71	Linear elastic/viscoelastic
Connectors	Plastic	$3.6E + 04$	0.23	4235	All	Linear elastic plastic
Board	FR4	$2.5E + 04$	0.15	1925	All	Orthotropic elasticity

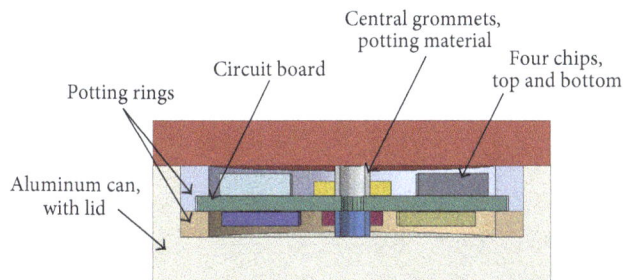

FIGURE 5: Geometry with potting rings supporting the circuit board.

As shown in Table 4, the Young's modulus for the potting varies by orders of magnitude over the temperature range. Analysis at the low and high ends of the temperature allows for assessing potting effects at different stiffness.

4. Finite Element Comparisons

4.1. Modeling and Simulation, Cold Temperatures (−40°C). Figure 8 shows the stresses in the chips at time = 0.004 second at cold temperature. The highest stress occurs at setback, roughly coincident with the largest axial acceleration. The stresses in the unpotted design were about 30% higher in the chips compared to the potted design. Figure 9 shows the displacements at setback. The relative deflection of the center of the board in the ring model is evident. Figure 10 shows the stresses at muzzle exit, around 0.0113 second. At muzzle exit, the boards tend to flex in the direction opposite setback. Moreover, the stresses in the fully potted design are observed to be higher than in the ring-only design. For both models, stresses were within the estimated strength of the chips (76 MPa [11]). In both cases the solder yielded during gun launch.

Table 5 compares the maximum stresses and maximum relative displacement in the chips over the time span at cold for the ringed geometry. In Table 5, the second column is the maximum equivalent Von Mises stress and column 3 is the time at which it occurs. Some of the chips reached a maximum stress in the gun around set back (around 0.0038 sec) and others during muzzle exit (around 0.011 sec). The Army uses 76-MPa as the allowable stress threshold for electronic components which corresponds to about a 10% chance of failure according to [11]. The relative displacement in chips is also noted, column 6 in Table 5. The Army considers maximum relative displacement in the same chip less than 0.10 mm as surviving. At cold, all chips in the ringed model are predicted to survive based on the Army criteria. The calculations were repeated at ambient and hot conditions.

Table 6 shows summary results for the fully potted model at cold temperatures. The allowable stress was slightly exceeded by Bottom-Chip-3. All maximum values for stress occurred at muzzle exit. Relative displacements, column 6, were well below the ringed geometry model.

TABLE 5: Results, ringed model at cold, maximum stresses.

Chips	Svm MPa	Time Sec	S1 MPa	Time Sec	Delta disp mm	Time Sec
Top-Chip-1	46.1	0.0026	24.2	0.0048	0.069	0.0113
Top-Chip-2	61.7	0.0111	18.9	0.0116	0.078	0.0042
Top-Chip-3	61.7	0.0116	14.6	0.0112	0.069	0.004
Top-Chip-4	22.7	0.0116	20.3	0.0116	0.024	0.0125
Bottom-Chip-1	40.7	0.0032	24.9	0.0032	0.065	0.0113
Bottom-Chip-2	38.2	0.004	28.5	0.004	0.065	0.004
Bottom-Chip-3	12.2	0.0032	10.0	0.0038	0.041	0.0129
Bottom-Chip-4	47.5	0.0032	36.5	0.0032	0.061	0.0042

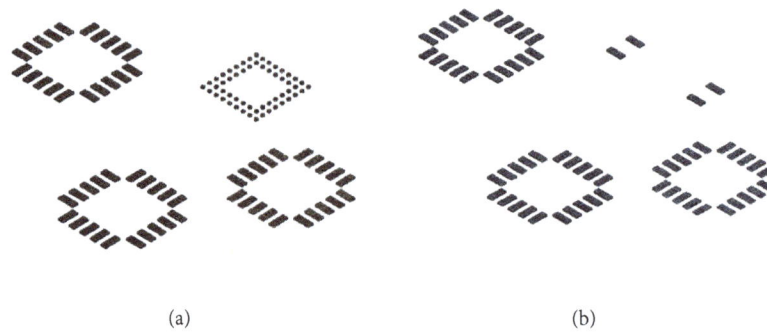

(a)   (b)

FIGURE 6: Solder configurations—Top (a) and Bottom Chips (b).

FIGURE 7: Material assumptions.

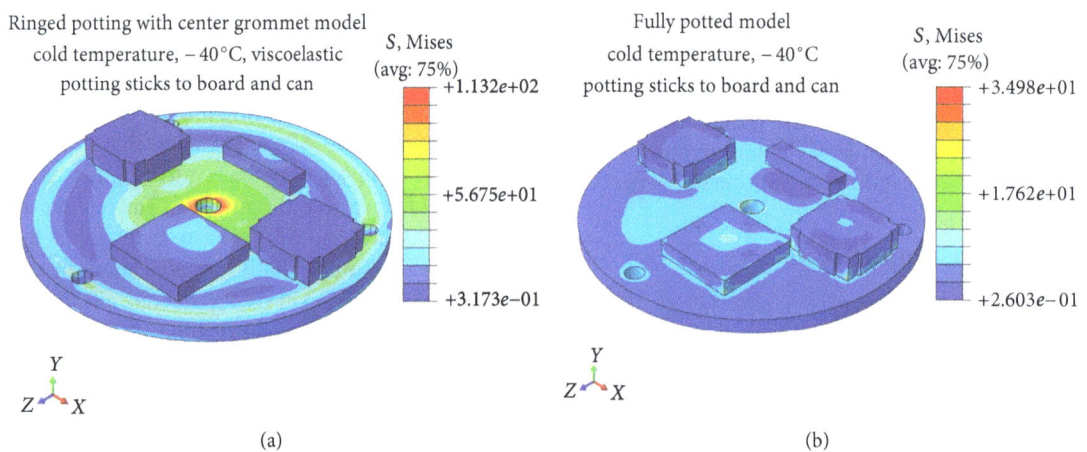

(a)   (b)

FIGURE 8: Comparison of stresses in chips without potting (a) and with potting (b) at cold temperature at set back.

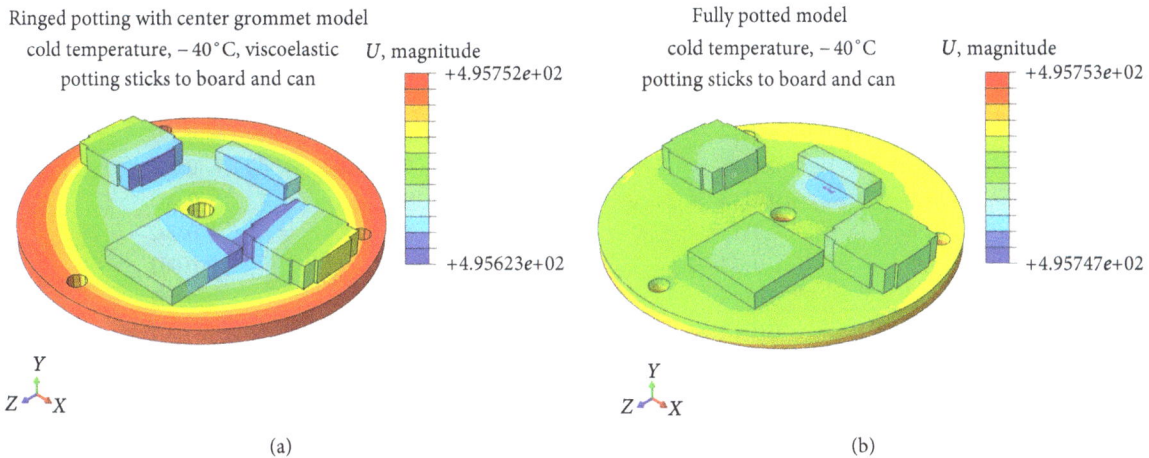

FIGURE 9: Comparison of displacement in chips without potting (a) and with potting (b) at cold temperature at setback.

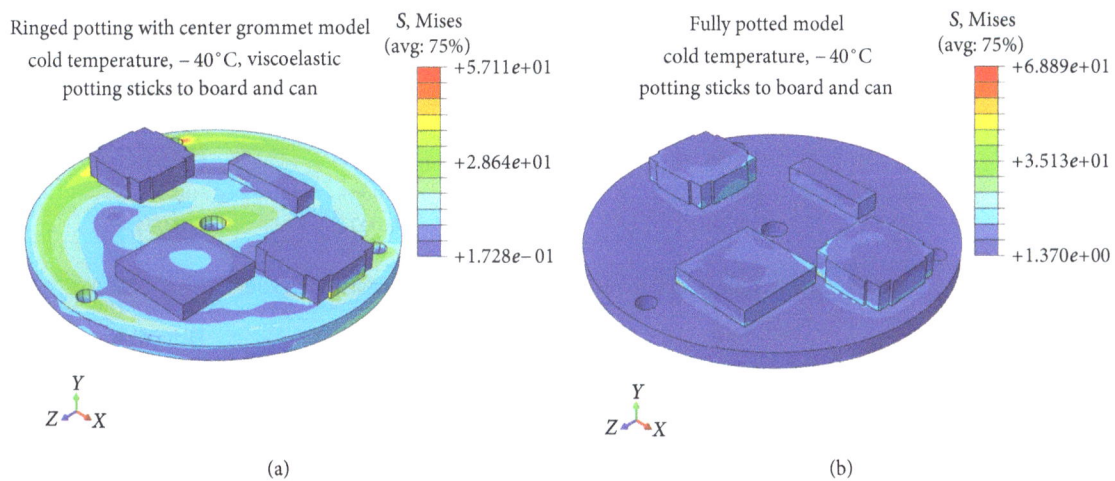

FIGURE 10: Comparison of stresses in chips without potting (a) and with potting (b) at cold temperature at muzzle exit.

The comparison between ringed potting and fully potting was repeated allowing the potting to be nonadhering in the fully potted model. In this case, Figure 11, the stresses in the fully potted model dropped below the ringed model when the potting does not adhere to chips. Stresses in nonadhering potting were lower than the sticky potting.

4.2. Modeling and Simulation, Hot Temperatures. Figure 12 compares stresses at hot conditions between the ringed and fully potted model at set back. For the ringed model, stress distributions were different and maxima were lower at hot temperatures. For the fully potted model, stresses were higher at high temperatures, and some chips exceeded allowable values.

Table 7 shows the stress in the rings-only model at high temperature. Comparing to the ringed analysis at cold in Table 5, stresses in the 1/2 of the top chips were higher than for the cold material data. Moreover, the chips on the bottom of the board are observed to have a slightly higher maximum stress under the high-temperature conditions. In the hot

model, most of the chips reached a maximum at set back. In the cold model, the bottom chips also reached maximum stress at set back. In the hot model, the relative displacement of the chips and the stresses were within acceptable values.

Table 8 shows results for the fully potted model at high temperatures. The stresses at the bottom of the board are roughly twice as high than for the ringed model. As with the ringed model, the stresses on the bottom of the board were higher than for the fully potted cold model.

5. Discussion and Comparisons

5.1. Errors and Omissions. The geometry in the finite element analysis was much simpler than actual electronics. It should be emphasized that the simplified models were used for comparisons to assess the effects of potting.

Solder configurations were simplified since (1) the study compares stresses in the chips, and (2) the solder serves as a nonlinear spring. If stress and fatigue within the solder joints are required, detailed solder models such as presented

TABLE 6: Results, fully potted model at cold, sticky potting, maximum stresses.

Chips	Svm MPa	Time Sec	S1 MPa	Time Sec	Delta disp mm	Time Sec
Top-Chip-1	26.3	0.0117	17.8	0.0113	0.002	0.0038
Top-Chip-2	33.8	0.0117	21.0	0.0113	0.002	0.0118
Top-Chip-3	25.0	0.0113	17.2	0.0118	0.003	0.0113
Top-Chip-4	16.9	0.0113	9.6	0.0115	0.002	0.0115
Bottom-Chip-1	61.8	0.0113	39.6	0.0115	0.002	0.0118
Bottom-Chip-2	61.7	0.0113	33.9	0.0115	0.002	0.0113
Bottom-Chip-3	77.6	0.0113	48.3	0.0118	0.002	0.0038
Bottom-Chip-4	60.3	0.0113	39.4	0.0112	0.002	0.0136

TABLE 7: Results, ringed model at hot, maximum stresses.

Chips	Svm MPa	Time Sec	S1 MPa	Time Sec	Delta disp mm	Time Sec
Top-Chip-1	51.0	0.0036	27.4	0.0036	0.046	0.0123
Top-Chip-2	44.7	0.0028	14.7	0.0132	0.096	0.0034
Top-Chip-3	41.5	0.0112	12.4	0.012	0.088	0.0038
Top-Chip-4	24.5	0.0034	16.1	0.0032	0.026	0.0038
Bottom-Chip-1	43.1	0.004	25.9	0.0048	0.056	0.0034
Bottom-Chip-2	41.6	0.0038	30.1	0.0038	0.088	0.0038
Bottom-Chip-3	12.2	0.0021	9.0	0.0034	0.043	0.0108
Bottom-Chip-4	54.1	0.0034	40.4	0.0034	0.074	0.0034

TABLE 8: Results, fully potted model at hot, sticky potting, maximum stresses.

Part	Svm MPa	Time Sec	S1 MPa	Time Sec	Delta disp mm	Time Sec
Top-Chip-1	16.4	0.0142	8.4	0.0124	0.024	0.0116
Top-Chip-2	24.3	0.0121	15.9	0.0122	0.021	0.0119
Top-Chip-3	38.0	0.012	23.7	0.012	0.019	0.0117
Top-Chip-4	22.1	0.0034	12.1	0.0112	0.015	0.0127
Bottom-Chip-1	107.5	0.0119	41.5	0.0124	0.022	0.0119
Bottom-Chip-2	82.6	0.0116	40.9	0.0127	0.020	0.0117
Bottom-Chip-3	80.6	0.0119	55.4	0.0121	0.017	0.0127
Bottom-Chip-4	77.7	0.0128	34.5	0.0128	0.021	0.0116

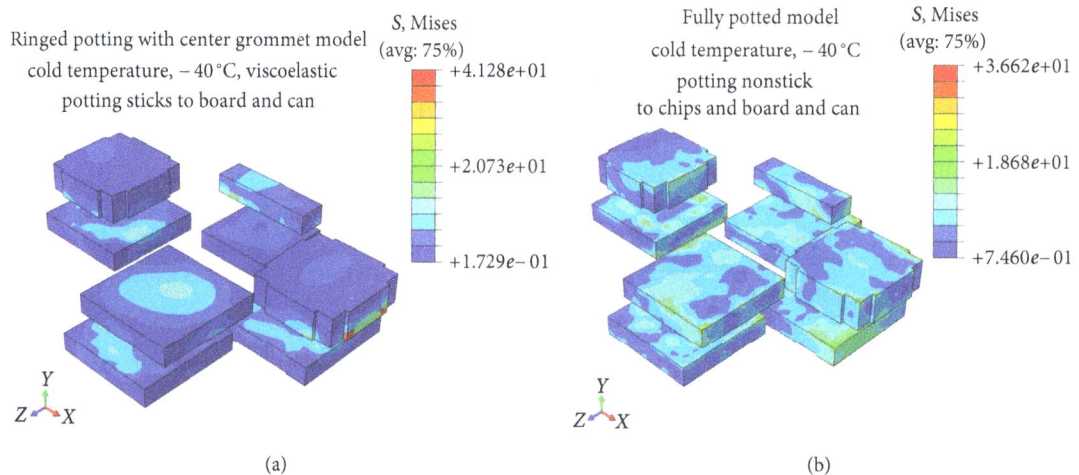

FIGURE 11: Comparison of stresses in chips with potting (a) and without potting (b) at cold temperature at set back, non-adhering potting.

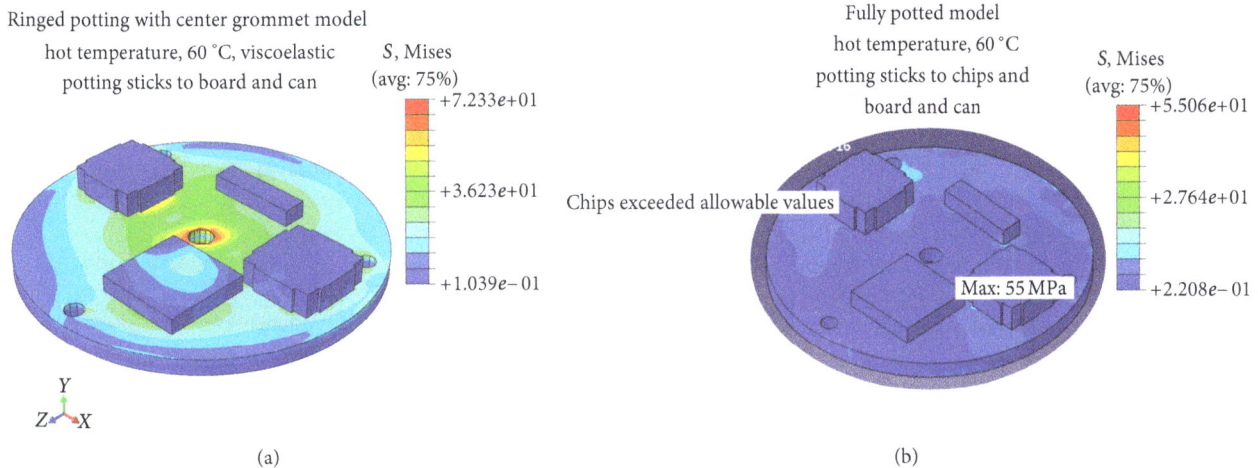

FIGURE 12: Comparison of stresses in chips with potting (a) and without potting (b) at hot temperature at set back, sticky potting.

by authors in [12] may be employed. Of course, high fidelity solder models would greatly increase the degrees of freedom of the finite element model. For these relative studies on chip behavior, detailed solder modeling was not required.

Like the solder, the boards and chips have been greatly simplified. The Army does model details of chips when failures occur. For this study, relative stresses and displacements were required for comparisons.

One drawback of this study is that the viscoelastic model employed is limited to a frequency range that falls below the vibratory frequencies the potted assembly is subjected to during gun launch (10–20 KHz). Further studies into the high-frequency material behavior are underway to determine the impact on the material behavior and structural stability of the electronic assembly. High-frequency viscoelastic properties should be incorporated into future models.

5.2. Estimating Failure in Chips and Electronic Components. It is recognized that the failure conditions for chips vary from chip to chip. Blattau et al. [11] described variations in the failure stress of capacitors with different materials for solder support. The reference also provides probability of failure as a function of calculated stress in chips. Failure is based on the stress required to cause cracking in capacitors, a well documented failure in capacitors [11, 13, 14]. The 76-MPa strength criterion is used by the Army as a rough guide for failure strength and should not be considered a hard failure condition. Modeling and simulation have predicted stresses as high as 101-MPa in chips that do not fail.

The relative deflection as another failure point is also a rough estimate. Flex failures have also been documented, by Prymak et al. [15]. The Army's use of 0.1 mm as a failure criterion on relative displacement within chips has worked well and is comparable to flexing magnitudes in the Prymak report.

5.3. Comparisons Computing Times. The current work was completed on a super computer at Picatinny Arsenal. The computer has 84 Total Cores and is a Linux-Based OS system

with 84 Cores on 7 Nodes/12 Cores per Node/48 GB RAM per Node Intel Xeon 5650, 2.67 GHz CPU, 6 Cores per CPU.

Analysis with full potting, Figure 4, was completed using 12 parallel processors. The cold temperature case took approximately 57 hours with the following statistics:

(i) number of elements is 376143,

(ii) number of nodes is 765110;

(iii) total number of variables in the model is 2295330.

The hot case analysis ran with three times the processors and took longer due to the smaller time step required. Because of the complexity of the potting geometrically, some of the elements were small. The element size coupled with the density/stiffness ratio at hot required a small time step.

In contrast, the ringed potting model, Figure 5, was completed using 12 parallel processors. The cold temperature case took approximately 5.7 hours with the following degrees of freedom:

(i) number of elements is 89246;

(ii) number of nodes is 119596;

(iii) total number of variables in the model 358788.

The hot and ambient cases took similar time because the finite elements were of reasonably uniform size. All elements were 8-node brick elements. The enhanced hourglass controls limited convergence and numerical difficulties. If multiple load cases need to be run, a simpler ringed design is easier to evaluate.

6. Recommendations

6.1. Stresses and Displacements at Different Temperatures. This study was conducted under multiple thermal mechanical loading conditions to identify the structural impact of potted electronics for gun launch applications. The mechanical test data acquired suggests that the material undergoes very large changes in strength and stiffness in the temperature range studied. Due to these changes, the stresses imparted by the potted vary in the temperature range from $-40°C$ to $60°C$. For reliability, small changes in a proven design are desirable. If potting is used, it is strongly suggested that a material should be selected with the glass transition temperature higher than the electronics operating range.

6.2. Thermal Effects. Effect of coefficient of thermal expansion mismatches between potting and electronic components was not addressed in this study. The expansion coefficient for potting in this study varies by a factor of two over the temperature range. The specific heat varies by about 50%. Chao et al. have shown that the potting would be expected to have significant thermal mismatch issues based on the coefficients of thermal expansion for the potting and the ceramic chips [16]. For thermal expansion issues, ringed potting should be considered over fully potted models.

The operating and storage range for munitions is $-42°C$ to $63°C$ and $-51°C$ to $71°C$, respectively. It is important to note that while the potted electronics assembly may protect the electronic components from large vibratory and shock loads; the material chosen must retain stable mechanical properties through the operating and storage temperature regimes. Material testing reveals that the mechanical properties of the potting material employed for this analysis change significantly with temperature. Because of this and as supported by [3, 6, 7, 16–18] the potting material potentially imposes additional risks to the mechanical stability of the electronics assembly. The material evaluated in this study is not an ideal potting solution because of its thermomechanical issues.

6.3. Interface Effects. Studies on the impact of interaction properties between potting and the electronics housing, solder connections, and electronic chips showed that sticky potting (potting tied to all surfaces) results in higher stresses than contact-only potting (nonadhering). Sticky potting can also lead to stress risers between components and the board and possible breaking of leads/wires. For this reason, non-sticky potting should be considered if fully potted electronics are required.

6.4. Ringed Geometry as Opposed to Fully Potted Model. As a result of this study and due to the thermal effects described by Chao et al. [16], a ringed assembly or similar is recommended over a fully potted design. The ringed design with a center support resulted in stresses and deflections in the acceptable range over the full temperature range.

7. Conclusion

Summarizing, although the stresses and displacements occurring during gun launch are adequately supported by a material with the hot and cold stiffness of the potting material, this type of potting is not recommended for gun-fired electronics. This is primarily because the glass transition temperature falls within the operating range and the material's structural integrity degrades with increasing temperature.

Further studies on the impact of high glass ($>71°C$) transition and low glass transition ($< -51°C$) materials are underway.

A comparison of a fully potted assembly to a rings-only-supported assembly indicated that full potting may not be required when adequate support is otherwise available. As discussed, ringed supports may add several other reliability benefits that relate to thermal, storage, and adhesion effects.

References

[1] M. Berman, "Electronic components for high-g hardened packaging," Tech. Rep. ARL-TR-3705, Army Research Laboratory, 2006.

[2] R. E. Keith, "Potting electronic modules," NASA Technical Report SP-5077, 1969.

[3] N.-H. Chao, J. A. Cordes, D. E. Carlucci et al., "The use of potting materials for electronic-packaging survivability in smart munitions," in *Proceedings of the International Mechanical Engineering Conference and Exposition*, 2010.

[4] A. R. Jeerfferie, M. Y. Yuhazri, O. Nooririnah et al., "Thermomechanical and morphological interelationship of polypropylene-mulitwalled carbon nanotubes (PP/MWCNTS) nanocomposites," *Internatioanl Journal of Basic and Applied Sciences*, vol. 10, no. 4, pp. 29–35, 2010.

[5] I. Baylakoglu, C. Hillman, and M. Pecht, "Characterization of some commercial thermally-cured potting materials," in *Proceedings of the International IEEE Conference on the Business of Electronic Product Reliability and Liability*, 2003.

[6] N. H. Chao, D. Carlucci, J. A. Cordes, M. E. DeAngelis, and J. Lee, "Implications of a fully-coupled thermal-stresses transient simulation in gun launch applications," Tech. Rep. ARMET-TR-10030, U.S. Army Armament Research Development and Engineering Center, 2010.

[7] A. S. Haynes and J. A. Cordes, "Characterization of a potting material for gun launch," in *Proceedings of the 26th International Symposium on Ballistics*, Miami, Fla, USA, September 2011.

[8] ABAQUS Users's Manual V. 6. 10. 1.

[9] J. A. Cordes, P. Vo, J. Lee, and D. Geissler, "Comparison of shock response spectrum for different gun tests," in *Proceedings of the 82nd Shock and Vibration Symposium (SAVIAC '11)*, Baltimore, Md, USA, October 2011.

[10] D. Carlucci, J. Cordes, S. Morris, and R. Gast, "Muzzle exit (set forward) effects on projectile dynamics," Tech. Rep. ARAET-TR-06003, U.S. Army Armament Research Development and Engineering Center, Dover, NJ, USA, 2006.

[11] N. Blattau, D. Barker, and C. Hillman, "Lead free solder and flex cracking failures in ceramic capacitors," in *Proceedings of the 24th Capacitor and Resistor Technology Symposium*, San Antonio, Tex, USA, March 2004.

[12] T. E. Wong, L. Suastegui, H. M. Cohen, and A. H. MAtsunaga, "Experimentally validated thermal fatigue life prediction model for leadleess chip carrier solder joint," in *Proceedings of the 10th Symposium on Mechanics of Surface Mount Assemblies*, ASME Winter Annual Meeting, Anaheim, Calif, USA, 1998.

[13] N. Blattau, P. Gormally, V. Iannaccone, L. Harvilchuck, and C. Hillman, "Robustness of surface mount multilayer ceramic capacitors assembled with pb-free solder," in *Proceedings of the CARTS Conference*, pp. 25–41, Orlando, FL, April 2006.

[14] G. F. Engel, G. Schlauer, V. Wischnat, and M. Pechloff, "Effective reduction of leakage failure mode after flex cracking events in X7R-type multilayer ceramic capacitors (MLCCS) by using

internal series connection," in *Proceedings of theCARTS Europe*, Bad Homburg, Germany, September 2006.

[15] J. Prymak, M. Prevallet, P. Blais, and B. Long, "New improvements in flex capabilities for MLC chip capacitors," in *Proceedings of the CARTS Conference*, pp. 251–255, Orlando, FL, April 2006.

[16] N. H. Chao, D. Carlucci, J. A. Cordes, M. E. DeAngelis, and J. Lee, "The use of potting materials for electronic-package survivability in smart munitions," *Journal of Electronic Packaging*, vol. 133, no. 4, Article ID 041003, 2011.

[17] J. A. Cordes, D. E. Carlucci, J. Kalinowski, and L. Reinhardt, "Design and development of reliability gun-fired structures," Tech. Rep. ARAET-TR-06009, U.S. Army Armament Research Development and Engineering Center, Dover, NJ, USA, 2006.

[18] P. Carlucci, J. A. Cordes, N. Payne, L. Reinhardt, and D. Troast, "Dynamic analysis of electronic components for precision munitions, a case study," Tech. Rep. ARMET-TR-09022, U.S. Army Armament Research, Development and Engineering Center, 2009.

4

Biosorption of Lead Ions from Aqueous Solution Using *Ficus benghalensis* L.

Venkateswara Rao Surisetty,[1] Janusz Kozinski,[1] and L. Rao Nageswara[2]

[1] *Faculty of Science & Engineering, York University, 4700 Keele Street, Toronto, ON, Canada M3J 1P3*
[2] *Department of Chemical Engineering, R.V.R & J.C College of Engineering, Guntur 522019, India*

Correspondence should be addressed to Venkateswara Rao Surisetty; vrsurisetty@gmail.com

Academic Editor: Dmitry Murzin

Ficus benghalensis L., a plant-based material leaf powder, is used as an adsorbent for the removal of lead ions from aqueous solution using the biosorption technique. The effects of process parameters such as contact time, adsorbent size and dosage, initial lead ion concentration, and pH of the aqueous solution on bio-sorption of lead by *Ficus benghalensis* L. were studied using batch process. The Langmuir isotherm was more suitable for biosorption followed by Freundlich and Temkin isotherms with a maximum adsorption capacity of 28.63 mg/g of lead ion on the biomass of *Ficus benghalensis* L. leaves.

1. Introduction

The removal of toxic metal ions and the recovery of valuable ions from wastewaters, soils, and waters are important in economic and environmental problems. Heavy metals and other metal ions exist as contaminants in aqueous waste streams of many industries, such as tanneries and mining. Some metals associated with these activities are Pb, Hg, Cr, and Cd. Toxic metals are released into the environment in a number of ways. Coal combustion, sewage wastewaters, automobile emissions, battery industry, mining activities, and the utilization of fossil fuels are just a few examples. Some of these metals accumulate in living organisms and cause various diseases and disorders.

Heavy metals such as lead, mercury, arsenic, copper, zinc, and cadmium are highly toxic when adsorbed into the body [1]. They can cause accumulative poisoning, cancer, brain damage, and so forth. Lead is a general metabolic poison and enzyme inhibitor. It can cause mental retardation and semipermanent brain damage in young children [2]. Lead has the ability to replace calcium in the bone to form sites for long-term release, hence, the imminent need to remove these toxic metals from waters and wastewaters. The permissible limit of lead in wastewater as set by the Environment protection Agency is 0.1 mg/L, whereas in drinking water it is 0.05 mg/L [3]. Among the various water-treatment techniques described, adsorption is generally preferred for the removal of heavy metal ions due to its high efficiency, easy handling, availability of different adsorbents, and cost effectiveness [4].

Earlier studies demonstrated that the living systems could often be unreliable because of many associated problems in maintaining active microbial populations under highly variable conditions of wastewaters [5]. Therefore, some types of biomass are often chosen to serve as biosorbents to remove heavy metals [5–8]. Biosorption, a term that describes the removal of heavy metals by passive binding to nonliving biomass from an aqueous solution, is considered as an alternative to conventional methods of metal recovery from solutions. Although freely suspended biomass may have better contact with adsorbate during biosorption process, the suspended biomass is normally not the practical form for direct use in the removal of heavy metals. Since cell immobilization can enhance its stability, mechanical strength, reusability, and the ease of treatment, the process has been well used to remove toxic heavy metals [5, 9–11].

The effectiveness of biosorption for the removal of heavy metals has been shown in a number of studies. However, only when the cost of the biosorption process can compete with the existing technologies will it be accepted commercially. Kuyucak [12] indicated that the cost of biomass production played an important role in determining the overall cost of a biosorption process. Therefore, low-cost biomass becomes a crucial factor when considering practical application of biosorption. Sari and Tuzen [13] performed biosorption characteristics of Cd(II) ions using the red alga (*Ceramium virgatum*) and found that the red alga is an effective and alternative biomass for the removal of Cd ions from aqueous solution. Anayurt et al. [14] performed equilibrium, thermodynamic, and kinetic studies on biosorption of Pb(II) and Cd(II) from aqueous solution using macrofungus (*Lactarius scrobiculatus*) biomass and concluded that the recovery of the metal ions from this biomass was found higher than 95% using 1 M HCl and 1 M HNO$_3$.

The present work investigated the potential use of untreated *Ficus benghalensis* L. biomass as metal sorbent for the removal of lead from aqueous solution. *Ficus benghalensis* L. (also known as Bodhi tree, peepul, and sacred fig) is native to southeast Asia, southwest China, India, and the Himalayan foothills. The tree is a large broadleaf evergreen tree with wide-spreading branching that grows to 60–100$'$ tall with trunk size as much as 9$'$ in diameter. This tree is semideciduous to deciduous in native monsoon climates. Tree seed (often deposited by birds) may germinate in upper tree crevices, producing dangling, nonparasitic, aerial roots that grow to the ground, root in the soil, and produce trunks. *Ficus benghalensis* L. was chosen as a biosorbent because of its availability if needed at a large scale and the relative lack of information about its sorption ability. Environmental parameters affecting the biosorption process such as pH, contact time, metal ion concentration, adsorbent concentration, and adsorbent size were evaluated. The equilibrium adsorption data were evaluated by Langmuir, Freundlich, and Temkin isotherm models. The kinetic data were correlated by first- and second-order kinetic models.

2. Materials and Methods

2.1. Preparation of Biosorbent. The *Ficus benghalensis* L. (Indian banyan tree) leaves were collected and washed with deionized water several times to remove dirt and they were dried. The dried leaves were powdered using a domestic grinder to the particle size of 75–212 μm and used as biosorbent without any pretreatment for lead adsorption.

2.2. Chemical. Pb(NO$_3$)$_2$, HCl, and NaOH were of AR grade and purchased from Merck (Mumbai, Maharashtra, India). Lead ions were prepared by dissolving their corresponding nitrate salt in distilled water. The pH of solutions was adjusted with 0.1 N HCl and NaOH.

2.3. Characterization of Biosorbent. The elemental analysis of the biosorbent was performed using PerkinElmer ELAN 5000 inductively coupled plasma-mass spectroscopy (ICP-MS) instrument. Surface area, pore volume, and average pore diameter of powdered biomass were measured using an

ASAP 2000 Micromeritics system. The sample was degassed for 2 h under vacuum, and the BET area, pore volume, and pore diameter were determined.

2.4. Biosorption Experiments. Biosorption experiments were performed at room temperature (30 ± 1°C) in a rotary shaker at 180 rpm containing 30 mL of different lead concentrations using 250 mL Erlenmeyer flasks. After one hour of contact (according to the preliminary sorption dynamics tests), with 0.1 g biomass of *Ficus benghalensis* L. leaves, equilibrium was reached and the reaction mixture was centrifuged for 5 min. The metal content in the supernatant was determined using Atomic Absorption Spectrophotometer (AAS) (GBC Avanta Ver 1.32, Australia) after filtering the adsorbent with 0.45 μm filter paper. The AAS was operated using air-acetylene flame at a wave length of 283.5 nm with slit width of 0.5 nm and cathode lamp current of 5 mA to determine the absorption of Pb metals. The amount of metal adsorbed by *Ficus benghalensis* L. leaves was calculated from the differences between metal quantity added to the biomass and metal content of the supernatant using the following equation:

$$q = \left(C_0 - C_f\right)\frac{V}{M}, \tag{1}$$

where q is the metal uptake (mg/g); C_0 and C_f, the initial and final metal concentrations in the solution (mg/L), respectively; V, the solution volume (mL); and M, the mass of biosorbent (g). The pH of the solution was adjusted by using 0.1 N HCl and 0.1 N NaOH. All the experiments were repeated five times, the average values have been recorded, and the statistical error was given in the figures. Also, blank experiments were conducted to ensure that no adsorption was taking place on the walls of the apparatus used.

The Langmuir [15] sorption model was chosen for the estimation of maximum lead sorption by the biosorbent. This model is based on the hypothesis that uptake occurs on a homogenous surface by monolayer sorption without interaction between adsorbed molecules and can be expressed as follows:

$$q = \frac{Q_{\mathrm{max}}bC_{\mathrm{eq}}}{1 + bC_{\mathrm{eq}}}, \tag{2}$$

where Q_{max} indicates the monolayer adsorption capacity of adsorbent (mg/g), and the Langmuir constant b (L/mg) is related to the energy of adsorption. For fitting the experimental data, the Langmuir model was linearized as follows:

$$\frac{1}{q} = \frac{1}{Q_{\mathrm{max}}} + \frac{1}{bQ_{\mathrm{max}}C_{\mathrm{eq}}}. \tag{3}$$

The Freundlich [16] model is represented by

$$q = KC_{\mathrm{eq}}^{1/n}, \tag{4}$$

where K (mg/g) is the Freundlich constant related to the adsorption capacity of adsorbent, and $1/n$ is the Freundlich exponent related to the adsorption intensity (dimensionless).

For fitting the experimental data, the Freundlich model was linearized as follows:

$$\ln q = \ln K + \frac{1}{n} \ln C_{eq}. \tag{5}$$

The Temkin isotherm [17] has generally been applied in the following form:

$$q = \frac{RT}{b_T} \ln \left(A_T C_{eq} \right), \tag{6}$$

where A_T (L/mg) is the Temkin isotherm energy constant, and b_T is Temkin constant related to the heat of sorption.

2.5. Biosorption Kinetics. The kinetic studies were carried out by conducting batch biosorption experiments with initial lead concentrations of 20, 40, 60, 80, and 100 mg/L. Biosorbent is separated by filtration, and the supernatant was used for analysis. Samples were taken at different time periods and analyzed for their lead concentration.

3. Results and Discussion

3.1. Characterization of the Biosorbent. Table 1 presents the elemental analysis of biosorbent *Ficus benghalensis* L. measured by ICP-MS. The biosorbent *Ficus benghalensis* L. mainly constitutes of carbon, oxygen, and hydrogen along with traces of other elements. The surface area of the biomass confirmed its porous nature. The surface area and elemental analysis of biosorbent are presented in Table 2.

3.2. Batch Biosorption Experiments. Batch biosorption experiments were carried out by varying the parameters, and the effect of contact time, pH, metal ion concentration, and the adsorbent dosage has been studied.

3.2.1. The Effect of Contact Time. The data obtained from the biosorption of lead ions on the *Ficus benghalensis* L. showed that a contact time of the 20 min was sufficient to achieve equilibrium, and the adsorption did not change significantly with further increase in contact time. Therefore, the uptake and unadsorbed lead concentrations at the end of 20 min were given as the equilibrium values, that is, q_e (mg/g) and C_{eq} (mg/L) (Figure 1). Other adsorption experiments were conducted at this contact time of 20 min. As shown in Figure 1, the biosorption process took place in two stages: the first rapid stage in which about 90% biosorption was achieved in 10 min, and a slower second stage, with equilibrium attained in another 5 to 10 min. During the first stage of biosorption, initial accumulation of metal takes place due to the availability of the large surface area. The penetration of metal molecules to the inner active sites of the biosorbent takes place in the second slower stage. This was confirmed as a decrease in pH was observed at the end of the experiments which might be due to the release of H^+ ions as result of ion exchange between Pb^{2+} and proton.

TABLE 1: Elemental analysis of biosorbent *Ficus benghalensis* L.

Element	Percentage (%)
C	49.22
O	37.85
H	8.14
N	0.98
S	0.11
Ca	0.75
Al	0.29
Na	0.22
K	0.14
Fe	0.03
Mg	0.02
Others	2.25

TABLE 2: Textual characteristics of biosorbent *Ficus benghalensis* L.

	BET surface area (m^2/g)	Total pore volume (cc/g)	Average pore diameter (nm)
Ficus benghalensis L.	7.21	0.07	120.7

3.2.2. Effect of pH. *Ficus benghalensis* L. presented a high content of ionizable groups (carboxyl groups from mannuronic and guluronic acids) on the cell wall of polysaccharides, which made it very susceptible to the influence of the pH [18]. The effect of pH on biosorption was shown in Figure 2, and it indicated that the uptake of lead increased with the increase in pH from 2.0 to 6.0. Similar results were also reported in the literature for different biomass systems [19–21]. At pH values lower than 2.0, lead removal was inhibited possibly as a result of the competition between hydrogen and lead ions on the sorption sites. This leads to an apparent preponderance of hydrogen ions, which restricted the approach of metal cations. This can be explained by the passive transport mechanism: metal ions diffuse to the surface of biomass where they bind to active sites on the cell surface formed with various chemical groups such as carboxylate, hydroxyl, amino, and phosphate that exhibit affinity for metal ions [22]. According to the surface complexation theory, the increase in metal removal as pH increases can be explained on the basis of a decrease in the competition between proton and metal species for surface sites and by the decrease in the surface charge [19]. As the pH increased, the ligands such as carboxylate groups in *Ficus benghalensis* L. would be exposed increasing the negative charge density of the biomass surface, which in turn increased the attraction of metallic ions with positive charge, and allowing the biosorption onto the cell surface. Based on the shift of pH value in the biosorption process, it may be concluded that ion exchange was the dominant mechanism during the sorption process.

In this study, the lead cations at pH 6 (approximately) were expected to interact more strongly with the negatively-charged binding sites in the adsorbent. As a result, the optimum pH for lead adsorption was found to be 6, and hence the other adsorption experiments were performed at this pH value.

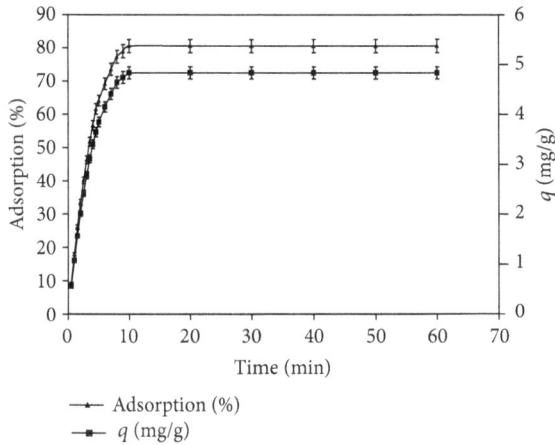

FIGURE 1: Effect of contact time on the biosorption of lead by *Ficus benghalensis* L. for 20 mg/L of metal, 0.1 g/30 mL of biosorbent concentration, and solution pH of 6.0.

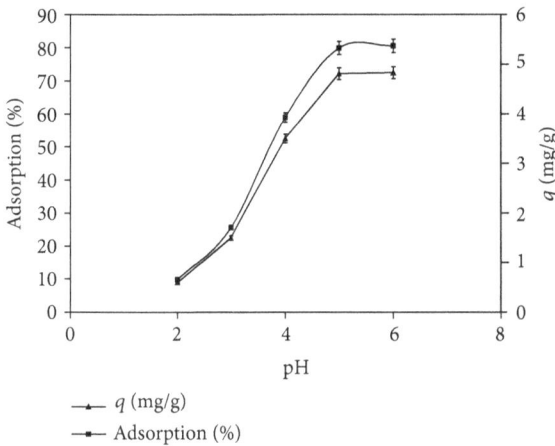

FIGURE 2: Effect of pH on lead biosorption by *Ficus benghalensis* L. for 20 mg/L of metal and 0.1 g/30 mL of biosorbent concentration.

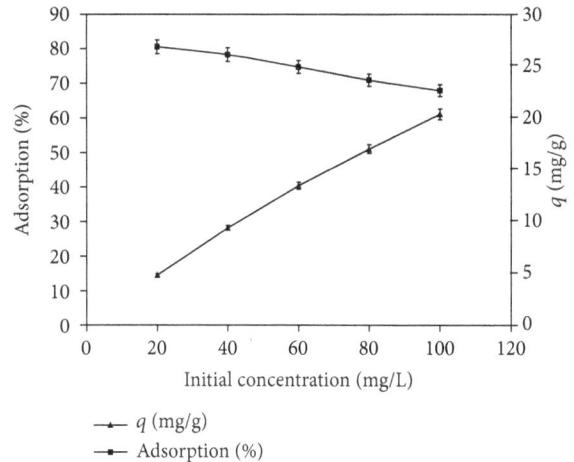

FIGURE 3: Effect of metal concentration on the biosorption of lead by *Ficus benghalensis* L. at 0.1 g/30 mL of biosorbent concentration and solution pH of 6.0.

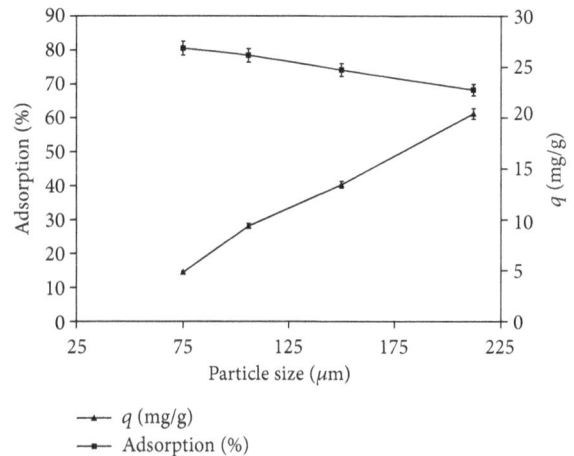

FIGURE 4: Effect of *Ficus benghalensis* L. particle size on biosorption of lead for 20 mg/L of metal, 0.1 g/30 mL of biosorbent concentration, and solution pH of 6.0.

3.2.3. Effect of Metal Ion Concentration. Figure 3 showed the effect of metal ion concentration on the adsorption of lead by *Ficus benghalensis* L. The metal uptake increased and the adsorption percentage of lead decreased with an increase in metal ion concentration. This increase (5.53–21.77 mg/g) was a result of the increase in the driving force that is a concentration gradient. However, the adsorption percentage of lead ions on *Ficus benghalensis* L. has decreased from 80.51 to 67.86%. Though an increase in metal uptake was observed, the decrease in adsorption percentage may be attributed to the lack of sufficient surface area to accommodate much more metal available in the solution. The percentage of adsorption at higher concentration levels showed a decreasing trend whereas the equilibrium uptake of lead showed an opposite trend. At lower concentrations, all lead ions present in the solution could interact with the binding sites and thus the percentage of adsorption was higher than that at higher lead ion concentrations. At higher concentrations, lower adsorption yield is due to the saturation of adsorption sites.

3.2.4. Effect of Adsorbent Size. The effect of different adsorbent particle sizes on removal percentage of lead was investigated, and the results were shown in Figure 4. It was noted that the adsorption of lead on *Ficus benghalensis* L. decreased from 80.51 to 68.32% with the increased particle size from 75 to 212 μm at an initial concentration of 20 mg/L. The smallest size obtained was 75 μm due to the limitation of available grinder configuration. This was due to the fact that decreasing the average particle size of the adsorbent increased the surface area, which in turn increased the number of sites available leading to an increase in adsorption.

3.2.5. Effect of Adsorbent Dosage. The effect of adsorbent dosage on the removal percentage at equilibrium conditions was studied, and it was observed (Figure 5) that the amount of lead adsorbed varied with varying adsorbent dosage. The amount of lead adsorbed increased with an increase in

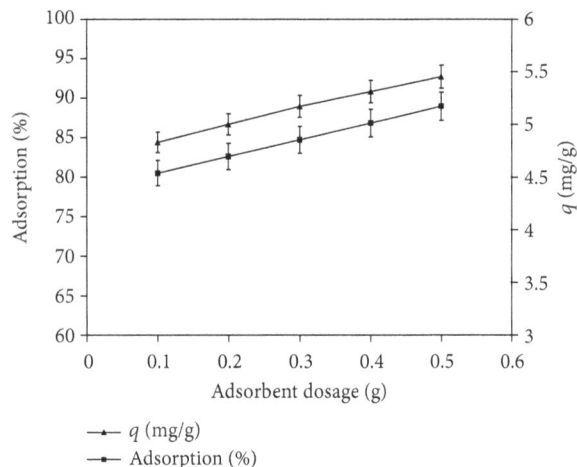

FIGURE 5: Effect of *Ficus benghalensis* L. dosage on biosorption of lead for 20 mg/L of metal concentration and solution pH of 6.0.

the adsorbent dosage from 0.1 to 0.5 g. The percentage of lead removal increased from 80.51 to 88.95% for an increase in adsorbent dosage from 0.1 to 0.5 g, respectively, at initial concentration of 20 mg/L. The increase in the adsorption of the amount of solute was obvious due to the increase in the amount of adsorbent, which in turn increased the number of sites available leading to an increase in adsorption. Similar trend was also observed by Bhattacharyya and Sharma [23] for the removal of lead using *Azadirachta indica* as adsorbent.

3.3. Biosorption Equilibrium. The equilibrium biosorption of lead on the *Ficus benghalensis* L. is established when the concentration of sorbate in bulk solution is in dynamic balance with that on the liquid-sorbent interface. The relationships between the concentrations of biosorbed metal and metal in solution at a given temperature are known as biosorption isotherms. In the present work, the Langmuir, Freundlich, and Temkin models described the equilibrium data. Equilibrium concentration, C_e, and equilibrium capacity, q_e, were determined in each case and displayed in Figure 6. The data obtained after linear regression analysis in each case was plotted and shown in Figures 7, 8, and 9. From these figures, it can be observed that the data is well fitted with the Langmuir model to the experimental data. This indicates that biosorption of lead on FRLs occurred in a monolayer. The values of the Langmuir, Freundlich, and Temkin models constants are listed in Table 3.

3.4. Kinetics of Adsorption. The prediction of adsorption rate gives important information for designing batch adsorption systems. Information on the kinetics of solute uptake is required for selecting optimum operating conditions for full-scale batch process. In Figure 1, the adsorption rate within the first 5 min was observed to be very high and thereafter the reaction proceeded at a slower rate till equilibrium, and finally the steady state was obtained after equilibrium. The saturation time was found to be 20 min based on the initial metal concentrations. The kinetics of the adsorption data was

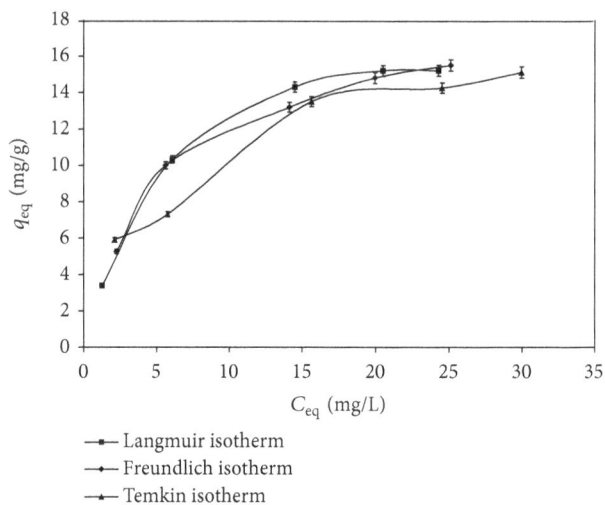

FIGURE 6: Langmuir, Freundlich, and Temkin isotherms plot for lead biosorption at 0.1 g in 30 mL of biosorbent concentration and solution pH of 6.0.

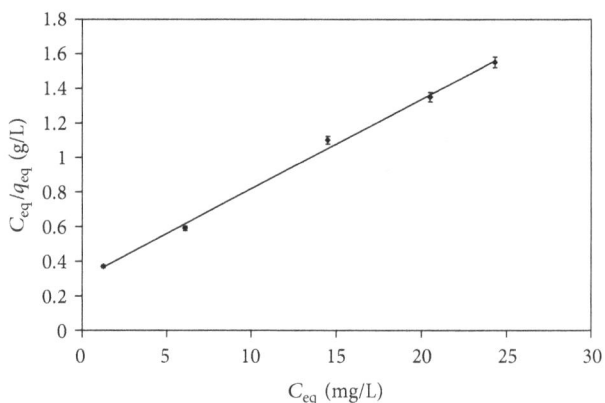

FIGURE 7: Langmuir biosorption isotherm for lead at 0.1 g in 30 mL of biosorbent concentration and solution pH of 6.0.

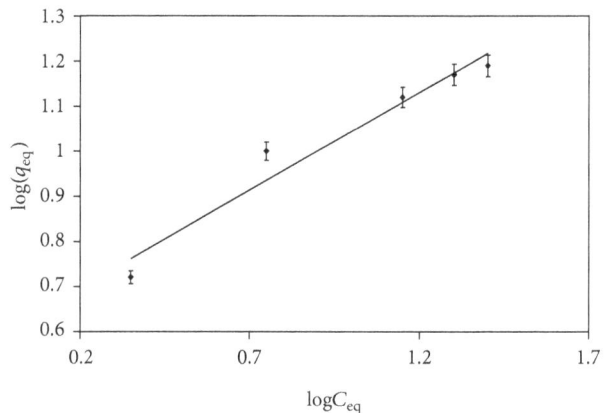

FIGURE 8: Freundlich biosorption isotherm for lead at 0.1 g in 30 mL of biosorbent concentration and solution pH of 6.0.

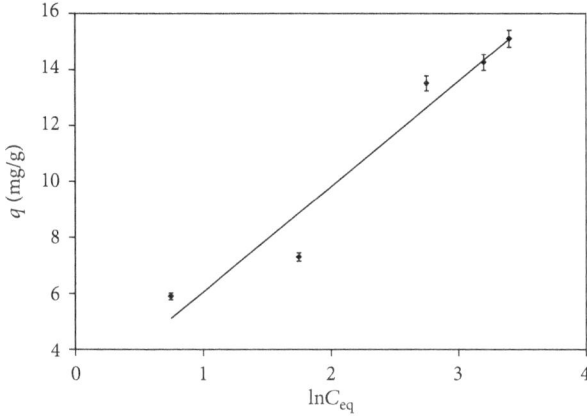

FIGURE 9: Temkin biosorption isotherm for lead at 0.1 g in 30 mL of biosorbent concentration and solution pH of 6.0.

TABLE 3: Langmuir, Freundlich, and Temkin isotherm constants and correlation coefficients.

Langmuir	
Q (mg/g)	28.636
b (L/mg)	0.1213
R^2	0.9979
Freundlich	
K_f	4.5552
n	2.0125
R^2	0.9545
Temkin	
A_T	1.4091
b_T	392.37
R^2	0.9473

analyzed using two kinetic models, pseudo-first- and pseudo-second-order kinetic models. These models correlated solute uptake, which was important in predicting the reactor volume.

3.4.1. The Pseudo-First-Order Equation. The pseudo-first-order equation of Lagergren is generally expressed as follows:

$$\frac{dq_t}{dt} = k_1 \left(q_e - q_t \right), \tag{7}$$

where q_e and q_t are the sorption capacities (mg/g) at equilibrium and at time t, respectively. k_1 is the rate constant of pseudo-first-order sorption (min^{-1}). Integration of (7) with the boundary conditions $q_t = 0$ when $t = 0$ and $q_t = q_t$ at $t = t$ resulted in the following:

$$\log \left(q_e - q_t \right) = \log \left(q_e \right) - \frac{k_1}{2.303} t. \tag{8}$$

The pseudo-first-order rate constant k_1 was obtained from the slope of the graph between $\log(q_e - q_t)$ versus time t (Figure 10). The calculated k_1 values and their corresponding linear regression correlation coefficients were shown in

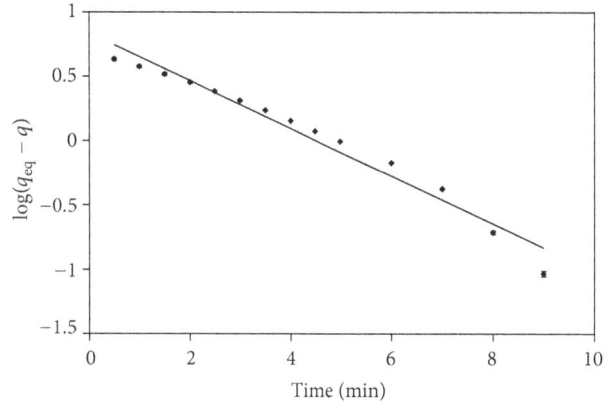

FIGURE 10: Pseudo-first-order biosorption of lead by *Ficus benghalensis* L. for 20 mg/L of metal, 0.1 g/30 mL of biosorbent concentration, and solution pH of 6.0.

Table 4. The linear regression correlation coefficient value R_1^2 was found to be 0.9695, which indicated that this model did not predict the adsorption data, and hence could not be considered as a true representative of the kinetic model.

3.4.2. The Pseudo-Second-Order Equation. If the rate of sorption follows second-order kinetics, the pseudo-second-order chemisorption kinetic rate equation is expressed as follows [28]:

$$\frac{dq_t}{dt} = k \left(q_e - q_t \right)^2, \tag{9}$$

where k is the rate constant of pseudo-second-order sorption (g/(mg min)). For the boundary conditions $q_t = 0$ when $t = 0$ and $q_t = q_t$ at $t = t$, the integrated form of (9) becomes

$$\frac{t}{q_t} = \frac{1}{kq_e^2} + \frac{1}{q_e} t, \tag{10}$$

where t is the contact time (min) and q_e (mg/g) is the amount of the solute adsorbed at equilibrium. Equation (10) does not require assigning any value for effective q_e. If pseudo-second-order kinetics is applicable, the graph t/q_t against t should show a linear relationship, from which q_e and k can be determined from the slope and interception of the plot. The data were plotted in Figure 11. The advantages of (10) were that there was no need to know any parameter beforehand.

The pseudo-second-order rate constant k_2, the calculated q_e value, and the corresponding linear regression correlation coefficient value R_2^2 were given in Table 4. At an initial lead concentration of 20 mg/L, the linear regression correlation coefficient R_2^2 value was higher. The higher R_2^2 value confirms that the adsorption data were well represented by pseudo-second-order kinetic model. The pseudo-second-order kinetics model applicability suggested that adsorption of lead on the *Ficus benghalensis* L. was based on chemical reaction involving the exchange of electrons between biosorbent and metal.

TABLE 4: Kinetic constants for lead onto *Ficus benghalensis* L.

Initial concentration (mg/L)	Pseudo-first-order			Pseudo-second-order		
	Rate constant k_1 (min^{-1})	Amount of lead absorbed on adsorbent, q_e (mg/g)	Correlation coefficient R_1^2	Rate constant k_2 (min^{-1})	Amount of lead absorbed on adsorbent, q_e (mg/g)	Correlation coefficient R_2^2
20	0.6032	9.532	0.9695	0.1165	5.535	0.9996

TABLE 5: Comparison of maximum adsorption capacities for lead adsorption to different adsorbents.

Adsorbent material	Adsorption capacity (mg/g)	pH	Reference
Na-Montmorillonite	3.6	5.0	Abollino et al. [8]
Crushed concrete fines	33.0	5.5	Coleman et al. [24]
Coir	8.6	5.5	Kathrine and Hansen [25]
Barley straw	5.3	5.5	Kathrine and Hansen [25]
Peat	11.7	5.5	Kathrine and Hansen [25]
Coniferous bark	7.4	5.5	Kathrine and Hansen [25]
Fontinalis antipyretica	14.7	5.0	Martins et al. [26]
Aspergillus niger 405	4.7	5.0	Filipovic-Kovacevic et al. [27]
Ficus benghalensis L.	28.6	6.0	Present study

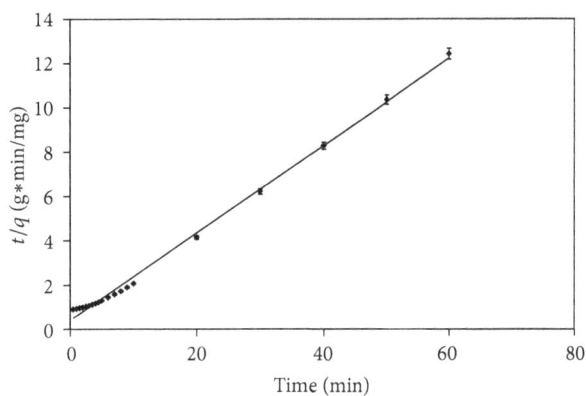

FIGURE 11: Pseudo-second-order biosorption of lead by *Ficus benghalensis* L. for 20 mg/L of metal, 0.1 g/30 mL of biosorbent concentration, and solution pH of 6.0.

3.5. Comparison of Ficus benghalensis L. with Other Sorbents. Table 5 summarizes the comparison of the maximum biosorption capacities of various sorbents. This comparison shows that *Ficus benghalensis* L. has higher biosorption capacities than Na-Montmorillonite [8], coir, barley straw, peat, coniferous bark [25], *Fontinalis antipyretica* [26], and *Aspergillus niger* 405 [27]. Crushed concrete fines [24] have higher biosorption capacity than *Ficus benghalensis* L. Cellulose, hemicelluloses, pectin, and lignin in the cell walls of *Ficus benghalensis* L. biomass are the most important sorption sites compared to other sorbents. Leaves have chlorophyll, carotene, anthocyanin, and tannin which contribute to metal biosorption as they contain hydroxyl, carboxylic, carbonyl,

amino, and nitro groups which are important sites for metal sorption [29]. The easy availability and cost effectiveness of *Ficus benghalensis* L. biomass are additional advantages which make it a better biosorbent for treatment of lead wastes.

4. Conclusions

The present study showed that the *Ficus benghalensis* L. was an effective biosorbent for the adsorption of lead ions from aqueous solution. The easy availability of *Ficus benghalensis* L. biomass and their cost effectiveness are additional advantages, which make it a better biosorbent for treatment of lead wastes. The effect of process parameters like pH, metal ion concentration, adsorbent dosage, and adsorbent size had considerable effect on the process equilibrium. The uptake of lead ions by *Ficus benghalensis* L. was found to increase by increasing the metal ion concentration and the adsorbent dosage and decrease by increase the adsorbent size. The metal uptake was also noted to increasing by increasing the pH up to 6. The adsorption isotherms could well represent the Langmuir equation followed by Freundlich equation. The biosorption process could be best described by pseudo-second-order kinetics.

References

[1] L. Friberg, G. F. Nordberg, and B. Vouk, Eds., *Handbook on the Toxicology of Metals*, Elsevier; Biomedical Press; North-Holland, Amsterdam, The Netherlands, 1979.

[2] A. Asghari, "Simultaneous determination of trace amounts of lead and zinc by adsorptive cathodic stripping voltammetry,"

The Malaysian Journal of Analytical Sciences, vol. 12, pp. 410–418, 2008.

[3] A. G. El-Said, "Biosorption of Pb(II) ions from aqueous solutions onto rice husk and its ash," *The Journal of American Science*, vol. 6, pp. 143–150, 2010.

[4] L. Guo, J. Liu, G. Xing, and Q. Wen, "Adsorption and desorption of zinc(II) on water-insoluble starch phosphates," *Journal of Applied Polymer Science*, vol. 111, no. 2, pp. 1110–1114, 2009.

[5] J. T. Matheickal and Q. Yu, "Biosorption of lead from aqueous solutions by marine algae *Ecklonia radiata*," *Water Science and Technology*, vol. 34, no. 9, pp. 1–7, 1996.

[6] J.-S. Chang and J. Hong, "Biosorption of mercury by the inactivated cells of Pseudomonas aeruginosa PU21 (Rip64)," *Biotechnology and Bioengineering*, vol. 44, no. 8, pp. 999–1006, 1994.

[7] K. H. Chu, M. A. Hashim, S. M. Phang, and V. B. Samuel, "Biosorption of cadmium by algal biomass: adsorption and desorption characteristics," *Water Science and Technology*, vol. 35, no. 7, pp. 115–122, 1997.

[8] O. Abollino, M. Aceto, M. Malandrino, C. Sarzanini, and E. Mentasti, "Adsorption of heavy metals on Na-montmorillonite. Effect of pH and organic substances," *Water Research*, vol. 37, no. 7, pp. 1619–1627, 2003.

[9] B. Volesky and Z. R. Holan, "Biosorption of heavy metals," *Biotechnology Progress*, vol. 11, no. 3, pp. 235–250, 1995.

[10] D. Kratochvil, B. Volesky, and G. Demopoulos, "Optimizing Cu removal/recovery in a biosorption column," *Water Research*, vol. 31, no. 9, pp. 2327–2339, 1997.

[11] J.-S. Chang, J.-C. Huang, C.-C. Chang, and T.-J. Tarn, "Removal and recovery of lead fixed-bed biosorption with immobilized bacterial biomass," *Water Science and Technology*, vol. 38, no. 4-5, pp. 171–178, 1998.

[12] N. Kuyucak, "Feasibility of biosorbents application," in *Biosorption of Heavy Metals*, B. Volesky, Ed., CRC Press, Boca Raton, Fla, USA, 1990.

[13] A. Sari and M. Tuzen, "Biosorption of cadmium(II) from aqueous solution by red algae (Ceramium virgatum): equilibrium, kinetic and thermodynamic studies," *Journal of Hazardous Materials*, vol. 157, no. 2-3, pp. 448–454, 2008.

[14] R. A. Anayurt, A. Sari, and M. Tuzen, "Equilibrium, thermodynamic and kinetic studies on biosorption of Pb(II) and Cd(II) from aqueous solution by macrofungus (Lactarius scrobiculatus) biomass," *Chemical Engineering Journal*, vol. 151, no. 1-3, pp. 255–261, 2009.

[15] I. Langmuir, "The adsorption of gases on plane surfaces of glass, mica and platinum," *The Journal of the American Chemical Society*, vol. 40, no. 9, pp. 1361–1403, 1918.

[16] H. M. F. Freundlich, "Over the adsorption in solution," *The Journal of Physical Chemistry*, vol. 57, pp. 385–470, 1906.

[17] C. Aharoni and M. Ungarish, "Kinetics of activated chemisorption. Part 2—Theoretical models," *Journal of the Chemical Society*, vol. 73, pp. 456–464, 1977.

[18] S. Qaiser, A. R. Saleemi, and M. Umar, "Biosorption of lead from aqueous solution by Ficus religiosa leaves: batch and column study," *Journal of Hazardous Materials*, vol. 166, no. 2-3, pp. 998–1005, 2009.

[19] Y. Sağ, A. Kaya, and T. Kutsal, "The simultaneous biosorption of Cu(II) and Zn on *Rhizopus arrhizus*: application of the adsorption models," *Hydrometallurgy*, vol. 50, no. 3, pp. 297–314, 1998.

[20] J. L. Zhou, P. L. Huang, and R. G. Lin, "Sorption and desorption of Cu and Cd by macroalgae and microalgae," *Environmental Pollution*, vol. 101, no. 1, pp. 67–75, 1998.

[21] J. T. Matheickal and Q. Yu, "Biosorption of lead(II) and copper(II) from aqueous solutions by pre-treated biomass of Australian marine algae," *Bioresource Technology*, vol. 69, no. 3, pp. 223–229, 1999.

[22] A. Özer and D. Özer, "Comparative study of the biosorption of Pb(II), Ni(II) and Cr(VI) ions onto S. cerevisiae: determination of biosorption heats," *Journal of Hazardous Materials*, vol. 100, no. 1-3, pp. 219–229, 2003.

[23] K. G. Bhattacharyya and A. Sharma, "Adsorption of Pb(II) from aqueous solution by *Azadirachta indica* (Neem) leaf powder," *Journal of Hazardous Materials B*, vol. 113, no. 1-3, pp. 97–109, 2004.

[24] N. J. Coleman, W. E. Lee, and I. J. Slipper, "Interactions of aqueous Cu2+, Zn2+ and Pb2+ ions with crushed concrete fines," *Journal of Hazardous Materials*, vol. 121, no. 1-3, pp. 203–213, 2005.

[25] K. Conrad and H. C. B. Hansen, "Sorption of zinc and lead on coir," *Bioresource Technology*, vol. 98, no. 1, pp. 89–97, 2007.

[26] R. J. E. Martins, R. Pardo, and R. A. R. Boaventura, "Cadmium(II) and zinc(II) adsorption by the aquatic moss Fontinalis antipyretica: effect of temperature, pH and water hardness," *Water Research*, vol. 38, no. 3, pp. 693–699, 2004.

[27] Ž. Filipovic-Kovacevic, L. Sipos, and F. Briški, "Biosorption of chromium, copper, nickel and zinc ions onto fungal pellets of aspergillus niger 405 from aqueous solutions," *Food Technology and Biotechnology*, vol. 38, no. 3, pp. 211–216, 2000.

[28] Y. S. Ho and G. McKay, "The kinetics of sorption of basic dyes from aqueous solution by sphagnum moss peat," *Canadian Journal of Chemical Engineering*, vol. 76, no. 4, pp. 822–827, 1998.

[29] S. Qaiser, A. R. Saleemi, and M. M. Ahmad, "Heavy metal uptake by agro based waste materials," *Electronic Journal of Biotechnology*, vol. 10, no. 3, pp. 409–416, 2007.

5

Multilanguage Semantic Interoperability in Distributed Applications

Agostino Poggi and Michele Tomaiuolo

Dipartimento di Ingegneria dell'Informazione, Università degli Studi di Parma, Viale U. P. Usberti 181/A, 43100 Parma, Italy

Correspondence should be addressed to Agostino Poggi; agostino.poggi@unipr.it

Academic Editor: Stavros Koubias

JOSI is a software framework that tries to simplify the development of such kinds of applications both by providing the possibility of working on models for representing such semantic information and by offering some implementations of such models that can be easily used by software developers without any knowledge about semantic models and languages. This software library allows the representation of domain models through Java interfaces and annotations and then to use such a representation for automatically generating an implementation of domain models in different programming languages (currently Java and C++). Moreover, JOSI supports the interoperability with other applications both by automatically mapping the domain model representations into ontologies and by providing an automatic translation of each object obtained from the domain model representations in an OWL string representation.

1. Introduction

Semantic information is assuming more and more importance both for the development of knowledge-based applications and for supporting the interoperability among different applications [1–4]. In particular, ontologies have been gaining interest for the representation of the application domain models and their use has been spreading in different applications fields [5–8].

Domain models are increasingly specified as formal ontologies through the use of a semantic Web language (e.g., OWL [9, 10]), but such models remain difficult to be utilized in applications developed through the used software languages and libraries. In fact, the mapping of such models into the code of a typical application development language often is not possible because of the different expressive power of the modeling and the implementation language. Moreover, when it is possible, the obtained implementation is too complex to be used by the large part of software developers.

However, the development of domain models that represent semantic information is very difficult without the use of a semantic language. To cope with this problem, a possible direction is to integrate usual programming techniques with some meta-programming techniques. In particular, the Java programming language supports meta-programming through annotations and reflection [11]. In fact, while annotations allow the decoration of the Java code with new concepts and idioms, reflection allows the retrieval of the information associated with annotations and then to use them for either modifying the usual execution of the Java code or for building new Java code.

In this paper, we present a software framework, called JOSI (Java and OWL for System Interoperability), whose goal is to simplify the development of the software libraries for managing the data that implement the domain models shared by the systems of a distributed enterprise application. The next section introduces related work on the use of annotations for the development of software and on the mapping between OWL ontologies and Java code. Section 3 describes the JOSI software framework. Section 4 describes how a domain model is represented. Section 5 presents how an implementation of a domain model is built starting from its JOSI representation. Section 6 introduces how domain model implementations are used in a software application. Sections 7 and 8 represent and discuss the experimentation of the JOSI software framework. Finally, Section 9 concludes the paper sketching some future research directions.

2. Related Work

The idea of using Java annotations for extending the Java language is not new and several research teams worked in that direction.

AspectJ [12] is probably the first important work that shows how Java annotation can provide a meta-programming layer on the top of Java programming structures. In particular, AspectJ is an aspect-oriented extension of the Java programming language that uses Java annotations for realizing declaring aspects, point-cuts, and advices.

Andreae et al. [13] proposed a software framework that supports pluggable type systems in the Java programming language by the definition of custom constraints on Java types through Java annotations.

AVal [14] is a software framework for the definition and checking of rules for programs written by using an attribute domain-specific built on the top of Java. This software framework allows the validation of such kinds of program through a set of predefined Java annotations. Moreover, it allows to the users of the framework to add new annotations to provide new kinds of validation.

Bordin and Vardanega [15] used Java annotations to embed in the source code a declarative specification of the required concurrent semantics and then for producing the source code that implements the declared concurrent semantics.

Cimadamore and Viroli [16] proposed a software framework that tried to simplify the seamless integration of Prolog code into Java applications taking advantage of Java annotations to incorporate the declarative features of Prolog into Java programs.

A lot of work has been done also towards the mapping of OWL ontologies into Java code and vice versa.

The first important work that shows the partial translation of OWL ontologies in Java code is the Protégé Bean Generator [17]. In particular, it transforms Protégé frame-based ontologies into Java source code for developing JADE agents [18, 19].

RDFReactor [20] is a toolkit for dynamically accessing an RDF model through domain-centric methods (getters and setters). In particular, it allows the access to the RDF model through a set of proxy objects that provide the methods for querying and updating the RDF elements.

A more sophisticated approach was presented by Kalyanpur et al. [21]. This approach deals with issues as multiple-inheritance by mapping OWL classes in Java interfaces. However, there is not a software tool which takes advantage of this approach for mapping OWL ontologies into Java code.

SeRiDA [22] is a methodology for enabling a three-tier mapping along ontologies, object-oriented java beans and relational database. In particular, it allows the generation of both an object-oriented and a relational model starting from a domain conceptualization expressed in OWL. This methodology has been experimented by realizing a software tool that generates programming interfaces as enterprise Java beans and Hibernate object-relational mappings from OWL ontologies.

Quasthoff and Meinel [23] presented a mechanism that allows application developers, with limited knowledge about RDF and OWL, to easily map arbitrary Java classes and interfaces to corresponding OWL concepts by using Java annotations. In particular, this mechanism has already been experimented in the development of a social network application testing new access control mechanisms on user-generated content with the help of Semantic Web rules [19].

OWLET [24] is a Java software environment based on an object-oriented model, which allows a simple and complete representation of ontologies defined by using OWL DL profile, and provides a complete set of reasoning functions together with a graphical editor for the creation and modification of ontologies. OWLET supports the development of heterogeneous and distributed semantic systems where nodes differ for their capabilities (i.e., CPU power, memory size, etc.). In fact, it offers a layered reasoning API that allows to deploy a system where high power nodes take advantages of all the OWLET reasoning capabilities, medium power nodes take advantages of a limited set of OWLET reasoning capabilities (e.g., reasoning about individuals) and low power nodes delegate reasoning tasks to the other nodes of the system.

Finally, the OWL API can be considered the reference Java API for managing ontologies [25, 26]. In fact, besides providing the manipulation of ontologies, it offers: a general purpose reasoner interfaces, the validators for the various OWL profiles, and the support for parsing and serializing ontologies in a variety of syntaxes. The API also has a very flexible design that allows third parties to provide alternative implementations for all major components.

Different works cope with the problem of defining models for integrating different data sources in enterprise information systems.

Astrova and Kalja [27] proposed an approach for system interoperability that maps relational database schemas into OWL ontologies and allows an improvement of database schemas by identifying "hidden" (implicit) semantic relationships and bad design solutions.

Lin and Harding [28] proposed a general manufacturing system engineering knowledge representation scheme to facilitate communication and information exchange in inter-enterprise, multidisciplinary engineering design teams. It has been developed and encoded in the standard semantic web language. The proposed approach focuses on how to support information autonomy that allows the individual team members to keep their own preferred languages or information models rather than requiring them all to adopt standardized terminology.

Salguero et al. [29] proposed a framework which encompasses the entire data integration process. The data source schemas as well as the integrated schema are expressed using an OWL extension which allows the incorporation of metadata to support the integration process.

3. Software Framework Overview

JOSI (Java and OWL for System Interoperability) is a software framework that tries to simplify the development of the

```
@Immutable
@Comparator ({"name", "domain"})
public interface Address extends Entity {
        @Getter ("name")
        @Cardinality (1)
        String getName ();

        @Getter ("domain")
        @Cardinality (1)
        Domain getDomain ();
}
@Immutable
@Comparator ({"name"})
public interface Domain extends Entity {
        @Getter ("name")
        @Cardinality (value = 1)
        String getName ();
}
@Name ("naming")
@Version ("1.0")
public interface Naming extends Factory {
        @Binding ({"name", "domain"})
        public Address getAddress (final String a, final Domain b);
        @Binding ({"name"})
        public Domain getDomain (final String a);
}
```

FIGURE 1: A simple naming domain model.

software libraries for managing the data that implement the domain models shared by the systems of a distributed enterprise application.

The main features of such software library are as follows: (i) a strict separation between the representation of a domain model and its implementation, (ii) an automatic generation of an implementation of the representation of the domain model in different programming languages; (iii) an automatic generation of an OWL ontology from the representation of a domain model and vice versa, and (iv) the possibility of using an OWL string representation of the domain model data to support the interoperability between systems implemented in different programming languages and so the possibility of translating domain model data to OWL string representations and vice versa.

JOSI is implemented in Java and takes advantage of Java interfaces and annotations to build a representation of a domain model and uses Java reflection to drive the processing of the information maintained by such interfaces and annotations for generating the source code of the classes that define the concrete implementation of the domain model.

The following sections will describe how a domain model is represented through Java interfaces and annotations, how the Java classes providing a concrete implementation of such a domain model are generated from such interfaces and annotations, and how such a software framework enables an application to use a concrete implementation of a domain model.

TABLE 1: Java annotations used in the representation of a domain model.

@Abstract	@Getter	@Name	@Symmetric
@AllValuesFrom	@HasValue	@Ordered	@SomeValuesFrom
@Binding	@Immutable	@Set	@Transitive
@Cardinality	@InverseOf	@Setter	@Version

4. Domain Model Representation

A domain model is represented by a set of Java interfaces. Each domain entity is represented by a Java interface (from here called entity interface) that defines the two methods for reading and modifying its attributes. Moreover, an additional Java interface (from here called factory interface) provides both some general information about the domain model and the factory methods for the creation of the Java classes which implement the different entity interfaces. Figures 1 and 2 show some entities of two domain models represented through the use of Java interfaces and annotations. Table 1 lists the Java annotations used in the representation of a domain model.

To support the creation of the implementation of such entities, each Java interface is enriched by some Java annotations and constant declarations.

The two annotations: *@Getter* and *@Setter* are applicable to the entity interface methods and define the reading and modifying methods of a specific attribute. The type of the

```
@Abstract
public interface Action extends Entity {
    @Getter ("resultTypes")
    @Cardinality (min = 1)
    String [] getResultTypes ();
    @Getter ("errorValues")
    Error [] getErrorValues ();
}
@Immutable
@Comparator ({"name"})
public interface Describe extends Action {
    public static final String DESCRIPTION = Description.class.getName ();
    public static final Error UNKNOWNACTION =
        ((Interaction) DataStore.getModel ("interaction")).getUnknownAction();
    public static final Error UNREACHABLEAGENT =
        ((Interaction) DataStore.getModel ("interaction")).getUnreachableAgent();
    @Getter ("name")
    @Cardinality (1)
    String getName ();
    @Override
    @Getter ("resultTypes")
    @HasValue (values={"DESCRIPTION"})
    @Cardinality (1)
    String [] getResultTypes ();
    @Override
    @Getter ("errorTypes")
    @HasValue (values = {"UNKNOWNACTION", "UNREACHABLEAGENT"})
    @Cardinality (2)
    Error [] getErrorValues ();
}
```

FIGURE 2: Two entities of a domain model describing the life-cycle of a software agent.

attribute is identified by both the return type of the reading method and the type of the argument of the modifying method (of course they need to identify the same type). In particular, the value of any attribute must be: a Java primitive data, an instance of the String class, an instance of a class implementing an entity interface, or an array of the previous kinds of value.

The four annotations: @Abstract, @Immutable, @OneOf, and @Singleton, are applicable to the entity interfaces. The first annotation identifies an abstract entity, that is, an entity that does not have any direct implementation. The second annotation identifies an entity that has an immutable implementation, that is, the interface cannot define methods that modify the value of its attributes and the implementation of its reading methods will be defined to return either the value of an attribute (if it is an immutable value) or a copy of the value (if it is a mutable value). The third annotation is used for identifying entities that have an extensional description (e.g., that can be defined through an enumeration). Finally, the forth annotation is used for the definition of some special entities that can be represented by a single class object.

Often the use of an implementation of a domain model inside an application needs the availability of operations for the comparison and ordering of their entities. In a Java implementation, such operations can be performed by implementing the *compareTo*, equals, and *hashCode* methods. The

annotation @Comparator is introduced for this scope. In fact, it identifies the sequence of attributes on which the previous three methods must work.

In a domain model often is necessary both to restrict the value that some attributes can assume and to establish a relationship between the attributes of some entities. It is done by associating some additional annotations to the reading methods of the entity interfaces.

The four annotations: @AllValueFrom, @SomeValues-From, @Cardinality, and @HasValue, define the most known constraints that OWL applies to the properties of an ontology. In particular, the first annotation constrains the values of an attribute to belong to specific type (of course, an implicit constraint of such a kind, is defined when the reading and modifying methods of an attributed are defined. However, an additional constraint can be added by imposing that the values of an attribute must belong to a subtype of the declared attribute type). The second annotation imposes that some of the values of an attribute must belong to a specific type (of course, such a type must be a subtype of the declared attribute type). The third annotation imposes that an attribute can have either a fixed number of values or a variable number of values defined by a minimum and/or a maximum value. Finally, the forth annotation imposes that an attribute must always contain some values (in this case, for the limited set of value types that can be associated with the attributes of an

```
public final class E1 implements Address {
    private String name;
    private Domain domain;
    // Class constructor.
    public E1 (final String a, final Domain b) {
        this.name = a;
        this.domain = b;
    }
    // Checks instance consistency.
    public boolean check () {
        if ((this.name != null) && (this.domain != null)) return true;
        return false;
    }
    @Override
    public String getName () {
        return this.name;
    }
    @Override
    public Domain getDomain () {
        return this.domain;
    }
}
public final class E2 implements Domain {
    private String name;
    // Class constructor.
    public E2 (final String a) {
        this.name = a;
    }
    // Checks instance consistency.
    public boolean check () {
        if (this.name != null) return true;
        return false;
    }
    @Override
    public String getName () {
    return this.name;
    }
}
```

FIGURE 3: Java implementation of the entities of the naming domain model.

annotation, the values of such constraints are defined through constant variables and the annotations refer to the names of such constant variables).

In some cases it can be necessary to impose that an attribute does not have duplicated values and that its values are maintained ordered: the two annotations: *@Set*, and *@Ordered,* impose the previous two constraints (in particular, the second constraint is implemented either by using the natural ordering between values or the ordering defined by the *compareTo* method built through the *@Comparator* annotation introduced above).

The three annotations: *@InverseOf*, *@Symmetric*, and *@Transitive*, define the most known constraints that OWL applies to the relationship between properties of an ontology. The first annotation defines an inverse relationship between attributes. The second annotation defines a symmetric relationship between the entities that have such kind of attribute. Finally, the third annotation defines a transitive relationship between the entities that have such kind of attribute.

Finally, the two annotations: *@Name* and *@Version,* are applicable to the factory interfaces: the first annotation indicates the name associated with the domain model and the second annotation identifies the version of the model. Lastly, the annotation *@Binding* is associated with a factory method of a model interface. This annotation identifies the attribute that each argument of the factory method will initialize.

5. Domain Model Implementation

A domain model representation, defined as described in the previous sections, contains all the information for building an implementation of such a domain model. This implementation is realized by an annotation processor that builds a Java class for each Java interface of the model. Figures 3 and 4 show the source code of the Java classes obtained through the naming domain model introduced in the previous section.

```
public final class F1 implements Naming {
    private static final String NAME = "naming";
    private static final String VERSION = "1.0";
    private static final String [] ENTITIES = {
        Address.class.getName (), Domain.class.getName ()};
    @Override
    public String getModelName () {
        return NAME;
    }
    @Override
    public String getModelVersion () {
        return VERSION;
    }
    @Override
    public String [] list () {
        return ENTITIES;
    }
    @Override
    public E1 getAddress (final String a, final Domain b) {
        E1 i = new E1 (a, b);
        if (i.check ()) return i;
        return null;
    }
    @Override
    public E2 getDomain (final String a) {
        E2 i = new E2 (a);
        if (i.check ()) return i;
        return null;
    }
}
```

FIGURE 4: Java implementation of the model of the naming domain model.

The result of such an annotation processor is a set of Java files. Each Java file contains the source code of a class that implement an interface of the domain model representation. Moreover, each class that implements an entity interface provides a method for building an OWL string representation of an entity class instance, and each class that implements a model interface provides a method for building an entity class instance from its OWL string representation.

The annotation processor used for generating the domain model implementation is composed by two software modules. The first module, called processing module, extracts the information from the domain model representation, generates an intermediate representation and then calls the second module. Then the second module, called generation module, builds the domain model implementation from the intermediate representation.

The intermediate representation is based on a two level tree where the root object maintains the information about the model interface and each leaf object maintains the information about an entity interface.

The processing module is independent from the implementation of the generation module because it calls a generation module by a Java interface and the generation module implementation is a parameter of the processing module constructor.

Therefore, it is very easy to provide different implementations of some domain model representations by defining new generation modules able to process in different ways the intermediate representation built by the processing module. In particular, the current version of the software framework provides another generation module which builds OWL ontologies from the domain model representations and stores them in RDF format [30].

6. Domain Model Application

After the creation of an implementation of a domain model, its use inside an application is very simple. In fact, the JOSI software framework provides a class, called *DataStore*, which has the duty of both maintaining the information about the different domain models available for the current application and providing the access to their implementation through the creation of an instance of the class that implements their domain interface. In particular, the *Datastore* instance can access to the list of the domain models used by the application through a property file.

Therefore, after the creation of an instance of the *DataStore* class, the code of the application can create instances of any class implementing the factory interface of a domain model and then use it for creating instances implementing

```
DataStore dt = Datastore.getInstance ();
Namning n = (Naming) dt.getFactory ("naming");
Domain d = n.getDomain ("localhost");
Address a1 = n.getAddress ("agent1", d);
Address a2 = n.getAddress ("agent2", d);
Lifecycle lc = (Lifecycle) dt.getFactory ("lifecycle");
Describe d1 = lc.getDescribe ("agent1");
```

FIGURE 5: Java code for creating instances of the entities of two domain models.

any entity interface of such a domain model. Figure 5 shows a sample of Java code performing the operations described above.

7. Experimentation

We are using the JOSI software framework for the development of the models and then the implementations of the data necessary for supporting the basic interactions among the components of a distributed system realized through the HDS software framework. Moreover, JOSI was experimented for defining the domain models of some applications in the fields of distributed information sharing and social networks.

HDS (Heterogeneous Distributed System) is a software framework that tries to simplify the realization of pervasive applications by merging the client-server and the peer-to-peer paradigms and by implementing all the interactions among the processes of a system through the exchange of typed messages and the use of composition filters for driving and dynamically adapting the behavior of the system [31].

Typed messages are one of the elements that mainly characterize such a software framework. In fact, typed messages can be considered an object-oriented "implementation" of the types of message defined by an agent communication language and so they are means that make HDS a suitable software framework both for the realization of multiagent systems and for the reuse of multiagent model and techniques in nonagent based systems.

In particular, the type of a message is defined by its content and its content is defined by an entity of a specific domain model defined with the JOSI software framework. Therefore, we used JOSI foe the definition of the domain models that support the basic interaction among HDS processes, that is, the managing of the processes themselves and of the resources that can they used in a distributed application. Moreover, we used JOSI for defining the domain models used for realizing the typical coordination algorithms of intelligent distributed systems.

RAIS (Remote Assistant for Information Sharing) is a peer-to-peer multiagent system supporting the sharing of information among a community of users connected through the Internet [32]. RAIS offers search facility similar to Web search engines, but it avoids the burden of publishing the information on the Web and it guaranties a controlled and dynamic access to information through the use of agents.

The use of agents in such a system is very important because it simplifies the realization of the three main services: (i) the filtering of the information coming from different users on the basis of the previous experience of the local user; (ii) the pushing of the new information that can be of possible interest for a user; and (iii) the delegation of access capabilities on the basis of a network of reputation built by the agents on the community of users.

RAIS is composed of a dynamic set of agent platforms connected through the Internet. In this case, JOSI has been used for the definition of the domain models supporting the definition of the interaction of agents for the retrieval and pushing of the information and for the management of the user profiles.

About the applications in the field of the social networks, we are starting the development a system for the study of the most known social networks and, in particular, of the social networks that provide semantic support for the management of both the profiles and the information published by the users [33].

In particular, we built a system that can simulate the behavior of some of the most known social networks and can compare them with some enhanced versions of such networks that provide semantic support through the use of JOSI domain models. In particular, we defined some domain models for representing the user profiles of different social networks and some domain models for supporting users in the publishing and retrieval of information related to some sample topics (e.g., computer science and music).

8. Experimental Results

The results of the experimentation of the software framework showed that the definition of a domain model can be done by any programmer with knowledge about the Java programming language, but does not require any knowledge about any knowledge engineering and semantic Web techniques and technologies. Moreover, if the entities of a domain model are defined as immutable objects, then the performance of managing such entities is similar to the one of managing JavaBean objects.

Other important results come from some tests that compared the result of the work of groups of students, which developed domain models using JOSI, with the work of other groups of students, which developed domain models without using it. In fact, while the first set of groups developed the domain model in few time spending a very limited part of it for code correction, the second set of groups developed the domain model in a very long time spending its large part for code correction. Moreover, the performance measures of the tests showed that the implementations of the domain model based on the JOSI framework provided better measures or at least similar to the ones provided by the "custom" implementations. Of course, while the use of JOSI guaranteed implementations in different programming languages (currently Java and C++) without additional costs, it was not true for "custom" implementations.

9. Conclusion

This paper presented a software framework, called JOSI (Java and OWL for System Interoperability), that has the goal of simplifying the development of the software libraries for managing the data that implement the domain models shared by the systems of a distributed enterprise application.

This software framework allows to represent a domain model through Java interfaces and annotations and then to use such a representation for automatically generating a Java implementation of the domain model. Moreover, it provides the interoperability with other kinds of systems both automatically mapping the Java domain representation in an OWL ontology and providing an automatic translation of each object defined by the domain model representation in an OWL string representation.

JOSI derived from O3L (Object-Oriented Ontology Library), a software library that provides a complete representation of ontologies compliant with OWL 2 W3C [34]. O3L has not the goal to be used for the creation and manipulation of ontologies, but provides a simplified and efficient API for the realization of applications, that interoperate through the use of shared ontologies, and allows: (i) the use of OWL individuals as data of the applications, (ii) the exchange of OWL individuals between applications, (iii) the reasoning about OWL individuals, and (iv) the classification of OWL classes and properties. The experimentation of O3L showed that it is a powerful means for developing applications but with two main limits: developers must have a good knowledge of semantic techniques and technologies and often applications cannot provide the required performances.

Current and future research activities are dedicated, besides to continue the experimentation of the current implementation of JOSI, to: (i) the development of a software generation module that allows the automatic generation of a C++ and Python implementation from a JOSI model representation, (ii) the generation of a JOSI model representation from an OWL ontology compliant with the JOSI domain model representation, (iii) the generation of OWL ontologies compliant with such a representation from OWL ontologies that contain classes and properties that cannot be defined through the annotations defined in the JOSI software framework, (iv) the introduction of new annotations for increasing the expressive power of the JOSI model representation.

References

[1] P. A. Bernstein and L. M. Haas, "Information integration in the enterprise," *Communications of the ACM*, vol. 51, no. 9, pp. 72–79, 2008.

[2] M. Ciocoiu, D. S. Nau, and M. Gruninger, "Ontologies for integrating engineering applications," *Journal of Computing and Information Science in Engineering*, vol. 1, no. 1, pp. 12–22, 2001.

[3] R. García-Castro and A. Gómez-Pérez, "Interoperability results for semantic web technologies using OWL as the interchange language," *Web Semantics*, vol. 8, no. 4, pp. 278–291, 2010.

[4] S. Heiler, "Semantic interoperability," *ACM Computing Surveys*, vol. 27, no. 2, pp. 271–273, 1995.

[5] D. Oberle, S. Staab, R. Studer, and R. Volz, "Supporting application development in the semantic web," *ACM Transactions on Internet Technology*, vol. 5, no. 2, pp. 328–358, 2005.

[6] M. Quasthoff, H. Sack, and C. Meinel, "Who reads and writes the social web? A security architecture for Web 2.0 applications," in *Proceedings of the 3rd International Conference on Internet and Web Applications and Services (ICIW '08)*, pp. 576–582, Athens, Greece, June 2008.

[7] M. Uschold, "Ontology-driven information systems: past, present and future," in *Proceedings of the 5th International Conference on Formal Ontology in Information Systems (FOIS '08)*, pp. 3–18, Amsterdam, The Netherlands, 2008.

[8] N. F. Noy, "Semantic integration: a survey of ontology-based approaches," *SIGMOD Record*, vol. 33, no. 4, pp. 65–70, 2004.

[9] B. C. Grau, I. Horrocks, B. Motik, B. Parsia, P. Patel-Schneider, and U. Sattler, "OWL 2: the next step for OWL," *Web Semantics*, vol. 6, no. 4, pp. 309–322, 2008.

[10] D. L. McGuinness and F. van Harmelen, OWL web ontology language overview, W3C Recommendation, 2004, http://www.w3.org/TR/owl-features/.

[11] B. Joy, J. Gosling, G. Steele, and G. Bracha, *The Java Language Specification*, Addison-Wesley, New York, NY, USA, 3rd edition, 2005.

[12] G. Kiczales, E. Hilsdale, J. Hugunin, M. Kersten, J. Palm, and W. G. Griswold, "An overview of AspectJ," in *Proceedings of the 15th European Conference on Object-Oriented Programming (ECOOP '01)*, vol. 2072 of *Lecture Notes in Computer Science*, pp. 327–354, Springer, Berlin, Germany, 2001.

[13] C. Andreae, J. Noble, S. Markstrum, and T. Millstein, "A framework for implementing pluggable type systems," in *Proceedings of the 21st Annual ACM SIGPLAN Conference on Object-Oriented Programming Systems, Languages, and Applications (OOPSLA '06)*, pp. 57–74, New York, NY, USA, October 2006.

[14] C. Noguera and R. Pawlak, "AVal: an extensible attribute-oriented programming validator for Java," *Journal of Software Maintenance and Evolution*, vol. 19, no. 4, pp. 253–275, 2007.

[15] M. Bordin and T. Vardanega, "Real-time Java from an automated code generation perspective," in *Proceedings of the 5th International Workshop on Java Technologies for Real-Time and Embedded Systems (JTRES '07)*, pp. 63–72, New York, NY, USA, September 2007.

[16] M. Cimadamore and M. Viroli, "A Prolog-oriented extension of Java programming based on generics and annotations," in *Proceedings of the 5th International Symposium on the Principles and Practice of Programming in Java (PPPJ '07)*, pp. 197–202, New York, NY, USA, September 2007.

[17] C. van Aart, R. Pels, G. Caire, and F. Bergenti, "Creating and using ontologies in agent communication," in *Proceedings of the*

Workshop on Ontologies in Agent Systems, Bologna, Italy, July 2002.

[18] F. Bellifemine, A. Poggi, and G. Rimassa, "Developing multi agent systems with a FIPA-compliant agent framework," *Software Practice & Experience*, vol. 31, no. 2, pp. 103–128, 2001.

[19] F. Bellifemine, G. Caire, A. Poggi, and G. Rimassa, "JADE: a software framework for developing multi-agent applications. Lessons learned," *Information and Software Technology*, vol. 50, no. 1-2, pp. 10–21, 2008.

[20] M. Volkel, "RDFReactor: from ontologies to programmatic data access," in *Proceedings of the Jena User Conference*, Bristol, UK, 2006.

[21] A. Kalyanpur, D. J. Pastor, S. Battle, and J. Padget, "Automatic mapping of OWL ontologies into Java," in *Proceedings of the 16th International Conference on Software Engineering and Knowledge Engineering*, pp. 98–103, Banff, Canada, 2004.

[22] I. N. Athanasiadis, F. Villa, and A. Rizzoli, "Enabling knowledge-based software engineering through semantic-object-relational mappings," in *Proceedings of the 3rd International Workshop on Semantic Web-Enabled Software Engineering, 4th European Semantic Web Conference*, pp. 16–30, Innsbruck, Austria, 2007.

[23] M. Quasthoff and C. Meinel, "Semantic web admission free—obtaining RDF and OWL data from application source code," in *Proceedings of the 4th International Workshop on Semantic Web Enabled Software Engineering*, pp. 17–25, Karlsruhe, Germany, 2008.

[24] A. Poggi, "OWLET: an object-oriented environment for OWL ontology," in *Proceedings of the 11th WSEAS International Conference on Computers (ICCOMP '07)*, pp. 44–49, Stevens Point, Wis, USA, 2007.

[25] S. Bechhofer, R. Volz, and P. Lord, "Cooking the semantic web with the OWL API," in *Proceedings of the 2nd International Semantic Web Conference (ISWC '03)*, D. Fensel, K. Sycara, and J. Mylopoulos, Eds., vol. 2870 of *Lecture Notes in Computer Science*, pp. 659–675, Springer, Berlin, Germany, 2003.

[26] M. Horridge and S. Bechhofer, "The OWL API: a Java API for working with OWL 2 ontologies," in *Proceedings of the 5th International Workshop on OWL: Experiences and Directions (OWLED '09)*, Chantilly, Va, USA, 2009.

[27] I. Astrova and A. Kalja, "Mapping of SQL relational schemata to OWL ontologies," in *Proceedings of the 6th WSEAS International Conference on Applied Informatics and Communications (AIC '06)*, pp. 376–380, Stevens Point, Wis, USA, August 2008.

[28] H. K. Lin and J. A. Harding, "A manufacturing system engineering ontology model on the semantic web for inter-enterprise collaboration," *Computers in Industry*, vol. 58, no. 5, pp. 428–437, 2007.

[29] A. Salguero, F. Araque, and C. Delgado, "Ontology based framework for data integration," *WSEAS Transactions on Information Science and Applications*, vol. 5, no. 6, pp. 953–962, 2008.

[30] J. Z. Pan, "Resource description framework," in *International Handbook on Ontologies*, S. Staab and R. Studer, Eds., Handbooks on Information Systems, Part 1, pp. 71–90, Springer, Berlin, Germany, 2009.

[31] A. Poggi, "HDS: a software framework for the realization of pervasive applications," *WSEAS Transactions on Computers*, vol. 9, no. 10, pp. 1149–1159, 2010.

[32] F. Bergenti and A. Poggi, "Building distributed and pervasive information management systems with HDS," in *Advances in Distributed Agent-Based Retrieval Tools*, V. Pallotta, A. Soro, and E. Vargiu, Eds., vol. 361 of *Studies in Computational Intelligence*, pp. 129–142, Springer, Berlin, Germany, 2011.

[33] F. Bergenti, E. Franchi, and A. Poggi, "Selected models for agent-based simulation of social networks," in *Proceedings of the Social Networks and MultiAgent Systems Symposium (SNAMAS '11)*, pp. 27–32, York, UK, 2011.

[34] A. Poggi, "Developing ontology based applications with O3L," *WSEAS Transactions on Computers*, vol. 8, no. 8, pp. 1286–1295, 2009.

6

Countermeasures Assessment of Liquefaction-Induced Lateral Deformation in a Slope Ground System

Davide Forcellini[1] and Angelo Marcello Tarantino[2]

[1] *Dipartimento Economia e Tecnologia, Università degli Studi della Repubblica di San Marino, Via Salita alla Rocca, 44. San Marino Rep., San Marino*

[2] *Dipartimento Ingegneria Meccanica e Civile DIMeC, Università degli Studi di Modena e Reggio Emilia, Via Vignolese 905, Modena, Italy*

Correspondence should be addressed to Davide Forcellini; davforc@omniway.sm

Academic Editor: Lucian Dascalescu

Liquefaction-induced lateral spreading may result in significant damage and disruption of functionality for structures and Slope Ground System. In this regard, finite-element simulations are increasingly providing a versatile environment in order to assess economical and effective countermeasures. Several systematic bidimensional FEM computations have been conducted to evaluate mitigation strategies under the action of an applied earthquake excitation. The presented study highlights the potential of computations in providing insights for analysis of liquefaction-induced lateral deformations. In the analysis, some specific assumptions are introduced and verified such as a nine-node quadrilateral elements, massive columns of soil with periodic boundary conditions, and a Lysmer-Kuhlemeyer dashpot used to model the finite rigidity of the underlying elastic medium. Moreover, the study aims to systematically explore the effectiveness of densification as a countermeasure and then evaluate the best extension comparing two scenarios.

1. Introduction

Lateral spreading refers to the development of large horizontal ground displacements due to earthquake-induced liquefaction. This phenomenon may result in significant damage and considerable replacement costs for existing buildings and civil engineering structures (quay walls, bridge piers, etc.) since it imposes notable lateral loads and may lead to widespread failures. Such adverse response is documented during several seismic events, such as the earthquakes of Niigata, Japan (1964, [1–4]), Dagupan City, Philippines (1990, [5–8]), Chi-Chi, Kocaeli, Turkey (1999, [9]), Taiwan (1999, [10]), and recent Tohoku, Japan (2011, [11, 12]). Particularly important is the dynamic slope stability under liquefaction lateral deformation into areas where potential structures can be of interest to landslides, such as dams and river or pier banks, especially if previously predisposing phenomena have acted in the static stress field [13, 14]. In this regards, several ground remediations have been developing, such as

vibroreplacement [15], solidification (cementation) [16, 17], gravel drains, or stone columns [18–21].

This paper aims to assess the seismic reliability and evaluate the ground improvement method of densification for an Italian real case study of a pier founded on a partially submerged layered ground slope strongly vulnerable to liquefaction. The earthquake response of the Ground System is simulated with a two-dimensional, advanced, nonlinear finite element model adopting the open-source computational platform OpenSees [22]. The performance of structural and geotechnical systems subjected to static and seismic loadings can be simulated thanks to the platform implementing a framework for saturated soil response as a two-phase materials, following the *U-P* formulation of Chan [23] and Zienkiewicz et al. [24].

The present study may be viewed as a further development of earlier and ongoing efforts [25–28] to generate appropriate numerical models for simulation prediction of liquefaction-induced ground response using OpenSees.

In particular, the main numerical challenge was modelling the boundary conditions to reproduce the elastic half-space under the soil system.

In the following sections, the employed computational formulation is described. The computational platform for slope-ground analysis is then presented focusing on the adopted materials, boundary conditions, and analysis assumptions in Section 3. Results of the conducted analysis for the free-field configuration are presented as well as the two scenarios countermeasures' response, respectively, in Sections 4 and 5. Finally, insight and conclusions based on the effectiveness of the proposed remediation are drawn.

2. Computational Formulation

All simulations were conducted using the open-source computational platform OpenSees [22]. This platform allows for developing applications to simulate the performance of structural and geotechnical systems subjected to static and seismic loadings. Implemented in OpenSees [25–28] is an analysis framework for saturated soil response as a two-phase material following the U-P formulation of Chan [23] and Zienkiewicz et al. [24], where U is displacement of the soil skeleton and P is pore pressure. This implementation is based on the following assumptions:

(i) small deformation and rotation;

(ii) solid and fluid density constant in both time and space;

(iii) porosity locally homogeneous and constant with time;

(iv) soil grains incompressible;

(v) solid and fluid phases equally accelerated.

The soil constitutive model, Figures 1 and 2 implemented in OpenSees [25–29], is based on the multisurface plasticity theory for frictional cohesionless soils proposed by Prevost [30], where p' is the effective confining pressure, t the octahedral shear stress, and g the octahedral shear strain. In particular, this constitutive model was developed for simulating the characteristics of cyclic mobility observed in saturated medium to dense cohesionless soil response [25–28]. Within a multisurface plasticity framework, the model incorporates shear-induced contractive, perfectly plastic and dilative response phases implemented through an appropriate nonassociative flow rule motivated by experimental observations as to capture the involved phenomena. Emphasis is placed on accurately reproducing the development and accumulation of shear deformations. The hardening rule was also introduced to enhance numerical robustness and to increase efficiency. Finally, a model calibration procedure based on monotonic and cyclic laboratory sample test data was conducted.

3. Computational Platform for Slope: Ground Analysis

The 2D Ground System is a typical slope 1000 m long, 60–130 m high as shown in Figure 3. Such model is built

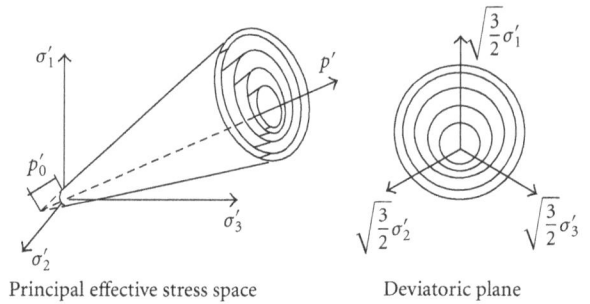

FIGURE 1: Conical yield surfaces in principal stress space and deviatoric plane [28, 30].

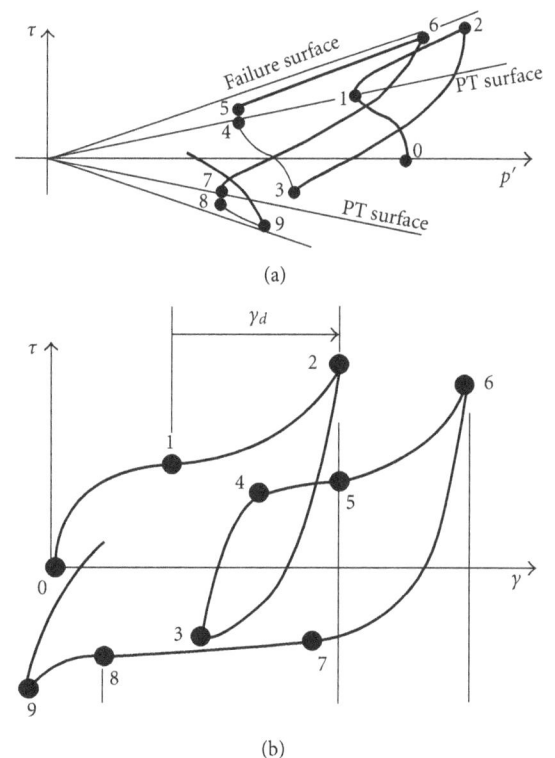

FIGURE 2: Shear stress-strain and effective-stress path under-undrained shear loading conditions [28].

with several layers of cohesive and cohesionless material, mainly of silty medium to fine sands (S), pliocene clays (C) as background layer, and a superficial layers of gravelly sands (G) located near the ground surface. OpenSees may implement a wise number of soil models including multiyield surface cohesionless (Drucker-Prager cone model) and cohesive (Mises or J2) ones. In order to model the different layers of the problem two models were adopted:

(i) Pressure-Independent Multiyield (Table 1) for pliocene clays (C) and

(ii) Pressure-Dependent Multiyield02 (Tables 2 and 3) for silty medium to fine sands (S).

TABLE 1: Characteristics adopted in the study for Pressure-Independent Multiyield model.

Unity	Description	Model	Mass (ton/m^3)	G (kPa)	B (kPa)	Cohesi on (kPa)	Peak shear strain (%)	Ref. press (kPa)	Pressure depend. coeff.
Unit 3	Pliocene clays	Pressure independ. multiyield	1.8	245090	408483	150	0.03	270	0

TABLE 2: Characteristics adopted in the study for Pressure-Dependent Multiyield02 model.

Unity	Description	Model	Mass density (ton/m^3)	Shear modulus (kPa)	Bulk modulus (kPa)	Friction angle	Peak shear strain (%)	Reference pressure (kPa)	Press. depend. coeff.	Pt angle
Fill	Artificial fill	Pressure depend. multiyield02	1.8	14580	24300	32	3	14	0.5	30
Unit 1 A	Gravelly sands	Pressure depend. multiyield02	1.8	43805	73008	36	3	53	0.5	34
Unit 1 B	Clean medium sands, clean coarse to medium sands	Pressure depend. multiyield02	1.8	97720	162867	35	3	154	0.5	33
Unit 1 C	Sandy silts + Clay sandy silts	Pressure depend. multiyield02	1.8	87914	146523	31.5	3	105	0.5	35
Unit 2	Slightly silty medium to fine sands	Pressure depend. multiyield02	1.8	146205	243675	33.5	3	200	0.5	31

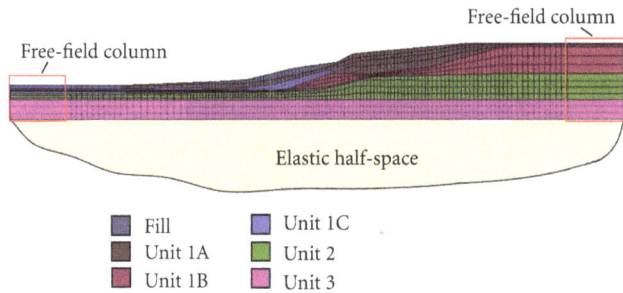

FIGURE 3: FEM model of Slope Ground System.

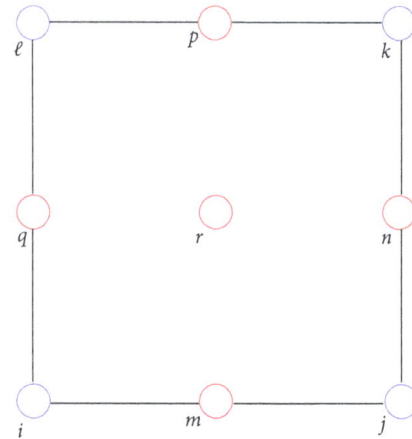

Node order $i\,j\,k\,\ell\,m\,n\,p\,q\,r$

FIGURE 4: 9_4_QuadUP element used in the study.

Model implementation is based on several computational assumptions (regarding finite elements, boundary conditions, and analysis), shown in next sections.

4. Finite Elements

The 9_4_QuadUP elements used (Figure 4) in this study were developed in plain-strain deformation conditions following the Biot Theory of porous medium. Such elements allow to take into consideration the solid skeleton of the soil (all 9 nodes) and also the fluid phase (4 corner nodes). In particular, the 4 corner nodes have 3 DOFs (2 displacements and 1 pore pressure) while the interior nodes have only 2 DOFs (two displacements).

5. Boundary Conditions

Numerical simulation of dynamic Slope Ground System problems requires many efforts to reproduce the real wave propagation adopting realistic boundary conditions. At this aim, several assumptions such as one elastic half-space under the soil system, massive columns of soil with periodic conditions, one Lysmer-Kuhlemeyer dashpot [31] at the base

TABLE 3: Liquefaction parameters for Pressure-Dependent Multiyield02 models.

Unity	Description	Model	Contraction and dilation model parameters				number of yield surfaces	Initial void ratio	Critical-state parameters		
			$c1$	$c3$	$d1$	$d3$			cs1	cs2	cs3
Fill	Artificial Fill	Pressure-Depend. Multiyield02	0.087	0.18	0	0	5	0.65	0.9	0.02	0.7
Unit 1 A	Gravelly sands	Pressure-Depend. Multiyield02	0.087	0.18	0	0	5	0.65	0.9	0.02	0.7
Unit 1 B	Clean medium sands, clean coarse to medium sands	Pressure-Depend. Multiyield02	0.045	0.15	0.06	0.15	5	0.75	0.9	0.02	0.7
Unit 1 C	Sandy silts + Clay sandy silts	Pressure-Depend. Multiyield02	0.013	0	0.3	0	5	0.8	0.9	0.02	0.7
Unit 2	Slightly silty medium to fine sands	Pressure-Depend. Multiyield02	0.013	0	0.3	0	5	0.8	0.9	0.02	0.7

of the model, and nodal mass simulating water conditions have been introduced. First of all, the entire site is underlain by an elastic half-space that was chosen to be consistent with the existence of bedrock below the slope site, as to allow the energy imparted by the seismic event to be removed from the site itself. The nodes on the base of the model are free to displace in horizontal directions and fixed against the vertical translation. Secondly, in the horizontal direction, the model represents a small section of a presumably infinite (or at least very large) soil domain. To ensure that free-field conditions exist at the horizontal boundaries of the model, the elements in these locations (indicated as red lines in Figure 1) are modelled significantly more massive than the interior mesh. For this purpose, their thickness was increased notably and the nodes on either side of these columns are tied together. Finally, to ensure that the critical portions of the model are not affected by the horizontal boundaries, the free-field columns were located sufficiently far away from the critical regions. Moreover, the dynamic excitation motion was assigned to a Lysmer-Kuhlemeyer [31] dashpot defined through a single zeroLength element. The first end of the dashpot element is fixed against all displacements, while the other end is connected to the soil node in the lower left-hand corner. The constitutive behaviour of the Lysmer-Kuhlemeyer dashpot in the horizontal direction is modelled by a viscous uniaxial material, that requires the dashpot coefficient, C. Following the method of Joyner and Chen [32], this coefficient is defined as the product of the mass density and shear wave velocity of the underlying medium (assumed to have the bedrock properties). In order to ensure that equivalent loading is applied to the model, the dashpot coefficient must be scaled by the area of the base of the model. Finally, the slope was considered completely submerged. Aimed to incorporate the dynamic effects of the water on the site without altering the effective stresses in the soil elements, a nodal mass is assigned manually to each node on the boundary of the mesh which is below water level. For the nodes on the level surface, the horizontal mass is set to zero and the vertical mass is set as the mass of the volume of water supported by the node.

6. Analysis

To control the various parts of the problems and to manage the wise quantities of results, the analysis was split into two consequent steps. The first one is gravity application and the second one is the dynamic analysis itself. Gravity application ensures that the distributions of pore pressure and effective stresses are appropriate for the site conditions prior to the application of a ground motion. Separate recorders were set up as to distinct the gravity analysis from any other postgravity results. Nodal displacements, accelerations, and pore pressures are recorded along with the elemental stresses and strains at each of the nine Gauss points. In order to achieve hydrostatic pore pressure conditions, gravity application analysis is divided into two parts: soil elements are considered to be firstly linear elastic and then elastoplastic. The elastic portion of the gravity analysis is run as a Transient Analysis with very large time steps, thus simulating a static analysis. Gravity is applied for 10 steps with a time step of 500, and 10 steps with a time step of 5000. Therefore, the plastic portion of the gravity analysis is run using smaller time steps to aid in convergence. Dynamic excitation analysis is developed using the method of Joyner and Chen [32]. The force time history was applied as a Plain Load Pattern at the Lysmer-Kuhlemeyer [31] dashpot. Two input motions (shown in Figure 5) are considered at the following hazard levels:

(i) *T-475*: 5% of probability of exceedance in 475 years, representing service limit conditions;

(ii) *T-5000*: 5% of probability of exceedance in 5000 years, both representing collapse limit conditions.

7. Free Field Response

Figures 6 and 7 show the results in terms of horizontal displacement at final time step. As expected, the site is strongly

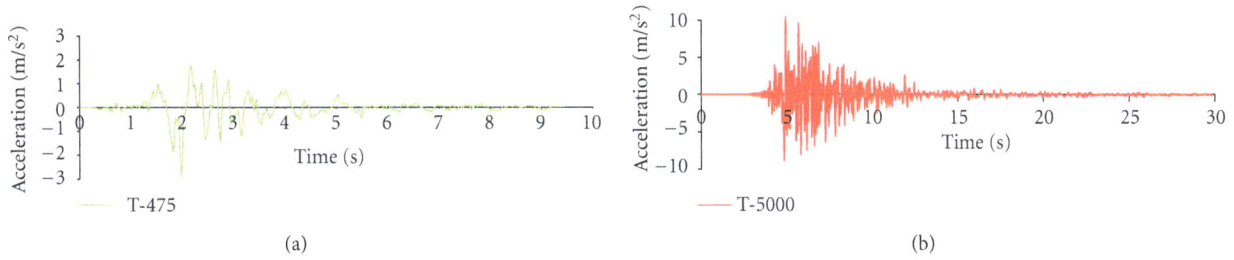

(a)   (b)

FIGURE 5: Input motions.

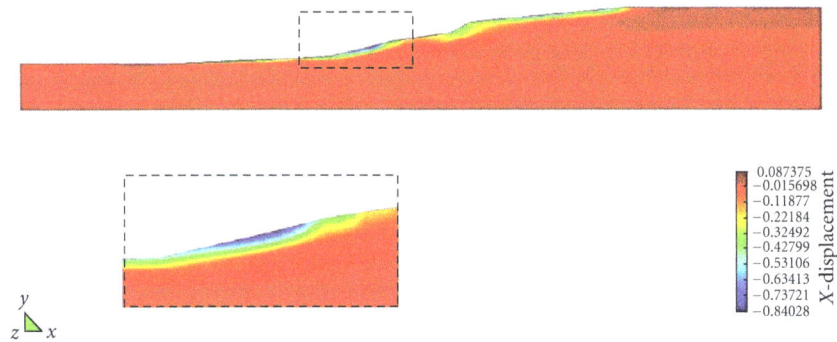

Contour fill of a. nodal displacements, X-displacement

FIGURE 6: T-475: contour deformation at final time step.

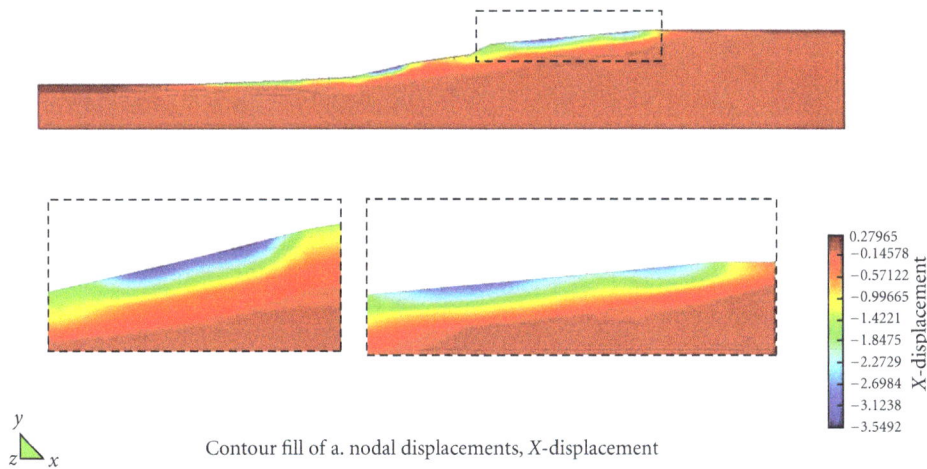

Contour fill of a. nodal displacements, X-displacement

FIGURE 7: T-5000: contour deformation at final time step.

FIGURE 8: Soil locations considered in the study.

subjected to soil liquefaction and consequent approach fill settlement and lateral spreading. The zoomed views show that the main values are reached in correspondence with the superficial layers that slide on the deeper ones. Under service limit condition (T-475 motion) the displacements are around 0.70 m, while for collapse conditions (T-5000 motion) they grow up to more than 2.50 m. On the superficial layer, three main locations (Figure 8) and respective time histories were considered (Figure 9). Even if the time histories have similar shape, locations 1 and 2 present different values. In particular, for T-475, the displacements at the final time step are around 0.70 m for location 1 and around 0.40 m for location 2. For T-5000, these values grow to around 2.50 m for location 1 and more than 1 m for location 2. Location 3, the upper side

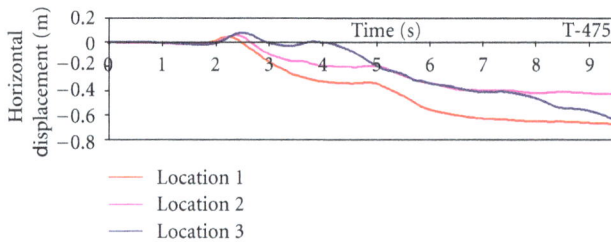

(a)

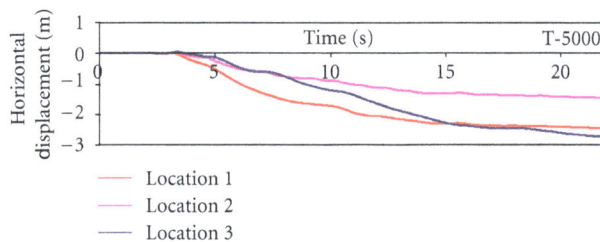

(b)

FIGURE 9: Time histories—three locations.

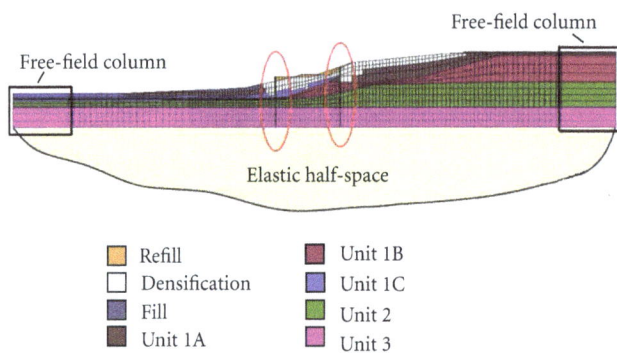

FIGURE 10: FEM model of Slope Ground System.

of the slope, shows different time histories. This behaviour evidences that the upper sides need more time to slip if compared to the other parts of the slope. In particular, for T-475, the displacement at the final time step is around 0.65 m, while for T-5000 is around 2.80 m.

8. Countermeasures Response

This study aims to systematically explore the effectiveness of densification (Figure 10) consisting of the following:

(i) construction of two series of steel 2.5 m diameter hammered piles (EI = $4.032 \cdot 10^8$ kNm^2) in the downhill and uphill of the central zone. The piles are modelled with elastic beam column elements with equivalent flexural characteristics;

(ii) densification of the areas that are resulted to be mainly subjected to liquefaction risk. In particular, such areas are evaluated with two different increasing extensions (named model 1 and model 2). Model 1: densification inside the two series of steel piles and for the first 5

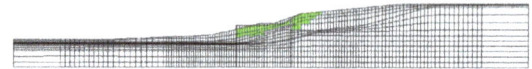

FIGURE 11: Countermeasure: model 1.

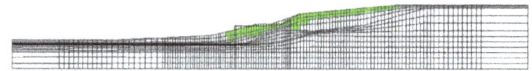

FIGURE 12: Countermeasure: model 2.

FIGURE 13: Section considered in the study.

meters out of section and 30 meters maximum depth (as painted in green colour in Figure 11).

Model 2: the model 1 densification is extended to the superficial layer (at maximum 15 meters depth) out of section 2 for 150 meters (as shown in green colour in Figure 12).

The densification technique consists of increasing the superficial layer (named as Fill in Figure 3) relative density to a value equal to 70% that considerably modifies the development of pore pressure. Such value was taken as a reference as the main goal to the technique itself in order to practically annulling liquefaction reproduction;

(iii) refill with the same soil material as the original one (named Fill) in order to create an horizontal platform.

The comparison between the two scenarios and the free field response is discussed in correspondence with the two sections pointed out in Figure 13. Longitudinal displacement time histories for the two scenarios are drawn in Figures 14, 15, 16, and 17 for T-475 and T-5000. Scenario 1 is seen to be effective only in correspondence with section 1; while extending the densification in the superficial layer (scenario 2), the displacements in both sections decrease sensitively. These considerations are expressed numerically in Tables 4 and 5, where the two scenario displacements are compared with the free field response values for both sections. In Figures 18 and 19 the displacement time histories at the top of the two piles are compared for both considered motions. Finally, Figures 20 and 21 show the displacement and contour deformation at final time step for the two scenarios for the collapse limit conditions (T-5000).

9. Conclusions

The study conducted in this paper demonstrates OpenSees high potentialities in performing appropriate numerical simulations for predicting liquefaction-induced lateral deformation. The results confirm the assumptions concerning the reproduction of boundary conditions such as the elastic half-space under the soil system, massive columns of soil with

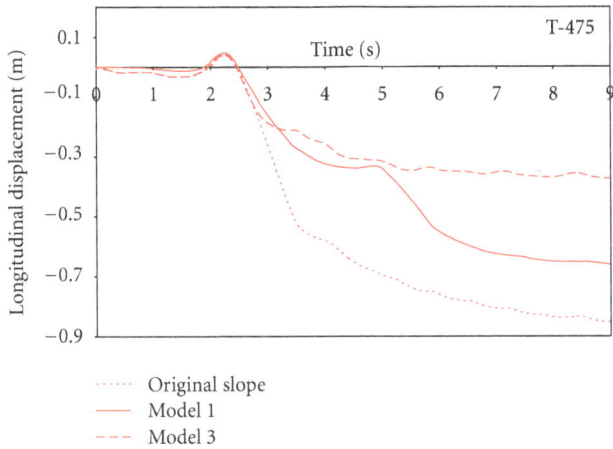

FIGURE 14: Comparison time histories—Section 1—acc: T-475.

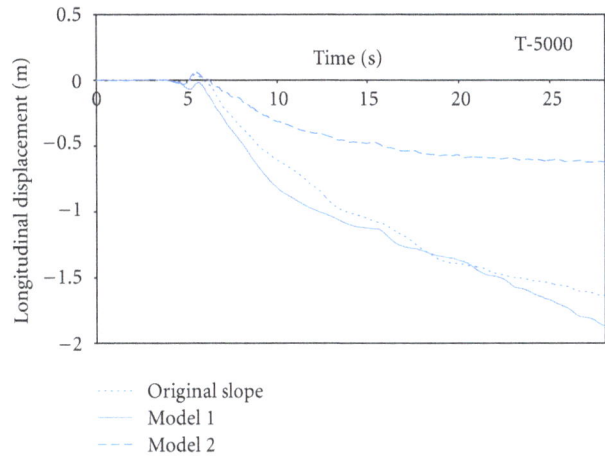

FIGURE 15: Comparison time histories—Section 2—acc: T-475.

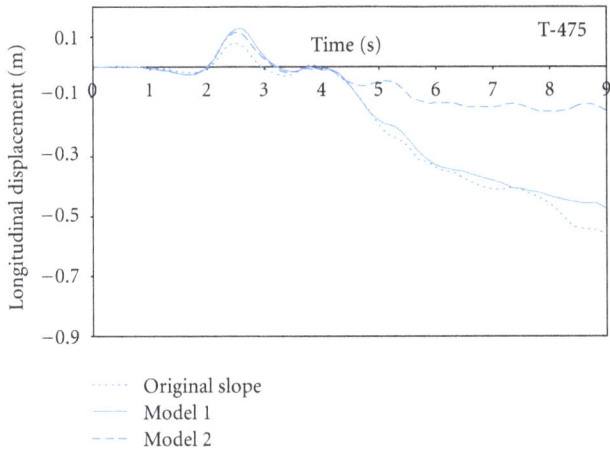

FIGURE 16: Comparison time histories—Section 1—acc: T-5000.

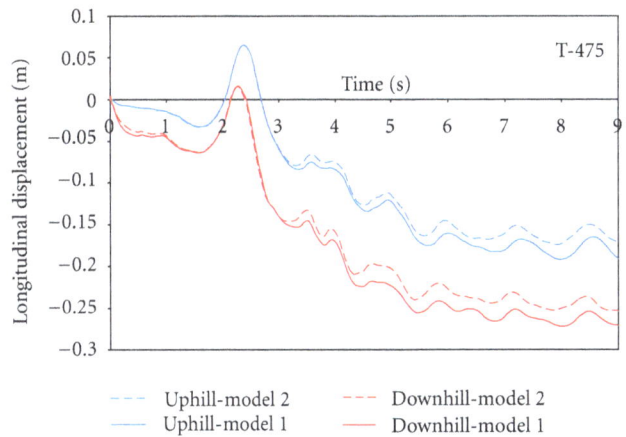

FIGURE 17: Comparison time histories—Section 2—acc: T-5000.

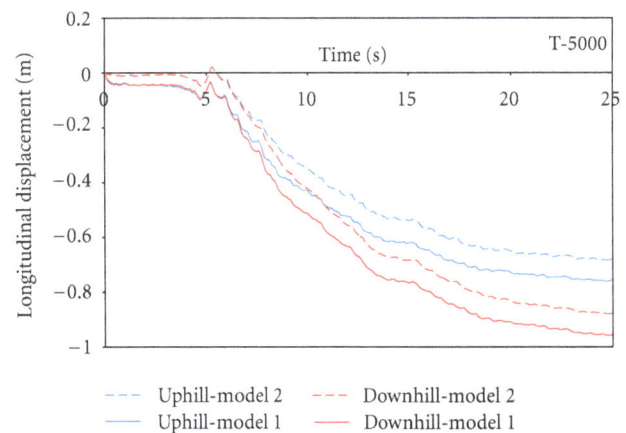

FIGURE 18: Comparison top displacement uphill and downhill wall—acc: T-475.

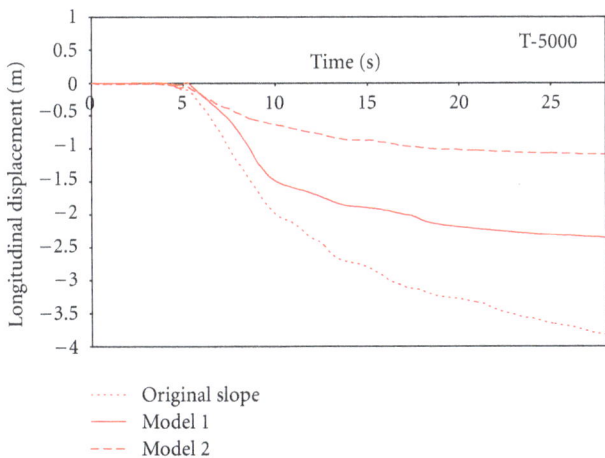

FIGURE 19: Comparison top displacement uphill and downhill wall—acc: T-5000.

periodic boundary conditions, and one Lysmer-Kuhlemeyer dashpot. Moreover, the assignment of nodal mass for each node on the boundary below water level allowed to reproduce the dynamic effects of the water on the site without altering the effective stresses themselves. Finally, splitting of the analysis into two consequent steps (gravity application and dynamic motions) with different computational choices in terms of time steps and force time history reveals its potentiality.

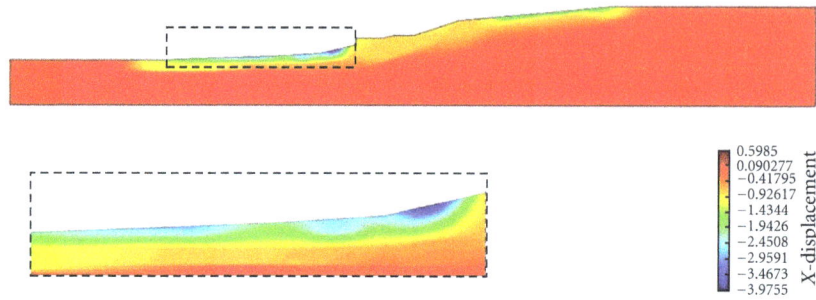

Contour fill of a. nodal displacements, X-displacement

FIGURE 20: Contour deformation at final time step—model 1—acc: T-5000.

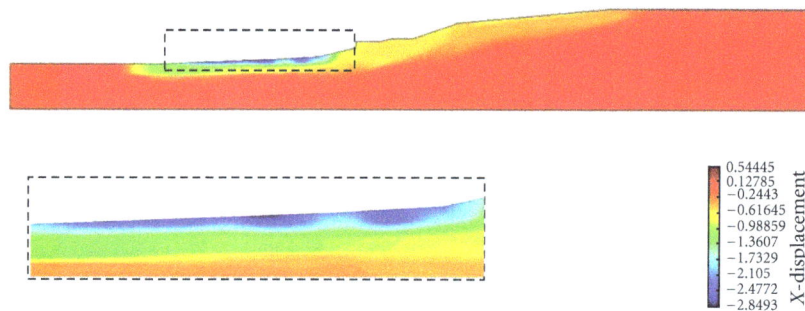

Contour fill of a. nodal displacements, X-displacement

FIGURE 21: Contour deformation at final time step—model 2—acc: T-5000.

TABLE 4: Comparison: longitudinal displacement of the models ratio—Section 1.

Section 1	T-475	T-5000
Model 1	0.78	0.62
Model 2	0.44	0.28

TABLE 5: Comparison: longitudinal displacement of the models ratio—Section 2.

Section 2	T-475	T-5000
Model 1	0.84	1.10
Model 2	0.27	0.38

The analyses confirm the vulnerability of such a partially submerged slope in terms of approach fill settlement and lateral spreading due to liquefaction. In particular, the results offer a main reference in the evaluation of the counter-measures design. The extensions for ground improvement densification are compared in terms of displacements in correspondence with significant positions for both service and collapse conditions.

The analyses evidence the effectiveness of densification in reducing the original fill settlement and lateral spreading due to liquefaction. In this regards, comparing these results with economic evaluation can help to quantify the performance and risk of liquefaction using metrics that are of immediate use to both engineers and stakeholders.

References

[1] H. Kishida, "Damage to reinforced concrete buildings in Niigata city with special reference to foundation engineering," *Soils Foundation*, vol. 6, no. 1, pp. 71–88, 1966.

[2] Y. Ohsaki, "Niigata earthquake, 1964 building damage and soil condition," *Soils Foundations*, vol. 6, no. 2, pp. 14–37, 1966.

[3] H. B. Seed and I. M. Idriss, "Analysis of soil liquefaction: Niigata earthquake," *Journal of Soil Mechanics and Foundations*, vol. 93, no. 3, pp. 83–108, 1967.

[4] Y. Yoshimi and K. Tokimatsu, "Settlement of buildings on saturated sand during earthquakes," *Soils and Foundations*, vol. 17, no. 1, pp. 23–38, 1977.

[5] K. Tokimatsu, S. Midorikawa, S. Tamura, S. Kuwayama, and A. Abe, "Preliminary report on the geotechnical aspects of the Philippine earthquake of July 16, 1990," in *Proceedings of the 2nd International Conference on Recent Advances in Geotechnical Earthquake Engineering and Soil Dynamics*, pp. 357–364, University of Missouri-Rolla, St. Louis, Mo, USA, 1991.

[6] T. Adachi, S. Iwai, M. Yasui, and Y. Sato, "Settlement of incli-nation of reinforced concrete buildings in Dagupan city due to liquefaction during 1990 Philippine earthquake," in *Proceedings of the 10th World Conference on Earthquake Engineering*, pp. 147–152, Rotterdam, The Netherlands, 1992.

[7] K. Ishihara, A. A. Acacio, and I. Towhata, "Liquefaction-induced ground damage in Dagupan in the July 16, 1990 Luzon earthquake," *Soils and Foundations*, vol. 33, no. 1, pp. 133–154, 1993.

[8] K. Tokimatsu, H. Kojimaa, S. Kuwayama, A. Abe, and S. Midorikawa, "Liquefaction-induced damage to buildings in 1990 Luzon Earthquake," *Journal of Geotechnical Engineering*, vol. 120, no. 2, pp. 290–307, 1994.

[9] Earthquake Engineering Research Institute (EERI), "Kocaeli, Turkey, earthquake of august 17, 1999 reconnaissance report," Earthquake Spectra, 2000.

[10] Earthquake Engineering Research Institute (EERI), "Chi-chi, Taiwan, earthquake of September 21, 1999, reconnaissance report," Earthquake Spectra, 2001.

[11] K. Irikura and S. Kurahashi, "Source model for generating strong ground motions during the 11 March 2011 off Tohoku, Japan earthquake," in *Proceedings of the Japan Geoscience Union International Symposium*, Makuhari, Chiba, Japan, 2011.

[12] S. Bhattacharya, M. Hyodo, K. Goda, T. Tazoh, and C. A. Taylor, "Liquefaction of soil in the Tokyo Bay area from the 2011 Tohoku (Japan) earthquake," *Soil Dynamics and Earthquake Engineering*, vol. 31, pp. 1618–1628, 2011.

[13] D. K. Keefer, "Landslides caused by earthquakes," *Geological Society of America Bulletin*, vol. 95, no. 4, pp. 406–421, 1984.

[14] G. Tosatti, D. Castaldini, M. Barbieri et al., "Additional causes of seismically-related landslides in the Northern Apennines," *Italy. Revista De Geomorfologie*, no. 10, pp. 5–21, 2008.

[15] H. J. Priebe, "The prevention of liquefaction by vibroreplacement," in *Proceedings of the 2nd International Conference on Earthquake Resistant Construction and Design*, S. A. Savidis, Ed., pp. 211–219, Balkema, Rotterdam, The Netherlands, 1991.

[16] J. K. Mitchell, C. D. P. Baxter, and T. C. Munson, "Performance of improved ground during earthquakes," in *Proceedings of the Conference of the Geotechnical Engineering Division of the ASCE in Conjunction with the ASCE Convention*, pp. 1–36, October 1995.

[17] R. Boulanger, I. Idriss, D. Stewart, Y. Hashash, and B. Schmidt, "Drainage capacity of stone columns or gravel drains for mitigating liquefaction," *Geotechique Special Publications*, vol. 1, no. 75, pp. 678–690, 1998.

[18] Japanese Geotechnical Society (JGS), *Special Issue on Geotechnical Aspects of the January 17, 1995 Hyogoken-Nanbu Earthquake*, Soils Foundations, 1998.

[19] S. Thevanayagam, G. R. Martin, T. Shenthan, and J. Liang, "Post-liquefaction pore pressure dissipation and densification in silty soils," in *Proceedings of the 4th International Conference on Recent Advances in Geotechnical Earthquake Engineering and Soil Dynamics*, S. Prakash, Ed., no. 4.28, San Diego, Calif, USA, 2001.

[20] A. Elgamal, E. Parra, Z. Yang, and K. Adalier, "Numerical analysis of embankment foundation liquefaction countermeasures," *Journal of Earthquake Engineering*, vol. 6, no. 4, pp. 447–471, 2002.

[21] A. Elgamal, J. Lu, and D. Forcellini, "Mitigation of liquefaction-induced lateral deformation in a sloping stratum: three-dimensional numerical simulation," *Journal of Geotechnical and Geoenvironmental Engineering*, vol. 135, no. 11, pp. 1672–1682, 2009.

[22] S. Mazzoni, F. McKenna, M. H. Scott, and G. L. Fenves, *OpenSystem For Earthquake Engineering Simulation User Manual*, University of California, Berkeley, Calif, USA, 2006, http://opensees.berkeley.edu/.

[23] A. H. C. Chan, *A unified finite element solution to static and dynamic problems in geomechanics [Ph.D. thesis]*, University College of Swansea, Swansea, UK, 1988.

[24] O. C. Zienkiewicz, A. H. C. Chan, M. Pastor, D. K. Paul, and T. Shiomi, "Static and dynamic behaviour of soils: a rational approach to quantitative solutions. I. Fully saturated problems," *Proceedings of the Royal Society of London A*, vol. 429, no. 1877, pp. 285–309, 1990.

[25] Z. Yang, *Numerical modeling of earthquake site response including dilation and liquefaction [Ph.D. thesis]*, Columbia University, New York, NY, USA, 2000.

[26] Z. Yang and A. Elgamal, "Influence of permeability on liquefaction-induced shear deformation," *Journal of Engineering Mechanics*, vol. 128, no. 7, pp. 720–729, 2002.

[27] A. Elgamal, Z. Yang, E. Parra, and A. Ragheb, "Modeling of cyclic mobility in saturated cohesionless soils," *International Journal of Plasticity*, vol. 19, no. 6, pp. 883–905, 2003.

[28] Z. Yang, A. Elgamal, and E. Parra, "Computational model for cyclic mobility and associated shear deformation," *Journal of Geotechnical and Geoenvironmental Engineering*, vol. 129, no. 12, pp. 1119–1127, 2003.

[29] E. Parra, *Numerical modelling of liquefaction and lateral ground deformation including cyclic mobility and dilation response in soil systems [Ph.D. thesis]*, Department of Civil Engineering, Renseealear Polytechnic Institute Troy, New York, NY, USA, 1996.

[30] J. H. Prevost, "A simple plasticity theory for frictional cohesionless soils," *International Journal of Soil Dynamics and Earthquake Engineering*, vol. 4, no. 1, pp. 9–17, 1985.

[31] J. Lysmer and A. M. Kuhlemeyer, "Finite dynamic model for infinite media," *Journal of the Engineering Mechanics Division*, vol. 95, pp. 859–877, 1969.

[32] W. B. Joyner and A. T. F. Chen, "Calculation of nonlinear ground response in earthquakes," *Bulletin of the Seismological Society of America*, vol. 65, no. 5, pp. 1315–1336, 1975.

7

Distal Placement of an End-to-Side Bypass Graft Anastomosis: A 3D Computational Study

John Di Cicco and Ayodeji Demuren

Department of Mechanical and Aerospace Engineering, Old Dominion University, Norfolk, VA 23529, USA

Correspondence should be addressed to Ayodeji Demuren; ademuren@odu.edu

Academic Editor: Alireza Khataee

A three-dimensional (3D) computational fluid dynamics study of shear rates around distal end-to-side anastomoses has been conducted. Three 51% and three 75% cross-sectional area-reduced 6 mm cylinders were modeled each with a bypass cylinder attached at a 30-degree angle at different placements distal to the constriction. Steady, incompressible, Newtonian blood flow was assumed, and the full Reynolds-averaged Navier-Stokes equations, turbulent kinetic energy, and specific dissipation rate equations were solved on a locally structured multiblock mesh with hexahedral elements. Consequently, distal placement of an end-to-side bypass graft anastomosis was found to have an influence on the shear rate magnitudes. For the 75% constriction, closer placements produced lower shear rates near the anastomosis. Hence, there is potential for new plaque formation and graft failure.

1. Introduction

The leading cause of death in the developed world is the cardiovascular disease, atherosclerosis. It is a progressive disease, in which atherosclerotic plaques, consisting of lipids and cholesterol, slowly develop over time to cause a narrowing of the arterial lumen. The narrowed arterial lumen is called a stenosis, which may grow to significantly reduce or completely obstruct the flow of blood. Turbulence and an adverse pressure gradient may be produced by a stenosis, where separated flow regions could exist and extend several diameters downstream in the poststenotic region.

Surgery is often needed to restore blood flow to tissues affected by a critical arterial stenosis, which is around a 70% cross-sectional area reduction. Commonly, stenotic arteries are repaired by vascular surgical procedures that bypass the stenosis with a conduit called a graft. Furthermore, the end-to-side anastomosis is a common technique used in bypass grafting, where anastomosis refers to a connection between two vessels, and in the case of an end-to-side anastomosis, the end of a graft is attached to the side of an artery with sutures. A significant occurrence associated with this major operation is that it frequently requires revision due to graft failure.

Late graft failure, thirty days after operation, frequently occurs because of normal cell proliferation that results in a thickening of the arterial inner wall called intimal hyperplasia [1]. The primary site for graft failure is the distal anastomosis [2, 3], where intimal hyperplasia is thicker along the floor of the host artery under the anastomosis and along the wall just distal to the toe of the anastomosis [4].

Researchers such as Bandyk et al. [5] and Dobrin et al. [6] have reported that intimal hyperplasia is found in regions of low wall shear stress. Furthermore, researchers such as Caro et al. [7, 8] have observed a possible connection between blood flow and arteriosclerosis by way of low wall shear stress, thereby, leading to a hypothesis correlating low wall shear stress and atherogenesis. In spite of this, it should be noted that atherogenesis may not be a direct result of low wall shear stress, but rather low wall shear stress may result in intimal thickening, which may in turn lead to atherogenesis [9].

Previous computational studies of the effect post-stenotic blood flow phenomena has on wall shear stress around a distal end-to-side anastomosis include Bertolotti and Deplano's [10] low Reynolds number work and a low resolution work by Kute and Vorp [11]. Further investigation is warranted due to the substantial research, indicated previously, that has shown the proliferation of intimal hyperplasia, which causes late distal graft failure, over a wide range of flow Reynolds numbers, in regions of low wall shear stress. Moreover, if such a relationship between distal placement and wall shear stress is determined, then to some degree vascular surgeons can control the influence of wall shear stress around the anastomosis, which may lead to a prolongation of bypass graft functionality. Accordingly, the present study is concerned with determining how distal placement of an end-to-side bypass graft anastomosis affects wall shear rates.

The following sections describe briefly governing flow equations and computational methods. Then computational results are presented followed by a summary and conclusion of the present study's main findings.

2. Mathematical Formulation

2.1. Flow Equations. Reynolds averaging was used to time-average the instantaneous full Navier-Stokes equations to produce Reynolds-averaged equations of fluid motion, which are better suited to predict the velocity field of a turbulent flow. Assuming the flow is steady and the gravity is negligible, these equations in Cartesian tensor notation are

$$\frac{\partial U_i}{\partial x_i} = 0, \tag{1}$$

$$U_j \frac{\partial U_i}{\partial x_j} = \nu \frac{\partial^2 U_i}{\partial x_j x_j} - \frac{1}{\rho} \frac{\partial P}{\partial x_i} + \frac{\partial \tau_{ij}}{\partial x_j}, \tag{2}$$

where $x_i = (x_1, x_2, x_3)$ represent the Cartesian coordinates, $U_i = (U_1, U_2, U_3)$ represent the Cartesian time-averaged velocity components, P represents the time-averaged pressure, and ν represents the kinematic viscosity, which is a ratio of absolute viscosity to density.

The Reynolds-stress tensor in (2) has introduced six more unknowns in addition to the four unknowns already present, but has not introduced any additional equations. Thus, there is a closure problem. A low Reynolds number form of the $k - \omega$ turbulence model [12] was used to achieve closure. The Boussinesq approach is incorporated relating the Reynolds stresses to the mean velocity gradients. This is represented as

$$\tau_{ij} = \nu_t \left(\frac{\partial U_i}{\partial x_j} + \frac{\partial U_j}{\partial x_i} \right), \tag{3}$$

where ν_t is the turbulent kinematic viscosity and is assumed to be an isotropic scalar quantity. It is expressed as

$$\nu_t = \alpha^* \frac{k}{\omega}, \tag{4}$$

where α^* is the low Reynolds number correction term, which is defined as

$$\alpha^* = \alpha_\infty^* \left(\frac{\alpha_0^* + \mathrm{Re}_t / R_k}{1 + \mathrm{Re}_t R_k} \right). \tag{5}$$

The transport equation for turbulent kinetic energy, k, in Cartesian tensor notation is

$$U_i \frac{\partial k}{\partial x_i} = \frac{\partial}{\partial x_j} \left(\Gamma_k \frac{\partial k}{\partial x_j} \right) + G_k - Y_k. \tag{6}$$

The transport equation for specific dissipation rate, ω, in Cartesian tensor notation is

$$U_i \frac{\partial \omega}{\partial x_i} = \frac{\partial}{\partial x_j} \left(\Gamma_\omega \frac{\partial \omega}{\partial x_j} \right) + G_\omega - Y_\omega. \tag{7}$$

2.2. Computational Details. FLUENT 6.0 is a robust commercial computational fluid dynamics (CFDs) software package that includes the programs GAMBIT 2.0, TGrid 3.4, and FLUENT (the solver). These programs were used in a process to create, to solve, and to interpret the computational tasks of this paper.

During the preprocessing stage, the computational model's coordinate system, geometry, mesh, and boundary conditions were created with GAMBIT, and the quality of the mesh was checked with TGrid. The geometry was based on an ideal femoral artery and bypass graft with 6 mm open diameters. The stenoses studied were 2D in length and smoothly constricted to either a 51% or a 75% area reduction at the throat. Both stenoses resembled an hour-glass shape and were created using cosine equations. In addition, fully developed laminar flow was chosen to flow into the stenosis and into the artery from the graft. Furthermore, the post-stenotic length was chosen to reestablish fully developed laminar flow. Lastly, the anastomotic junction angle was 30 degrees, which is taught to vascular surgeons as an optimal angle (or less) for graft attachment [13].

Six different geometries are presented for this study: three 51% and three 75% arterial models each with a different bypass graft attachment site. Table 1 summarizes the distal bypass graft attachment sites distal to the arterial stenosis, where L_G is the length measured from the throat of the stenosis to the toe of the graft. For case one, the graft was placed in the separation region, and for case two, the graft was placed in the reattachment region of separated flow. Both were also subjected to post-stenotic turbulence. For case three, the graft was placed far downstream to allow a confluence of fully developed laminar flow from the artery and graft. A typical 3D mesh generated in this study is shown in Figure 1, which corresponds to case one.

In addition, a locally structured multiblock mesh was chosen for the present study. A grid sensitivity study was performed on simulations of flow configurations, which correspond to experiments of Ahmed and Giddens [14] and Keynton et al. [15]. Unstructured meshes generated with tetrahedral elements were compared to structured multiblock meshes with hexahedral elements. Mesh quality improved in the latter, with skewness factors going down from 0.3 to below 0.1. Corresponding results were also more accurate. Some mesh sensitivity results are shown in Figure 3.

Geometry of an artery with one end fully occluded and with a graft end-to-side anastomosis at a 30 degree angle was created and meshed. The meshes included: (1) a boundary layer mesh (BL), where cells were clustered toward the walls; (2) a longitudinal mesh (Long), where cells were clustered toward the junction along the x-axis; (3) a refined mesh; and (4) a coarse mesh. Shear rates calculated at the outer wall were normalized with the corresponding shear rate of the graft indicated as "b" in Figure 2. Figure 3 shows the outer

TABLE 1: Summary of graft placements.

	L_G/D	
	51% Stenosis	75% Stenosis
Case 1	3	3
Case 2	~4	~5.7
Case 3	67	67

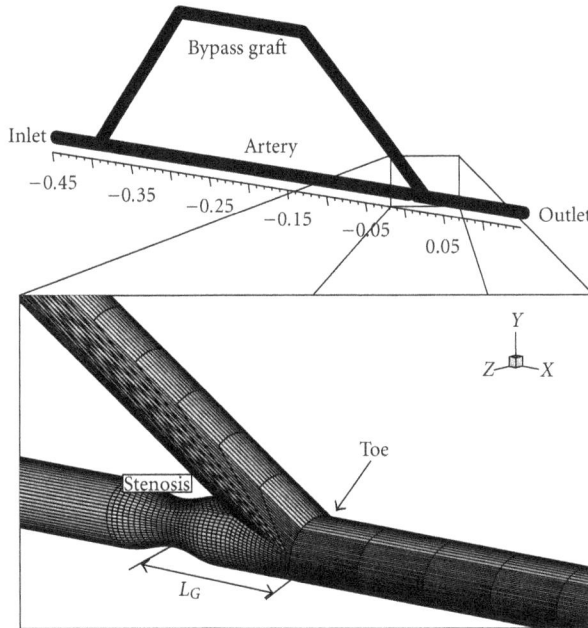

FIGURE 2: Axial Measurements Reference Schematic.

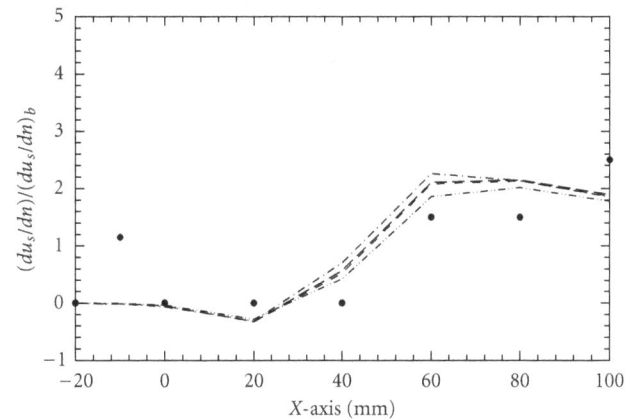

FIGURE 1: Three-dimensional computational domain and typical mesh of the present study.

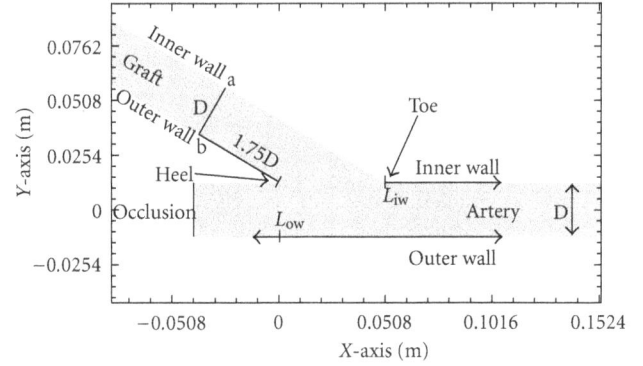

- - - - Multi-block (BL)-280.000 cells
- - - - Multi-block (long)-176.000 cells
- - - - Multi-block-176.000 cells
- - - - Multi-block-72.000 cells
●   Keynton et al. (1991)

FIGURE 3: Normalized Shear Rate at the Outer Wall at Re = 100.

wall normalized computational shear rates compared with Keynton et al.'s [15] experiment.

Some additional benefits of using a multi-block mesh include full control of edge mesh density, production of minimally skewed 3D elements for complicated geometries, and accurate and efficient computations.

Conversely, multi-block mesh generation can be much more time consuming than unstructured mesh generation. For instance, the present study's multi-block meshes involved creating and meshing separate blocks that fit together to form the global meshed geometry. After geometry creation, the individual blocks were meshed with a mapping concept that generated eight node hexahedral elements. Moreover, node density was controlled by an interval count spacing function that created $i + 1$ uniformly spaced nodes on an edge, where i is a user defined interval ($i = 0, 1, 2 \ldots$).

The solving stage involved the use of the FLUENT solver to define the model's scale, fluid properties, flow physics, solver, and initial conditions. In addition, numerical schemes, such as SIMPLE and second-order upwind-difference approximation, were defined to control the stability and accuracy of the solution of the governing nonlinear partial differential equations. For each model, the mesh files were imported into a double-precision solver. Subsequently, the mesh was checked and then scaled to meters. Then the

steady, segregated solver was chosen along with the default algebraic multigrid solver. In addition, the low Reynolds number formed of the standard $k - \omega$ turbulence model, with default model constants, was selected to predict turbulent flow. Furthermore, Newtonian blood was defined with $\mu = 0.0035$ kg/ms and $\rho = 1060$ kg/m^3. The solution was then initialized using the velocity inlet boundary condition, where the Reynolds number was 1,100. Note: first-order accuracy was utilized initially to aid in convergence, and then the models were solved with second-order accuracy with the residual monitor set to 10^{-5}. Experience showed that such residual level was adequate for convergence of the solution in most cases. Furthermore, convergence was assumed when residuals were stable and not decreasing after approximately 1,000 iterations. Under-relaxation factors for pressure, momentum, turbulent kinetic energy, and specific dissipation rate were adjusted as needed. Computations were performed on a single workstation with 2.5 GHz CPU, and typical results were obtained for each case in 10 to 30 hours, depending on the mesh density and the Reynolds number.

Moreover, flow visualization and quantitative analysis of results obtained from solving were carried out during postprocessing and are described next.

3. Results

3.1. Flow Visualization

(a) *51% Stenosis.* The flow field around the distal bypass graft anastomosis (DBGA) placed just distal to the 51% stenosis in a region of separated flow is shown in Figure 4(a). Flow visualization indicates that the flow's velocity increases as it passes through the stenosis and accelerates again just before the anastomotic toe. Moreover, the separation region at the inner wall is truncated by the graft flow. Also, the inner wall separation region appears to be thicker than the outer wall separation region. In addition, the graft flow's momentum causes the arterial flow to skew toward the outer wall under the toe region. Also, graft flow along the outer wall near the heel appears to be detached from the core flow.

For case two, the DBGA was placed in the reattachment region of the separated flow Figure 4(b). Flow visualization shows the flow accelerating through the stenosis followed by a deceleration, where it accelerates once again just before the toe region. Also, the separation region at the inner wall just distal to the stenosis is truncated, and the arterial flow is skewed toward the outer wall by the graft flow beginning under the anastomosis. In addition, flow detachment from the graft outer wall near the anastomotic heel can be seen; however, the graft flow detachment is more conspicuous than for case one.

For case three, placing the DBGA far downstream subjected the region around the anastomosis to fully developed laminar flow and still allowed the influence of the upstream stenosis to be a factor in flow calculations Figure 4(c). Furthermore, the post-stenotic flow separation and turbulence do not directly affect the region around the DBGA, as with the previous two cases. Comparing case three with cases one and two shows that its velocity magnitude is overall slightly less. Similarly, though, the flow from the graft skews the arterial flow toward the outer wall. In addition, flow detachment from the graft outer wall near the anastomotic heel appears to occur at around the same place as for case two; however, graft streamlines are not skewed as much toward the toe.

(b) *75% Stenosis.* The flow field around the DBGA placed in a separation region just distal to the 75% stenosis is shown in Figure 5(a). Similar to Figure 4(a), the separation region at the inner wall is truncated by the graft flow; however, the inner wall separation region is thicker, where its distance from the inner wall is slightly past the inner wall of the throat of the stenosis. Also, flow from the graft is seen causing the arterial flow to skew toward the outer wall in Figure 5(a); however, the impinging graft flow does not cause a noticeable deceleration of the arterial flow under the anastomosis, as was seen in Figure 4(a). Moreover, the graft flow in Figure 5(a) does not appear to detach from the graft outer wall near the heel, which appears to have happened in Figure 4(a).

For case two, the DBGA was placed over the reattachment point of separated flow Figure 5(b). The decrease in cross-sectional area has increased the extent of the separation region. Thus, placement of the DBGA is farther from the stenosis than case two with a 51% stenosis, and as a result, a fairly developed recirculating zone at the post-stenotic inner wall is evident. Also, Figure 5(b) shows a slight increase in overall velocity magnitude from the previous case. Particular to this placement, a confluence of the graft flow and the arterial flow has formed a thin core distal to the anastomotic toe; whereas, for the other placements, the graft flow has noticeably skewed the arterial flow toward the outer wall leaving a thicker higher velocity inner core confluence. Also, Figure 5(b) shows flow detachment from the graft outer wall near the heel.

For case three, the DBGA was placed far downstream in fully developed laminar flow Figure 5(c). Although it is not apparent from the contour coloring, the velocity magnitude of the confluence of arterial flow and graft flow for both Figures 4(c) and 5(c) is similar. In addition, similar flow phenomena are observed, which includes skewing of the arterial flow toward the outer wall caused by the impinging graft flow and flow detachment from the graft outer wall near the anastomotic heel.

3.2. Shear Rates. Velocity gradients were measured at the inner wall and outer wall for all six cases. Figure 2, for example, shows how shear rates were measured for computational cases simulating Keynton et al.'s [15] experiment. For the present computational cases, the toe was considered the zero axial reference point for both inner wall and outer wall shear rate measurements. Inner wall measurements were made in 0.5D increments from the toe of the graft to 3D. For the outer wall, measurements were taken in 0.5D increments from −1.5D to 3D. In addition, velocity gradients were normalized with velocity gradients measured at 5D before the throat of the stenosis, where "a" represents the shear rate at the inner wall and "b" represents the shear rate at the outer wall.

The normalized velocity gradients at the inner wall and at the outer wall for the six cases are shown in Figures 6 and 7, respectively. Figure 6 indicates that case three (75%) produced the overall highest inner wall shear rates, and out of the 51% cases studied, case three also showed the highest shear rates. Beginning at the anastomotic toe, shear rates for the 51% cases, as well as the 75% cases, did not differ much from each other. However, further downstream the 75% cases differed significantly. For instance, case one (DBGA placement in a region of separated flow) differed from case three (DBGA placement far downstream) by approximately one normalized magnitude.

For the 51% cases measured beyond the toe, the differentiation among the cases was not as large. For instance, shear rate measurements for case one and case two (DBGA placement in a reattachment region of separated flow) insignificantly differed. Moreover, they differed at most by approximately 0.4 of a normalized magnitude from case three. Interestingly, as axial distance increased along the inner wall, case one (75%) had a normalized shear rate value

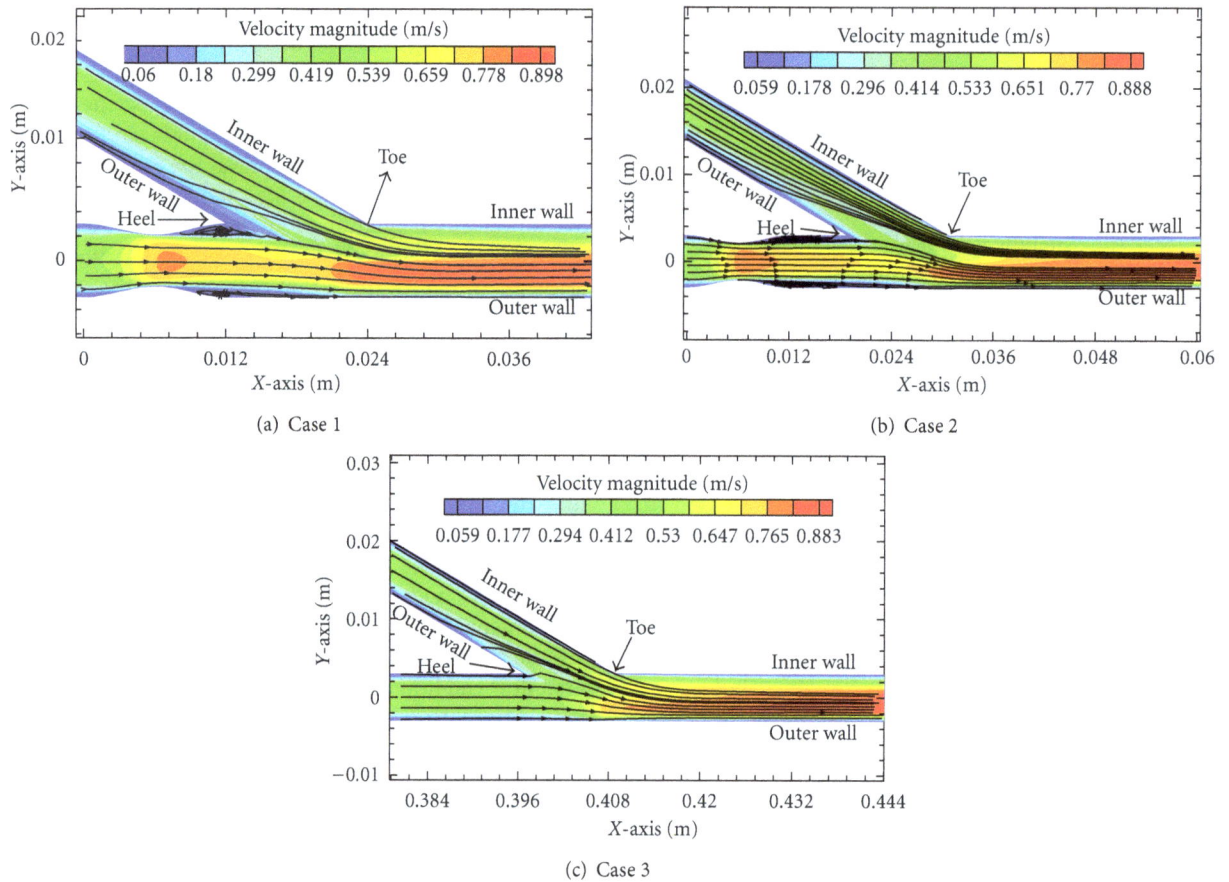

(a) Case 1

(b) Case 2

(c) Case 3

FIGURE 4: Velocity magnitude contours with streamlines of the 51% stenosis cases.

similar to cases one and two (51%), and at $L_{iw}/D = 2$ and 2.5, it was lower than these other two cases.

Figure 7 also shows case three (75%) as having the highest overall shear rates, and out of the 51% cases, case three also showed the highest shear rates. For outer wall measurements, cases one and two for both the 51% and 75% stenoses did not vary in magnitude from each other significantly. In addition, before the anastomotic toe, cases one and two (75%) differed on average by approximately two normalized magnitudes from case three, and after the toe, they different from case three on average by approximately 1.5 normalized magnitudes. Moreover, before the anastomotic toe, normalized shear rates for cases one and two (51%) on average differed by approximately one magnitude, and after the toe, cases one and two (51%) on average differed by approximately 0.6 normalized magnitudes. Interestingly, the highest and lowest outer wall normalized shear rates measured throughout the range of L_{ow}/D studied are observed for 75% stenosis cases.

4. Summary and Conclusion

In summary, the leading cause of death in the developed world is the cardiovascular disease, atherosclerosis. This disease stifles blood flow leading to symptoms that require surgical intervention. The femoral artery is a common site requiring a bypass graft operation usually involving the end-to-side anastomosis technique. However, late graft failure is likely to occur due to intimal hyperplasia and requires surgical revision. Although the present understanding of the development of atherosclerosis and intimal hyperplasia is limited, hemodynamic theories of vascular pathogenesis offers a promising start into understanding the initiation and proliferation mechanisms for these types of diseases that cause bypass graft failure.

Therefore, different graft attachment locations were studied using a 51% and a 75% stenotic arterial model. Steady, incompressible, Newtonian flow was assumed and simulated on a locally structured multiblock mesh with hexahedral elements. Furthermore, the finite-volume method and a pressure correction scheme, SIMPLE, were used in addition to implicit Gauss-Siedel iteration to solve the full Navier-Stokes equations, turbulent kinetic energy, and specific dissipation rate equations with second-order accuracy. A preliminary study was conducted and confirmed acceptable prediction accuracy of the present computational code and technique. Consequently, the highest shear rates were recorded for the bypass graft anastomosis placed farthest downstream from the stenosis for both the 51% and the 75% stenotic models. Moreover, the lowest inner wall normalized shear rate was

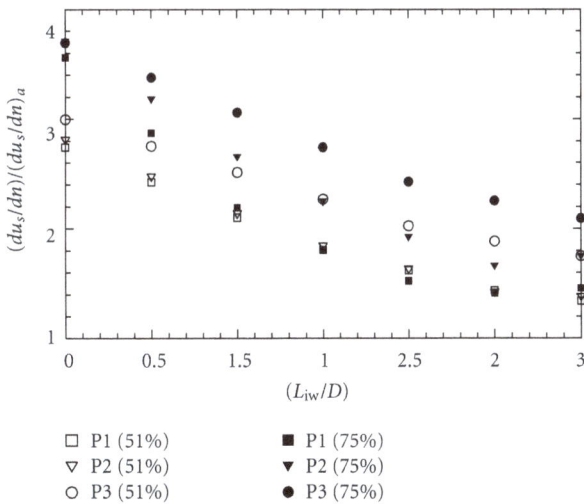

(a) Case 1

(b) Case 2

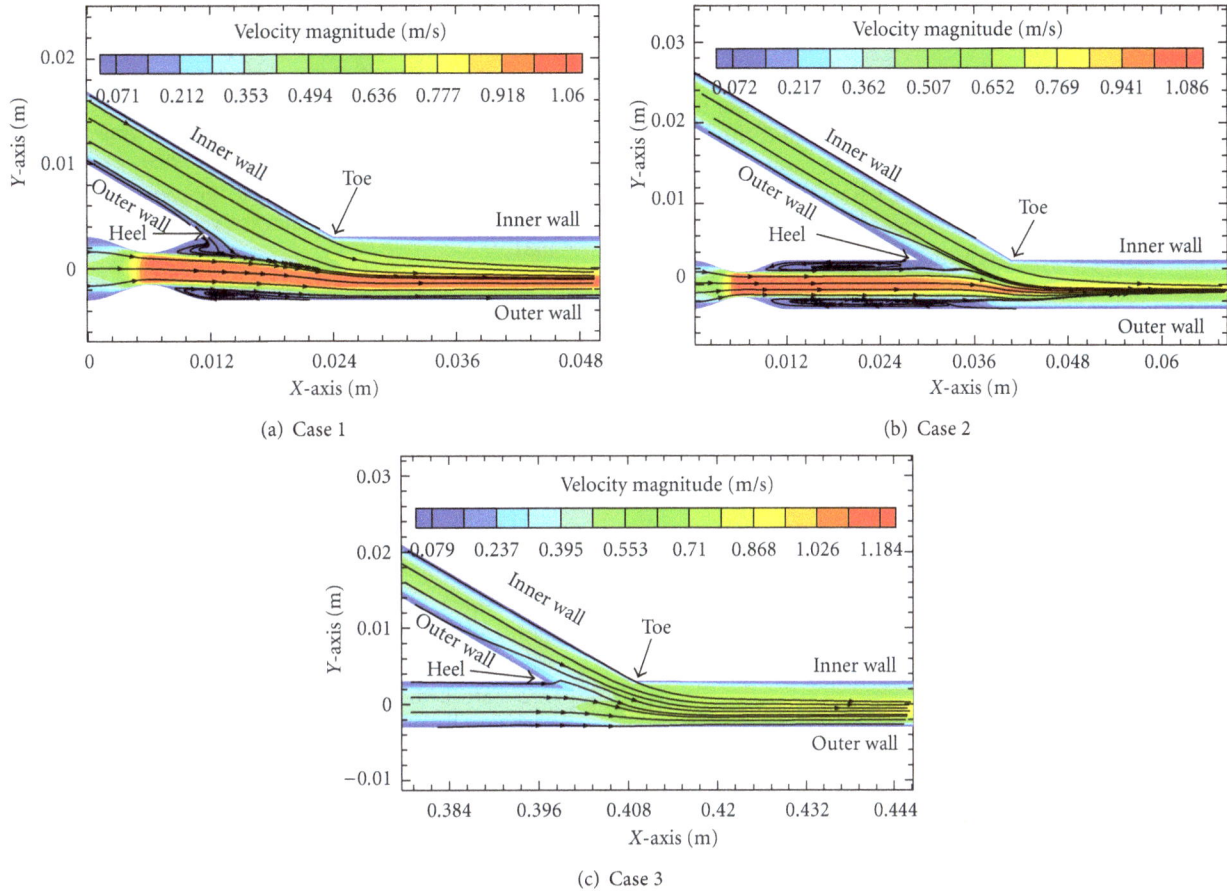

(c) Case 3

FIGURE 5: Velocity magnitude contours with streamlines of the 75% stenosis cases.

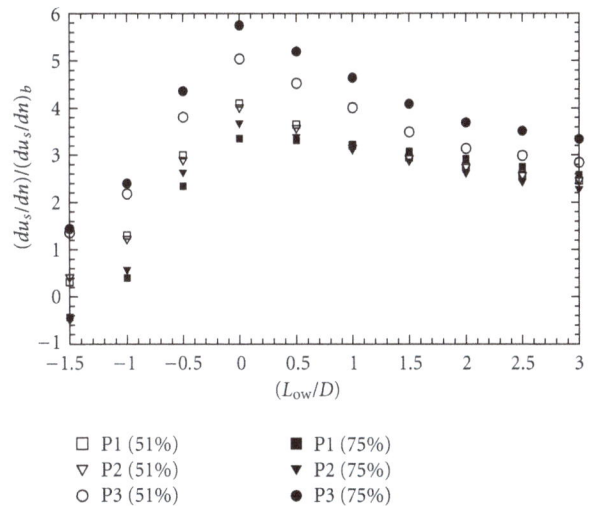

FIGURE 6: Comparison of axial variation of inner wall shear rates for all DBGA cases.

FIGURE 7: Comparison of axial variation of outer wall shear rates for all DBGA cases.

recorded for case one (51%), and the lowest outer wall normalized shear rate was recorded for case two (75%). From Figures 6 and 7, the difference in normalized shear rate magnitudes among the cases indicates that distal placement of an end-to-side bypass graft anastomosis influences shear rate magnitudes around the distal anastomosis, and with further study, an optimal placement could be determined that possibly extends patent blood flow to afflicted tissues.

As a final note, it is not clear exactly what effect these values of shear rates will have on an *in vivo* distal bypass graft anastomosis; however, previous studies indicate a connection between low wall shear rates and intimal hyperplasia development. The present computational study found the lowest wall shear rates occurring around the DBGA cases close to the stenosis for both 51% and 75% cases, where the 75% stenosis on average produced lower shear rates. Therefore, it appears that placement distance from the stenosis and the degree of the stenosis influence shear rates around the distal bypass graft anastomosis.

Nomenclature

D: Diameter
G: Production term
i: User-defined interval
k: Turbulent kinetic energy
L: Length
n: Normal to streamline
P: Time-averaged static pressure
R_k: Turbulence model constant
Re_t: Turbulence model term
u_s: Velocity along streamline
U: Time-averaged velocity component
x: Cartesian coordinate
Y: Dissipation term.

Greek Letters

α^*:   Low Reynolds number correction term
α_0^*:   Turbulence model constant
α_∞^*:   Turbulence model constant
Γ: Effective diffusivity
ν: Kinematic viscosity
ρ: Density
τ_{ij}:   Reynolds stress tensor
ω: Specific dissipation rate.

Subscripts

G: Graft
i, j: Einstein notation (counting index)
iw: Inner wall
k: Turbulent kinetic energy
ow: Outer wall
t: Turbulent
ω: Specific dissipation rate.

Acknowledgment

This work was supported by the National Science Foundation under Grant no. 0139336.

References

[1] D. E. Szilagyi, J. P. Elliott, J. H. Hageman, R. F. Smith, and C. A. Dall'olmo, "Biologic fate of autogenous vein implants as arterial substitutes: clinical, angiographic and histopathologic observations in femoro popliteal operations for atherosclerosis," *Annals of Surgery*, vol. 178, no. 3, pp. 232–246, 1973.

[2] A. W. Clowes, A. M. Gown, S. R. Hanson, and M. A. Reidy, "Mechanisms of arterial graft failure. 1. Role of cellular proliferation in early healing of PTFE prostheses," *American Journal of Pathology*, vol. 118, no. 1, pp. 43–54, 1985.

[3] P. N. Madras, C. A. Ward, W. R. Johnson, and P. I. Singh, "Anastomotic hyperplasia," *Surgery*, vol. 90, no. 5, pp. 922–923, 1981.

[4] V. S. Sottiurai, J. S. T. Yao, R. C. Batson, S. L. Sue, R. Jones, and Y. A. Nakamura, "Distal anastomotic intimal hyperplasia: histopathologic character and biogenesis," *Annals of Vascular Surgery*, vol. 3, no. 1, pp. 26–33, 1989.

[5] D. F. Bandyk, G. R. Seabrook, P. Moldenhauer et al., "Hemodynamics of vein graft stenosis," *Journal of Vascular Surgery*, vol. 8, no. 6, pp. 688–695, 1988.

[6] P. B. Dobrin, F. N. Littooy, and E. D. Endean, "Mechanical factors predisposing to intimal hyperplasia and medial thickening in autogenous vein grafts," *Surgery*, vol. 105, no. 3, pp. 393–400, 1989.

[7] C. G. Caro, J. M. Fitz-Gerald, and R. C. Schroter, "Arterial wall shear and distribution of early atheroma in man," *Nature*, vol. 223, no. 5211, pp. 1159–1161, 1969.

[8] C. G. Caro, J. M. Fitz-Gerald, and R. C. Schroter, "Atheroma and arterial wall shear. Observation, correlation and proposal of a shear dependent mass transfer mechanism for atherogenesis," *Proceedings of the Royal Society of London B*, vol. 177, no. 46, pp. 109–159, 1971.

[9] R. M. Nerem, "Hemodynamics and the vascular endothelium," *Journal of Biomechanical Engineering*, vol. 115, no. 4, pp. 510–514, 1993.

[10] C. Bertolotti and V. Deplano, "Three-dimensional numerical simulations of flow through a stenosed coronary bypass," *Journal of Biomechanics*, vol. 33, no. 8, pp. 1011–1022, 2000.

[11] S. M. Kute and D. A. Vorp, "The effect of proximal artery flow on the hemodynamics at the distal anastomosis of a vascular bypass graft," *Journal of Biomechanical Engineering*, vol. 123, no. 3, pp. 277–283, 2001.

[12] D. C. Wilcox, *Turbulence Modeling for CFD*, DCW Industries, Los Angeles, Calif, USA, 1998.

[13] R. B. Rutherford, "Basic vascular surgical techniques," in *Vascular Surgery*, R. B. Rutherford, Ed., pp. 476–486, W. B. Saunders, New York, NY, USA, 5th edition, 2000.

[14] S. A. Ahmed and D. P. Giddens, "Velocity measurements in steady flow through axisymmetric stenoses at moderate reynolds numbers," *Journal of Biomechanics*, vol. 16, no. 7, pp. 505–516, 1983.

[15] R. S. Keynton, S. E. Rittgers, and M. C. S. Shu, "The effect of angle and flow rate upon hemodynamics in distal vascular graft anastomoses: an *in vitro* model study," *Journal of Biomechanical Engineering*, vol. 113, no. 4, pp. 458–463, 1991.

8

Study of Dynamic Behavior of Multilayered Clamped Composite Skewed Hypar Shell Roofs under Impact Load

Sanjoy Das Neogi,[1] Amit Karmakar,[2] and Dipankar Chakravorty[3]

[1] Department of Civil Engineering, Meghnad Saha Institute of Technology, Kolkata 700150, India
[2] Department of Mechanical Engineering, Jadavpur University, Kolkata 700032, India
[3] Department of Civil Engineering, Jadavpur University, Kolkata 700032, India

Correspondence should be addressed to Sanjoy Das Neogi; sanjoy_civil08@yahoo.com

Academic Editor: Mickaël Lallart

With advancement in the field of structural engineering, hunt for smarter materials has channelised the research towards the application of composite material. It is the high specific weight and specific stiffness of this material that have drawn the interest of different industrial sectors. Civil engineers also picked up composites to use it as a roofing material. Laminated composite shells, which can cover large column-free area and reduces dead weight of structure, show vulnerability under sudden impact due to their low transverse shear resistances. This study utilises finite element tool to investigate the dynamic response of a multilayered laminated composite hypar shells for fully clamped boundary condition. This class of shells is unique in a sense that the curvature has only the radius of cross curvature and these shells do not admit easy closed form solution particularly when the boundary conditions are complicated. Contact behavior of impactor and impacted mass has been modeled by modified Hertzian contact law and time-dependent equations are solved using Newmark's time integration technique. Basic aim is to analyse the shell for symmetrically placed multilayered angle and cross ply lamination under different impact velocities.

1. Introduction

Composite material has gained popularity in different industrial sectors. It is the high specific weight and stiffness of this material that have gained the popularity for composite. With other sectors, civil engineering industries also started using the laminated composites as roofing material in the situation where large column-free space is required. The high specific weight also causes reduction in dead weight of structure and thus reduces seismic weight and foundation cost. In spite of these advantages, low transverse shear strength makes them vulnerable under sudden impact. It is obvious that an accurate modeling of contact behavior is one of the most important steps to analyze an impact response problem. The classical contact law between elastic solids derived by Hertz [1] was employed by different authors from time to time. The problem gets more complicated in case of composite materials and the Hertzian contact law which was derived for homogeneous isotropic materials may not be adequate.

Tan and Sun [2] undertook an experimental program on a graphite/epoxy laminated plate to establish an empirical indentation law. The theoretical basis of their experiment was checked by a nine-noded plate finite element. Time histories of contact force and displacement were reported by Sun and Chen [3] for simply supported initially stressed plate under impact. The contact law as proposed by Tan and Sun [2] was utilized. They used a steel ball as an impactor and a ten-layered simply supported composite plate as the impacted mass. Impact analysis of shell structure was first reported by Toh et al. [4] for an orthotropic laminated cylindrical shell under low velocity impact generated by a solid striker. Shim et al. [5] studied the elastic response of glass/epoxy laminated composite ogival shells subjected to low velocity impact at any arbitrary location by a solid striker.

They reported an analytic biharmonic polynomial solution. A finite element model, with and without geometric nonlinearity, was presented by Kistler and Waas [6] for a laminated composite cylindrical shell subjected to transverse

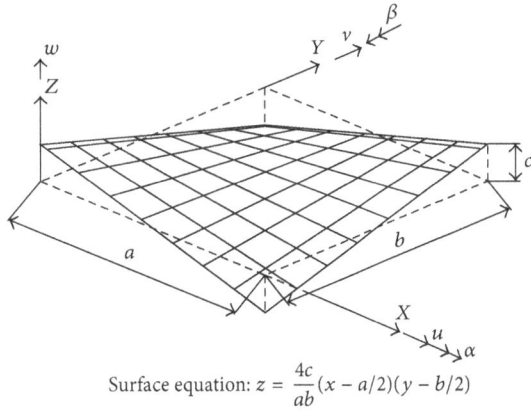

Surface equation: $z = \dfrac{4c}{ab}(x - a/2)(y - b/2)$

FIGURE 1: Surface of a skewed hypar shell and degrees of freedom.

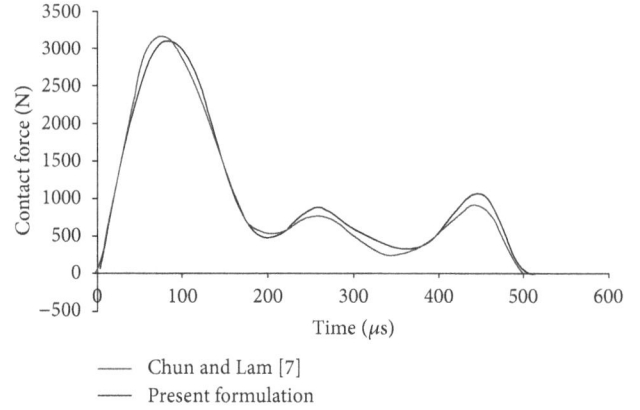

— Chun and Lam [7]
— Present formulation

FIGURE 2: Contact force history of clamped plate. $E_{11} = 142.73$ GPa, $E_{22} = 13.79$ GPa, $G_{12} = 4.64$ GPa, $\nu_{12} = 0.30$, $\rho = 1.61 \times 10^3$ kg/m^3, $a = b = 0.14$ m, $h = 3.81 \times 10^{-3}$ m, mass of striker = 0.014175 kg, velocity of striker = 22.6 m/s, contact stiffness $(k_c) = 1 \times 10^8$ N/m$^{1.5}$.

central impact. Chun and Lam [7] proposed a numerical formulation to calculate the dynamic response of fully clamped laminated composite plates subjected to low velocity impact. They utilised Lagrange's principle to derive the nonlinear, second order governing differential equation and Hertzian contact law. A centrally impacted three-layered laminated graphite-epoxy composite plate, of different ply orientation was investigated. Their study excluded any problem on shell forms. Karmakar and Kishimoto [8] undertook a transient dynamic finite element analysis to study the response of centrally impacted delaminated composite pretwisted rotating cylindrical shells under low velocity impact. Effects of transverse shear deformation and rotary inertia were included in their study. A transient dynamic response of rotating shallow shells subjected to low velocity impact was reported by Karmakar and Kishimoto [9].

A look through the literature reveals the fact that impact response of civil engineering shell structures has not received due attention though such shells may often be subjected to impact loads. Impact forces on shell roofs are encountered due to snow fall, due to air borne debris in cyclone prone areas and in other similar situations. So the necessity to study impact response of composite shell roofs is felt. A parallel review in the area of composite shells reveals that the industrially important hypar shell needs a lot more attention although some important aspects of these shell forms were reported recently by Sahoo and Chakravorty [10–12]. The only report on impact response of composite hypar shell was due to Das Neogi et al. [13] where they discussed some aspects of such two-layered simply supported shells. Hence, this paper aims to carry out impact response of composite hypar shells for fully clamped boundary condition.

2. Mathematical Formulation

The shell surface considered for the present study is hypar shell (Figure 1) with doubly curved and anticlastic surface having cross curvature $1/R_{xy}$ only and $1/R_x = 1/R_y = 0$. A doubly curved thin shallow shell of uniform thickness h made of homogeneous, laminated composite, linearly elastic material is considered. A shell is characterized as shallow

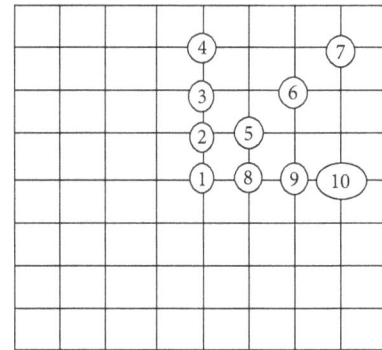

FIGURE 3: Points at which deflection is measured.

if any infinitesimal line element of its middle surface is approximated by the length of its projection on the XY plane. This implies that

$$\left(\frac{\partial z}{\partial x}\right)^2 \ll 1, \qquad \left(\frac{\partial z}{\partial y}\right)^2 \ll 1, \qquad \left(\frac{\partial z}{\partial x}\right)\left(\frac{\partial z}{\partial y}\right) \ll 1. \quad (1)$$

Moreover, the lateral boundary of a shallow shell is approximated by its projection on the XY plane with regard to its boundary conditions. According to Vlasov [14], the above conditions are practically satisfied for shells with a rise to span ratio less than 1/5 and cross curvature is approximately represented as

$$\frac{1}{R_{xy}} = \frac{\partial^2 z}{\partial x \partial y}. \quad (2)$$

TABLE 1: Nondimensional natural frequencies (ϖ) for a three-layer graphite epoxy twisted plate.

Angle of twist θ (deg)	0	15	30	45	60	75	90
$\phi = 15°$							
Qatu and Leissa [17]	1.0035	0.9296	0.7465	0.5286	0.3545	0.2723	0.2555
Present formulation	0.9990	0.9257	0.7445	0.5279	0.3542	0.2720	0.2551
$\phi = 30°$							
Qatu and Leissa [17]	0.9566	0.8914	0.7205	0.5149	0.3443	0.2606	0.2436
Present formulation	0.9490	0.8842	0.7181	0.5142	0.3447	0.2613	0.2444

$E_{11} = 138\,\text{GPa}$, $E_{22} = 8.96\,\text{GPa}$, $G_{12} = 7.1\,\text{GPa}$, $\nu_{12} = 0.3$, $a/b = 1$, $a/h = 100$.

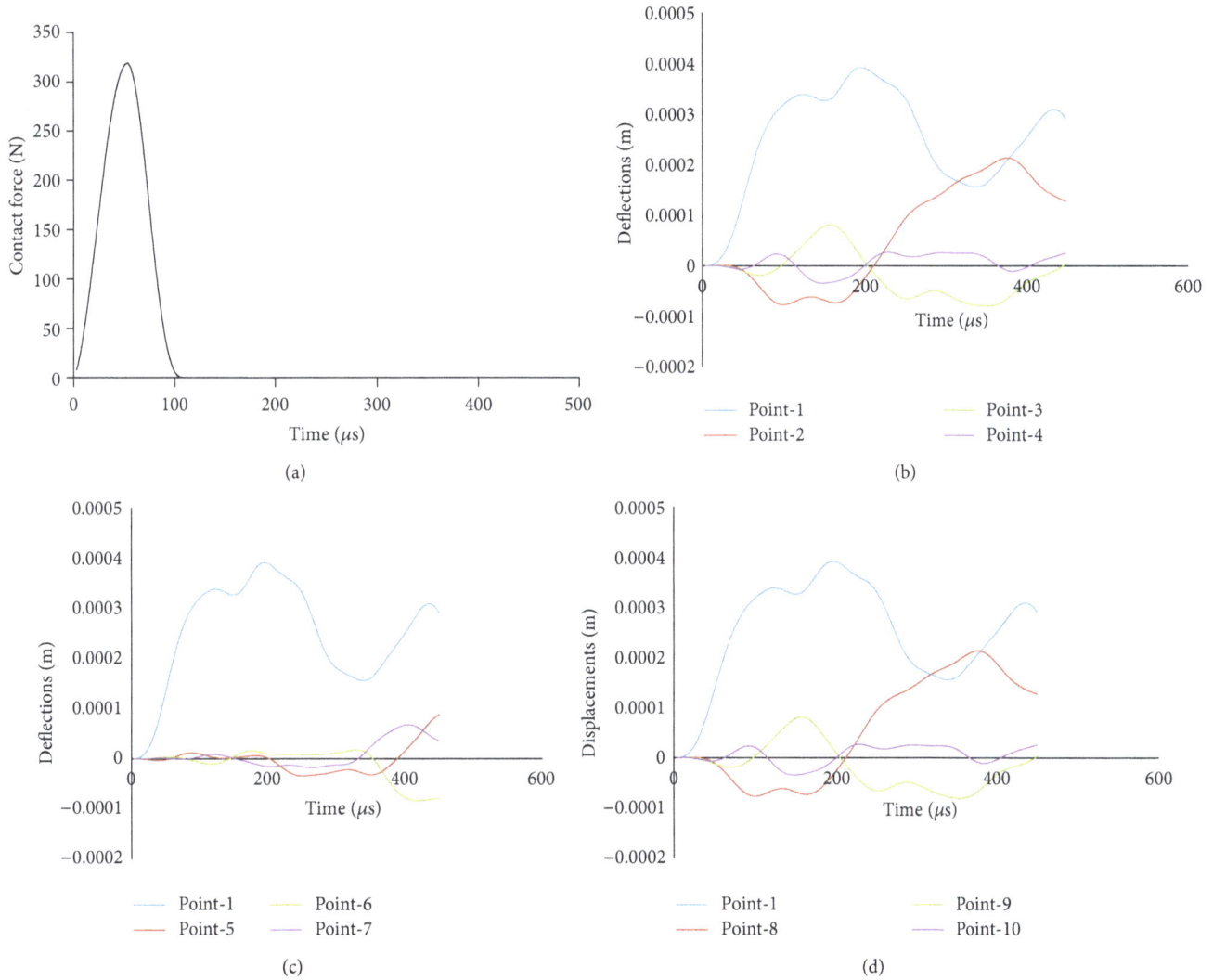

(a)

(b)

(c)

(d)

FIGURE 4: Impact response of clamped cross ply (CL/CP) composite hypar shells for impact velocity 1 m/s.

The generalised displacement vector of an element is expressed in terms of the shape functions and nodal degrees of freedom as

$$[u] = [N]\{d_e\},\tag{3}$$

that is,

$$\{u\} = \begin{Bmatrix} u \\ v \\ w \\ \alpha \\ \beta \end{Bmatrix} = \begin{bmatrix} [N_i] & & & & \\ & [N_i] & & & \\ & & [N_i] & & \\ & & & [N_i] & \\ & & & & [N_i] \end{bmatrix} \begin{Bmatrix} u_i \\ v_i \\ w_i \\ \alpha_i \\ \beta_i \end{Bmatrix}.\tag{4}$$

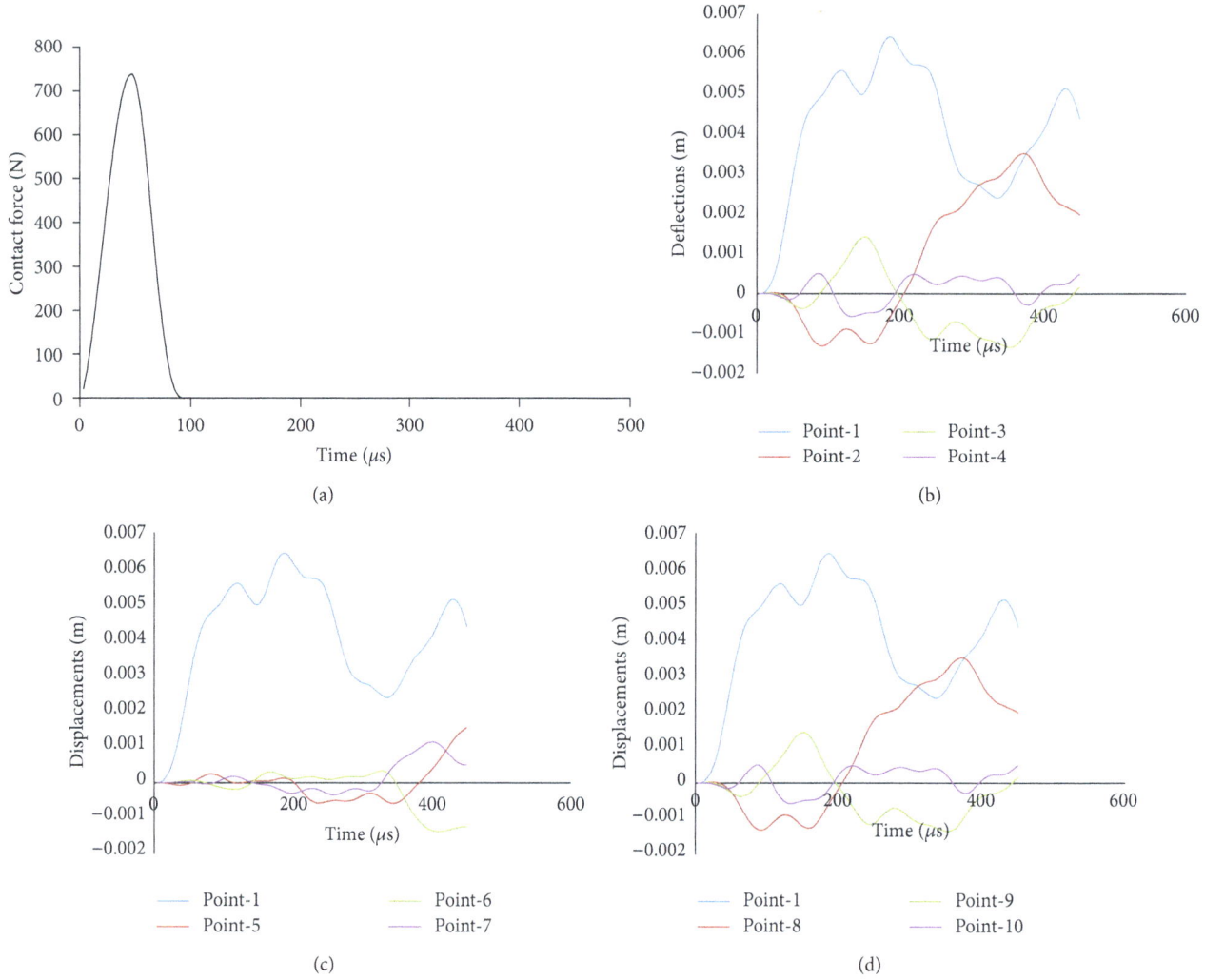

FIGURE 5: Impact response of clamped cross ply (CL/CP) composite hypar shells for impact velocity 2 m/s.

The element stiffness and mass matrices are derived by using the minimum energy principle. The element stiffness matrix is

$$[K_{she}] = \iint [B]^T [D] [B] \, dx \, dy. \tag{5}$$

The generalised inertia matrix per unit area includes the translatory and rotatory inertia terms. Incorporating both the translatory and rotatory inertia terms, the generalised inertia matrix takes the following form:

$$[M] = \iint [N]^T [\rho] [N] \, dx \, dy. \tag{6}$$

The dynamic equilibrium equation of the target shell for low velocity impact is given by the following equation:

$$[M] \{\ddot{\delta}\} + k \{\delta\} = \{F\}, \tag{7}$$

TABLE 2: Convergence study for time step.

Time step (μs)	Maximum contact force (N) $0°/90°/0°$
2	291.38795
2.5	301.58181
2.75	311.0179
3	318.5276
3.25	324.95893
3.5	329.38339

where $[M]$ and $[K]$ are global mass and elastic stiffness matrices, respectively. $\{\delta\}$ is the global displacement vector. For the impact problem, $\{F\}$ is given as

$$\{F\} = \{0 \; 0 \; 0 \cdots F_c \cdots 0 \; 0 \; 0\}^T. \tag{8}$$

Here F_c is the contact force given by the indentation law and the equation of motion of the rigid impactor is given as

$$\ddot{m}_i \ddot{\omega}_i + F_c = 0, \tag{9}$$

TABLE 3: Maximum contact force, maximum dynamic displacement, equivalent static load, dynamic magnification factor for different velocities.

Boundary condition and ply orientation	Velocity (m/s)	Maximum impact load (N)	Maximum displacement (m)	Equivalent static load (N)	Dynamic magnification factor
Clamped 0°/90°/0°	1	318.5276	0.0000534	1880	3.75
	2	736.1553	0.000111	3.910	3.46
	3	1204.515	0.00017	5.980	2.90
	5	2242.325	0.000292	10300	2.75
	7	3373.736	0.000417	14700	2.50
	10	5198.173	0.000612	21500	2.24

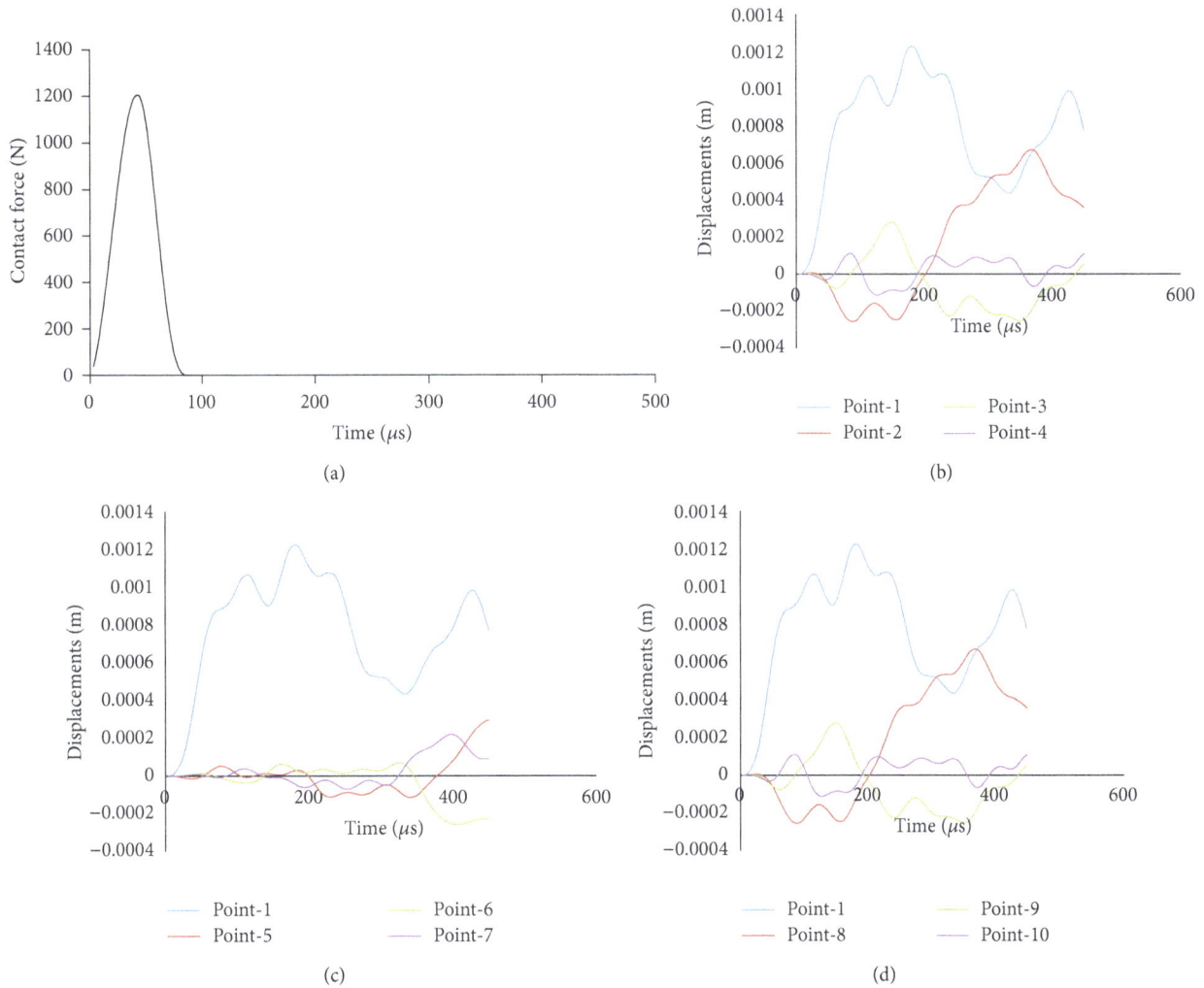

(a)

(b)

(c)

(d)

FIGURE 6: Impact response of clamped cross ply (CL/CP) composite hypar shells for impact velocity 3 m/s.

where $\ddot{m}_i$ and $\ddot{w}_i$ are the mass and acceleration of the impactor, respectively. A power law was proposed by Yang and Sun [15] based on static indentation tests using steel ball as an indentor. This contact law accounted for the permanent indentation after unloading cycle; that is, collisions upon the rebound of the target structure after the first period of contact were considered. The modified version of the above mentioned contact law was proposed by Tan and Sun [2] and was utilised by Sun and Chen [3]. The contact force model following Sun and Chen [3] is incorporated in the present finite element formulation. If k is the contact stiffness and α_m is the maximum local indentation, the contact force F_c during loading is given by

$$F_c = k\alpha^{1.5}, \quad 0 < \alpha \le \alpha_m. \quad (10)$$

The indentation parameter α depends on the difference of the displacements of the impactor and the target structure at

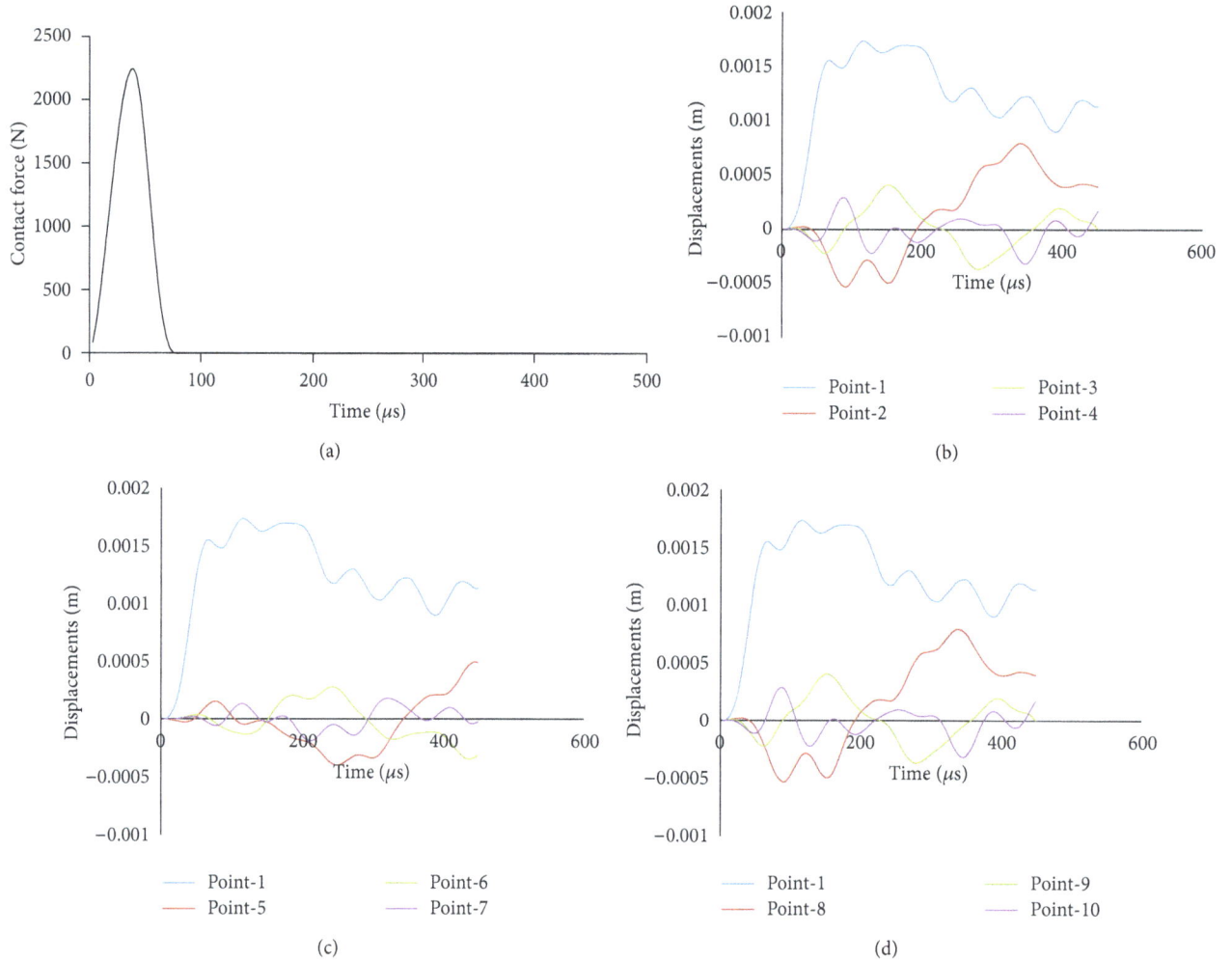

(a)

(b)

(c)

(d)

FIGURE 7: Impact response of clamped cross ply (CL/CP) composite hypar shells for impact velocity 5 m/s.

any instant of time, and so also the contact force. The values of α are changing with time because of time varying displacements of both the rigid impactor and the target structure. At an instant, the maximum indentation takes place and as a result maximum contact force is also obtained. At this instant, displacement of the impactor also attains the maximum value [16]. There after the displacement of the impactor gradually decreases, but the target point displacement keeps on changing and finally increases to a maximum value and some point of time these two displacements become equal [16]. This leads to zero value of indentation and eventually the contact force becomes zero. At this instant, the impactor loses the contact with the target. The process after attaining the maximum contact force till the reduction of contact force to zero value is essentially referred to as unloading [3]. If the mass of the impactor is not very small, a second impact may occur upon the rebound of the target structure leading to a same phenomenon of contact deformation and attainment of maximum contact force. This process is known as reloading. If F_m is the maximum contact force at the beginning of unloading and α_m is the maximum indentation during

loading, the contact force F_c for unloading and reloading is expressed as [3]

unloading phase:

$$F_c = F_m \left[\frac{\alpha - \alpha_0}{\alpha_m - \alpha_0} \right]^{2.5};$$ (11)

reloading phase:

$$F_c = F_m \left[\frac{\alpha - \alpha_0}{\alpha_m - \alpha_0} \right]^{1.5},$$ (12)

where α_0 denotes the permanent indentation in a loading-unloading cycle.

$$\alpha_0 = \beta_c \left(\alpha_m - \alpha_p \right) \quad \text{if } \alpha_m > \alpha_{cr}$$
$$\alpha_0 = 0 \quad \text{if } \alpha_m < \alpha_{cr},$$ (13)

where β_c is a material dependent constant and α_{cr} is the critical indentation beyond which permanent indentation

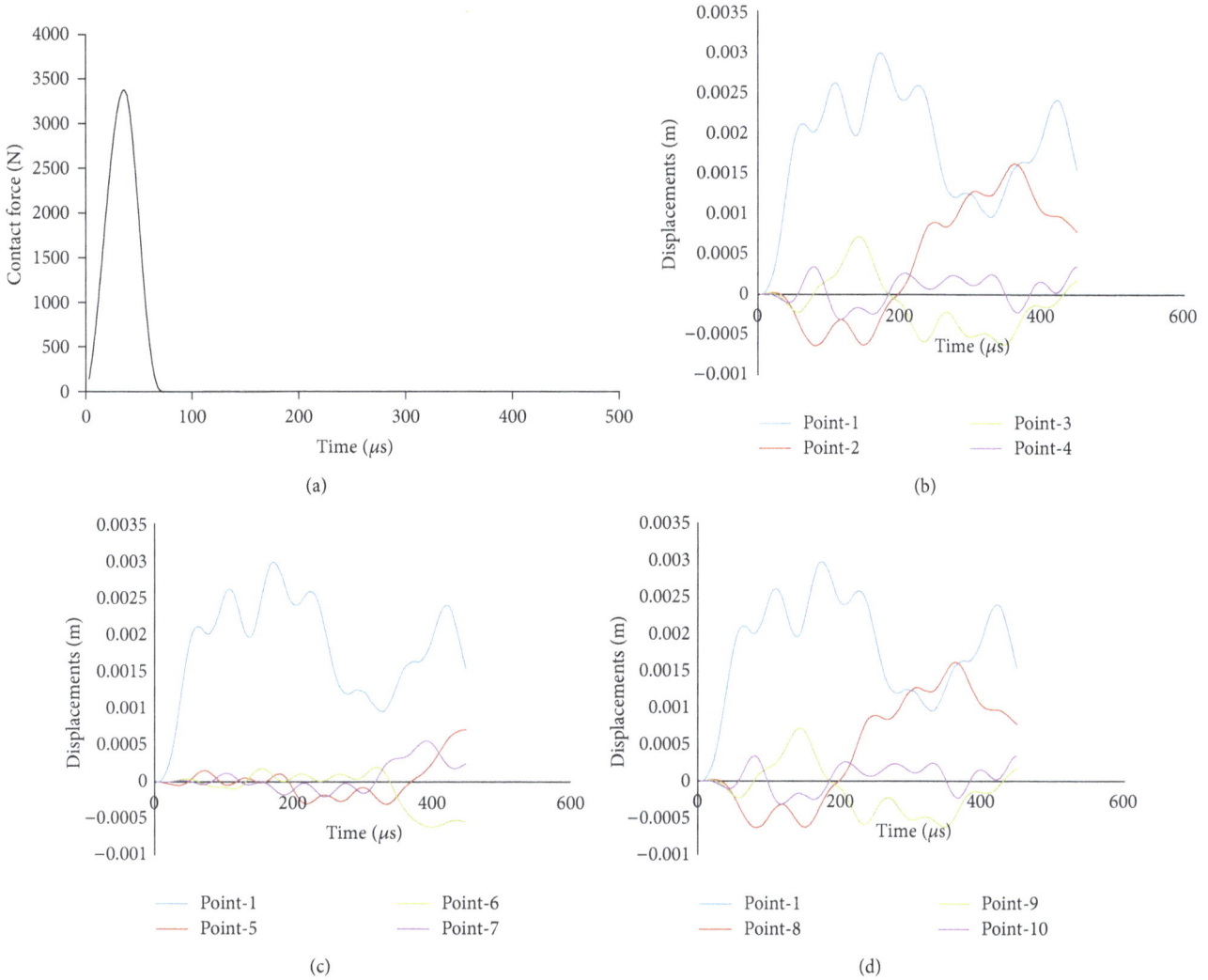

FIGURE 8: Impact response of clamped cross ply (CL/CP) composite hypar shells for impact velocity 7 m/s.

occurs, and the values are 0.094 and .01667 cm, respectively for graphite-epoxy composite [3]. Equations (7) and (9) are solved using Newmark constant-acceleration time integration algorithm in the present analysis. Equation (7) may be expressed in iteration form at each time step as follows:.

$$\left[\overline{K}\right]\{\Delta\}_{t+\Delta t}^{i+1} = \frac{\Delta t^2}{4}\{F\}_{t+\Delta t}^i + [M]\{b\}_i, \quad (14)$$

where

$$\left[\overline{K}\right] = \frac{\Delta t^2}{4}[K] + [M] \quad (15)$$

$$\{b\}_t = \{\Delta\}_t + \Delta t\{\dot{\Delta}\}_t + \frac{\Delta t^2}{4}\{\ddot{\Delta}\}_t. \quad (16)$$

In the above equations (14) and (15), $[K]$ and $[M]$ are the global stiffness and mass matrices obtained after assembling the matrices obtained at element level and imposing the boundary conditions. The degrees of freedom that are locked are selected through an input matrix and the corresponding

terms in the global stiffness and mass matrices and load vectors are eliminated. The approach enables imposition of arbitrary boundary conditions in the problem although the present results are obtained only for clamped edges. The same solution scheme is also utilized for solving the equation of motion of the impactor, that is, (7). In (14), i is the number of iterations within a time step. It is to be noted that a modified contact force $F_{t+\Delta t}^i$ obtained from the previous iteration is used to solve the current response $\{\Delta\}_{t+\Delta t}^{i+1}$. The iteration procedure is continued until the equilibrium criterion is met.

3. Numerical Examples

Numerical problems are solved with two different objectives. The present approach is applied to solve natural frequencies of graphite-epoxy twisted plates which are structurally similar to skewed hypar shells. This is to validate both the stiffness and mass matrix formulations in present finite element code. The problem details are furnished with Table 1. Another problem solved earlier by Chun and Lam [7] related to the

(a)

(b)

(c)

(d)

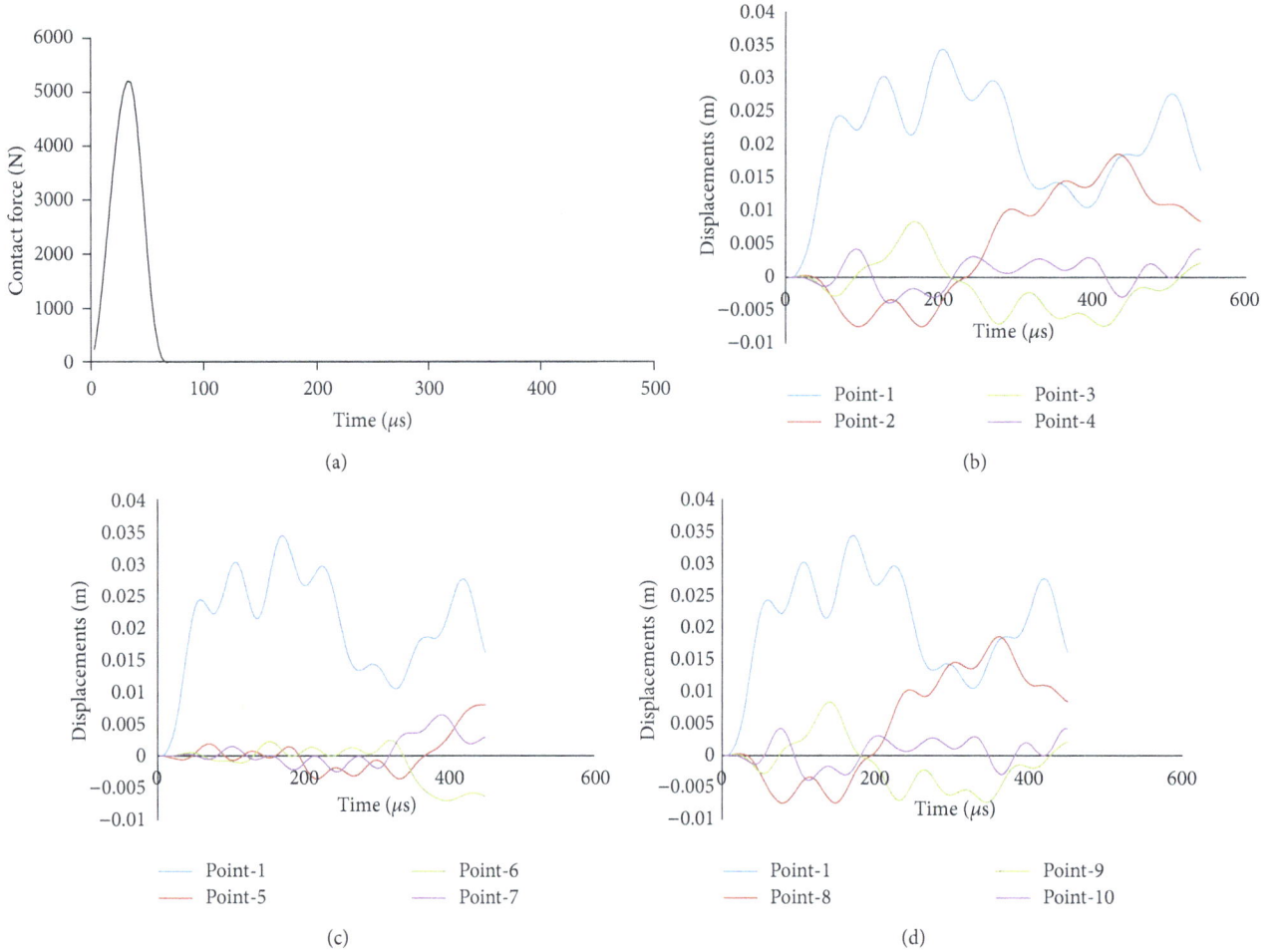

FIGURE 9: Impact response of clamped cross ply (CL/CP) composite hypar shells for impact velocity 10 m/s.

impact response of clamped composite plate, is taken up as the second benchmark to validate the impact formulation and the results are furnished as Figure 2.

Apart from the problems mentioned above, impact response of skewed hypar shells being impacted at the central point is also studied for clamped boundary condition, different laminations, and impact velocities. The details of the problems which are the authors' own are given below.

(i) Boundary condition: Clamped (CL).

(ii) Lamination: $0^0/90^0/0^0$(CP).

(iii) Velocity of impact in (m/s): 1, 2, 3, 5, 7, 10.

(iv) Details of shell geometry: $a = 1.0$ m, $b = 1.0$ m, $h = 0.02$ m, $c = 0.2$ m.

(v) Material details: $E_{11} = 120$ GPa, $E_{22} = 7.9$ GPa, $G_{12} = G_{23} = G_{13} = 5.5$ GPa $\nu_{12} = 0.30$, $\rho = 1.58 \times 10^{-5}$ N-sec^2/cm^4.

The converged value (Table 2) for a time step $\Delta t = 3 \, \mu s$ is adopted for the present analysis.

4. Results and Discussions

The results of Table 1 show that the fundamental frequency values of the twisted plates obtained by the present formulation agree closely to those reported by Qatu and Lessia [17]. Thus the correct incorporation of stiffness and mass matrix formulation in the present code is established. Figure 2 shows the time variation of the contact force induced in a composite plate under low velocity impact as obtained by Chun and Lam [7]. The values obtained by the present formulation are also presented graphically in the same figure in a different style. Here again excellent agreement of results is observed which establishes the correctness of impact formulation. The displacement history and dynamic magnification of several points (Figure 3) are studied along with the node of impact. For studying the impact response of clamped (CL) cross ply (CP) shell Figures 4, 5, 6, 7, 8, 9, and 10 and Table 3 are to be studied. All the results of contact force and displacement that are presented, in either graphical or tabular form, are arrived at after a study of time step convergence.

The finite element mesh adopted is also based on force and displacement convergence criteria. When low velocity normal impact response of clamped cross ply shell is studied

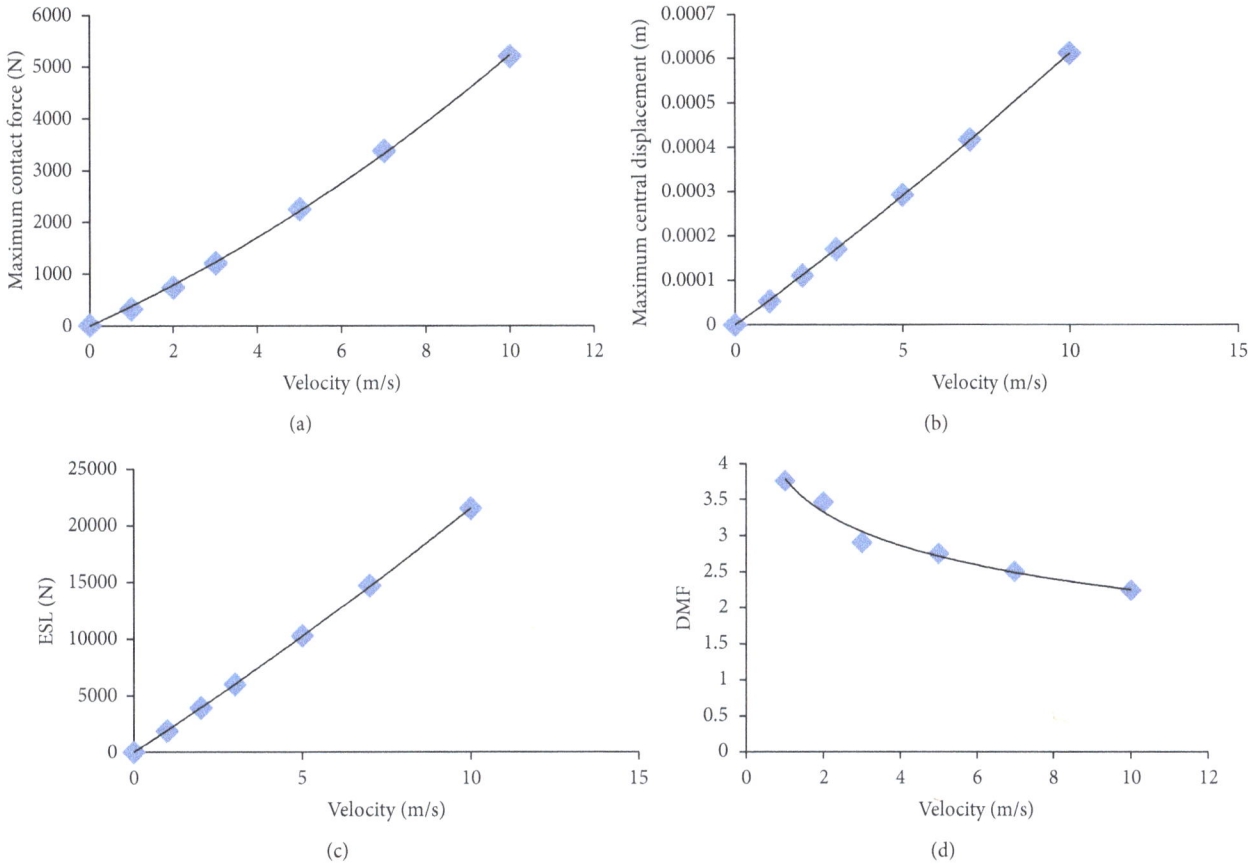

FIGURE 10: Variation of maximum impact load, maximum displacement, equivalent static load, and dynamic magnification factor with velocity for simply supported angle ply (CL/CP) composite hypar shells.

being struck by the spherical impactor centrally it is observed that the contact force shows a sort of parabolic variation with a single peak. After a given time span of $100 \, \mu$s or less, the contact force dies down to a null value. It is interesting to note that higher the impactor velocity higher is the contact force as expected, but the force dies down to a null value earlier. This behavior may be attributed to the fact that higher the velocity, more rapid is the elastic rebound of the impactor followed by detachment which causes the contact force to decay out. It is also very interesting to observe that the time instant corresponding to peak contact force and that for peak displacement do not match. This is because the resultant displacement at any time instant is a cumulative effect of the instantaneous contact force value and the inertia effect of the previous instant. The figures showing the transient displacements reflect the fact that vibration continues even after the force dies down with successively occurring peaks though, the peak values are less in magnitude than the highest peak which occurs a bit after the instant of maximum contact force but before the full decay of it.

To estimate the equivalent static load (ESL) corresponding to a particular impactor velocity, a concentrated load at the centre (point of impact) is applied and adjusted to yield a central displacement equal to the maximum dynamic displacement. The magnitude of the central displacement is calculated with the peak contact force applied at the point of

impact as a static concentrated load. The central displacement obtained under such a load when divides the maximum dynamic displacements yields dynamic magnification factor (DMF). The variations of maximum contact force, maximum dynamic displacement, and equivalent static load (ESL) with impact velocity are almost linear and all of the three above mentioned values are increasing functions of the impact velocity. However the dynamic magnification-factor (DMF) and the impact velocity show a logarithmic relation and the DMF is a decreasing function of velocity of impact.

It may also be noted that at some of the points, for example, point-2, the displacement at some instant may exceed the same occurring at the point of impact. But such displacements will never govern the design as they are always less than the absolute maximum displacement occurring at the point of contact.

5. Conclusions

The following conclusions may be derived from the present study.

(1) The close agreement of the results obtained by the present method with those available in the published literature establishes the correctness of the approach used here.

(2) Under the influence of normal low velocity impact, the contact force shows a parabolic combined loading and unloading curve with a single peak for the practical class of shells considered here. Higher magnitude of impact velocity results in higher value of the peak contact force but due to a sharp elastic rebound, the total duration of contact force is less for higher velocity of impactor.

(3) The time instants at which the maximum contact force and the maximum dynamic displacement occur show a phase difference and interestingly in some cases, the maximum displacement and hence stresses may occur even after the contact force dies down totally. Thus it is concluded that the study may be stopped only after when the major peaks of the dynamic displacement die down and not after the full decay of the contact force only.

(4) The maximum contact force, the peak dynamic displacement, and the equivalent static load are all increasing functions of impactor velocity, the relations being almost linear. However, the dynamic magnification factor shows a logarithmically decreasing tendency with increase of the velocity of impact.

(5) At some of the points, the displacement at some instant may exceed the same occurring at the point of impact. But such displacements will never govern the design as they are always less than the absolute maximum displacement occurring at the point of contact.

References

[1] H. Hertz, "On the contact of elastic solids," *Journal fur Die Reine und Angewandte Mathematik*, vol. 92, pp. 156–171, 1881.

[2] T. M. Tan and C. T. Sun, "Use of statical indentation laws in the impact analysis of laminated composite plate," *Journal of Applied Mechanics, Transactions ASME*, vol. 52, no. 1, pp. 6–12, 1985.

[3] C. T. Sun and J. K. Chen, "On the impact of initially stressed laminates," *Journal of Composite Materials*, vol. 19, no. 6, pp. 490–504, 1985.

[4] S. L. Toh, S. W. Gong, and V. P. W. Shim, "Transient stresses generated by low velocity impact on orthotropic laminated cylindrical shells," *Composite Structures*, vol. 31, no. 3, pp. 213–228, 1995.

[5] V. P. W. Shim, S. L. Toh, and S. W. Gong, "The elastic impact response of glass/epoxy laminated ogival shells," *International Journal of Impact Engineering*, vol. 18, no. 6, pp. 633–655, 1996.

[6] L. S. Kistler and A. M. Waas, "Impact response of cylindrically curved laminates including a large deformation scaling study," *International Journal of Impact Engineering*, vol. 21, no. 1-2, pp. 61–75, 1998.

[7] L. U. Chun and K. Y. Lam, "Dynamic response of fully-clamped laminated composite plates subjected to low-velocity impact of a mass," *International Journal of Solids and Structures*, vol. 35, no. 11, pp. 963–979, 1998.

[8] A. Karmakar and K. Kishimoto, "Transient dynamic response of delaminated composite twisted cylindrical shells subjected to low velocity impact," *Key Engineering Materials*, vol. 297–300, pp. 1285–1290, 2005.

[9] A. Karmakar and K. Kishimoto, "Transient dynamic response of delaminated composite rotating shallow shells subjected to impact," *Shock and Vibration*, vol. 13, no. 6, pp. 619–628, 2006.

[10] S. Sahoo and D. Chakravorty, "Finite element bending behaviour of composite hyperbolic paraboloidal shells with various edge conditions," *Journal of Strain Analysis for Engineering Design*, vol. 39, no. 5, pp. 499–513, 2004.

[11] S. Sahoo and D. Chakravorty, "Finite element vibration characteristics of composite hypar shallow shells with various edge supports," *Journal of Vibration and Control*, vol. 11, no. 10, pp. 1291–1309, 2005.

[12] S. Sahoo and D. Chakravorty, "Free vibration characteristics of point supported hypar shells-a finite element approach," *Advances in Vibration Engineering*, vol. 7, no. 2, pp. 197–205, 2008.

[13] S. Das Neogi, A. Karmakar, and D. Chakravorty, "Impact response of simply supported skewed hypar shell roofs by finite element," *Journal of Reinforced Plastics and Composites*, vol. 30, no. 21, pp. 1795–1805.

[14] V. Z. Vlasov, *Allegemeine Schalen Theorie und Ihre Anwendung in Dar Technik*, Akademie-Verlag GmBH, Berlin, Germany, 1958.

[15] S. H. Yang and C. T. Sun, "Indentation law for composite laminates," *Composite Materials*, vol. 787, pp. 425–446, 1985.

[16] W. Goldmith, *IMPACT the Theory and Physical Behavior of Colliding Solids*, Dover, New York, NY, USA, 2001.

[17] M. S. Qatu and A. W. Leissa, "Natural frequencies for cantilevered doubly-curved laminated composite shallow shells," *Composite Structures*, vol. 17, no. 3, pp. 227–255, 1991.

9

Bandwidth Allocation Based on Traffic Load and Interference in IEEE 802.16 Mesh Networks

Sanjeev Jain,[1] Vijay Shanker Tripathi,[2] and Sudarshan Tiwari[2]

[1] Department of Electronics & Communication Engineering, Motilal Nehru National Institute of Technology, Allahabad, India
[2] Department of Electronics & Communication Engineering, NIT, Raipur, India

Correspondence should be addressed to Sanjeev Jain; snjece@gmail.com

Academic Editor: Daniele Tarchi

This paper introduces a traffic load and interference based bandwidth allocation (TLIBA) scheme for wireless mesh network (WMN) that improves the delay and throughput performance by proper utilization of assigned bandwidth. The bandwidth is allocated based jointly on traffic load and interference. Then a suitable path is selected based upon the least routing metric (RM) value. Simulation results are presented to demonstrate the effectiveness of the proposed approach which indicates higher bandwidth utilization and throughput as compared with existing fair end-to-end bandwidth allocation (FEBA).

1. Introduction

Wireless mesh networking is an emerging hot topic and is still in infancy. Key features of WMN are being dynamically self-organized, self-configured, self-healing, scalable, reliable, easy to deploy, and it can establish adhoc network automatically and maintain connectivity. WMNs are activated in the industrial standard groups, such as IEEE 802.11, IEEE 802.15, and IEEE 802.16. [1]. Few applications of WMN are to access broadband internet, indoor WLAN, mobile user access and connectivity. WMNs are specifically constructed by the Firetide for providing connectivity [2].

Backhaul connectivity of the mesh networks is provided by the mesh base station in the IEEE 802.16 and controlling one or more subscriber stations is also provided. Collection of bandwidth request from subscriber station and management of resource allocation are the responsibilities of the mesh base station (BS) when a centralized scheduling scheme is used [3]. There are two types of routing in WMNs, namely, centralized scheduling and distributed scheduling. The IEEE 802.16 standard provides a centralized scheduling mechanism that supports contention-free and resource-guarantee transmission services in mesh mode. Research is going on towards designing an efficient way to realize centralized or distributed schedule by maximizing channel utilization. The designs are divided into two phases: routing and scheduling. First, a routing tree topology is constructed from a given mesh topology. Secondly, channel resource is allocated to the edges in the routing tree by a scheduling algorithm [4].

The channel resource is bandwidth, which is allocated on the basis of fundamental performance parameters. Generally delay, throughput, fairness, or interference is considered for bandwidth allocation. The wireless network has experienced significant growth to meet the increasing bandwidth demands of network users and support the emerging bandwidth-intensive applications such as videoconferencing and video on demand (VoD). In the IEEE 802.16 mesh networks the bandwidth negotiation is implicit which is based on the assumption that only the one-hop neighbors of a receiver can interfere with its ongoing data reception, which is also referred to as "protocol-model." In 802.16 mesh networks in order to satisfy the QoS in routing packets, it is very important to reserve sufficient bandwidth for the transmission of the individual links on a particular route. Because, in wireless mesh networks, the end-to-end throughput of traffic flows depends on the path length, that is, the higher the number of hops, the lower the throughput becomes.

Organization of this paper is as follows. In Section 2 we have presented related work. In Section 3 details of proposed work and algorithm are presented. Section 4 presents the simulation results using NS2 simulator. Section 5 presents conclusions and outlines directions for the future work.

2. Related Work

Cicconetti et al. [5] have proposed a fair end-to-end bandwidth allocation (FEBA) algorithm, in order to provide a maximum throughput in the end-to-end traffic flow. FEBA is implemented at the medium access control (MAC) layer of single-radio, multiple channels IEEE 802.16 mesh nodes, operated in a distributed coordinated scheduling mode. The advantage of this approach is that it negotiates bandwidth among neighbors to assign a fair share proportional to a specified weight to each end-to-end traffic flow. This way traffic flows are served in a differentiated manner, with higher priority traffic flows being allocated more bandwidths on average than the lower priority traffic flows.

Peng and Cao [6] have presented a dynamic programming based resource allocation and scheduling algorithm to address the problem of resource allocation with the goal of providing fairness access to channels in IEEE 802.16 mesh networks. They defined node's unsatisfactory index and throughput function. Then, a multiobjective programming formulation was proposed for optimizing network performance.

Zhang et al. [7] have proposed a novel QoS guarantee mechanism which includes protocol process and minislot allocation algorithm. It uses existing service classes in original standard. Protocol processes were defined to manage the dynamic service flow and minislot allocation algorithm was used to support data scheduling of various services. WiMAX MAC layer was redesigned to support service classification in mesh mode. Using extended distributed scheduling messages, the delivery method of dynamic service management messages in WiMAX mesh networks was implemented.

Mogre et al. [8] have proposed a CORE, which addresses the problem of jointly optimizing the routing, scheduling, and bandwidth savings via network coding. Prior solutions are either not applicable in the 802.16 MeSH mode or computationally too costly to be of practical use in the WMN under realistic scenarios. CORE's heuristics are able to compute solutions for the previous problem within an operator definable maximum computational cost, thereby enabling the computation and near real-time deployment of the computed solutions. And the advantage of this approach is that CORE is able to increase the number of flows admitted considerably and with minimal computational costs. We also see that CORE successfully increases the number of network coding sessions which can be established in the WMN.

De Rango et al. [9] have proposed a GCAD-CAC (greedy choice with bandwidth availability aware defragmentation) algorithm which is able to guarantee a respect for data flow delay constraints defined by three different traffic classes. By this approach it is possible to achieve good results and try to accept all the new requests, but when a higher priority request is received, a lower priority admitted request is preempted. This preemption can leave some small gaps which are not

sufficient for new connection admission; these gaps can be collected by the GCAD algorithm by activating a bandwidth availability based defragmentation process.

Yang et al. [10] have proposed zone-based bandwidth allocation for mobile users in the IEEE 802.16j multihop relay network (IEEE 802.16-MR). The main focus of the work was adaptive selection of the zone size fit for user mobility. The zone of a mobile user includes the current relay station and its neighboring relay stations within the zone size in hop count. Bandwidth allocation was done for the mobile user roaming within the zone, and calculation of the required bandwidth was also presented.

Shakeri and Khazaei [11] have presented a novel scheduling scheme in WMNs. This technique is a multiple gateway fair scheduling scheme. This scheme consists of distributed requirement table and requirement propagation algorithm for scheduling at the gateways. The requirement propagation algorithm allows each gateway to distribute the requirements and routing table for scheduling into the network.

Delay aware load balancing routing (DLBR) [12] was proposed for WMN by introducing combined RM. The same was compared with existing Load balancing metric (LBM) [13]. The projected simulation results showed considerable reduction in delay and overhead thereby increasing the overall packet delivery. However, this paper does not consider bandwidth allocation and to the best of our knowledge till date there is a lack of a systematic study of distributed bandwidth reservation strategies for the mesh networks [1, 2, 14, 15].

3. Methodology

Our proposed metric differs from prior methods in several ways; that is, an efficient route is established with least delay and load which is considered in bandwidth allocation in WMN. We consider the traffic load in the interfering neighbors as the metric of traffic interference. Initially, a combined routing metric (RM) is defined for efficient route selection using the metrics traffic interference (TIM) and end-to-end service delay (EDM) [12, 16]. The suitable path is selected based upon the least routing metric value. Next, bandwidth allocation is performed for the selected path using fair end-to-end bandwidth allocation (FEBA) [5]. The basic idea of FEBA is that each node assigns bandwidth requests and grants in a round-robin manner where the amount of allocated bandwidth in bytes is proportional to the number of traffic flows weighted on their priorities. In this, FEBA approach each active queue, both requesting and granting, is assigned a weight value which is used by the bandwidth request/grant procedure. So, the amount of service is proportional to the number of traffic flows under service. We are considering the traffic interference (TIM) [12, 17] metrics along with the traffic load in the request/grant procedure to make it possible to provide an efficient bandwidth.

3.1. Calculation of Traffic Interference Metric (TIM). We consider the traffic load in the interfering neighbors as the metric of traffic interference. Here both interflow interference and intraflow interference are considered. When the neighboring nodes transmit on the same channel, they compete with each other for channel bandwidth. The number of interfering nodes is not considered for degree of interference; instead the load generated by the interfering node is taken into account. This metric considers the traffic of interfering node to capture the interflow interference.

Let $\eta l(D)$ be the set of interfering neighbors of nodes a and node b, over channel D.

ETT_{ab} captures the difference in transmission rate and loss ratio of links.

Then the TIM metric is defined as follows:

$$\text{TIM} = \text{ETT}_{ab}(D) \times \text{Lavg}_{ab}(D), \quad \eta l(D) \neq 0$$
$$\text{TIM} = \text{ETT}_{ab}(D), \quad \eta l(D) = 0, \tag{1}$$

where Lavg_{ab} is the average load of $\eta l(D)$, given by

$$\text{Lavg}_{ab}(D) = \frac{\sum_{\eta 1} L_{\text{int}}(D)}{\eta 1(D)}, \tag{2}$$
$$\eta l(D) = \eta_a(D) \cup \eta_b(D).$$

$L_{\text{int}}(D)$ is the load of the interfering neighbors.

When there are no interfering neighbors, TIM metric selects the path with high transmission rate and low loss ratio. In the presence of interfering neighbors, TIM metric selects the path with minimum traffic load and minimum interference [17].

3.2. Calculation of End-to-End Service Delay Metric (EDM). The expected end-to-end service delay metric (EDM) is used [16] to allow any shortest path based routing protocol to select a route with lowest end-to-end latency.

The EDM is defined as "network load-aware and radio-aware service delay" which is the end-to-end latency spent in transmitting a packet from source to destination. In order to estimate the EDM value, the expected link transmission time (ELT^2) which is used for successfully transmitting a packet on each link is computed. Then this value is multiplied with mean number of backlogged packets in output queue at each relay node.

It is assumed that each node is serviced with a first-in-first-out (FIFO) interface queue. The per-hop service delay T_k is given by the expected time spent in transmitting all packets waiting for transmission through a link i at node k.

T_k considers the expected service delay, of any node such as queue delay, contention delay and transmission time of link i between node k and any neighbor node in the transmission range.

With a given T_k, the EDM of path p with h-hops, between source and destination, is estimated as follows:

$$\text{EDM}(p) = \sum_{j=1}^{h} T_k. \tag{3}$$

3.2.1. Estimation of T_k. Let there be M neighbor nodes in transmission range of node n. Let $\eta_{k,i}$ be the mean number of packets waiting for transmission on link i at node k

to successfully transmit through link I; T_k is estimated as follows:

$$T_k = \sum \left(\eta_{k,i} \times \left(dc_{k,w} + \text{ELT}^2(k,i) \right) \right) + \text{ELT}^2(k,i), \tag{4}$$

where the $\text{ELT}^2(k,i)$ is the ELT^2 of link i at node k and $dc_{k,w}$ is the mean contention delay at node k. As a result, route selection using the EDM finds the path with the lowest end-to-end service delay in terms of current network load. In addition, a routing protocol using this metric can simultaneously perform traffic load balancing.

3.2.2. Estimation of ELT^2. $\text{ELT}^2(k,i)$ is defined as the link transmission time spent by sending a packet over link i at node k. This measure is approximated and designed for ease in implementation and interoperability.

The ELT^2 for each link is calculated as

$$\text{ELT}^2(k) = \left[O_{\text{cnt}} + \frac{F_s}{t} \right] \times \frac{1}{(1 - F_r)}, \tag{5}$$

where O_{cnt} is the control overhead, $dc_{k,w}$ is the mean contention delay, and the input parameters t and F_r are the bit rate in Mbs and the frame error rate of link i for frame size F_s, respectively.

3.3. Route Metric. A combined route metric (RM) [12] is proposed which includes both TIM and EDM metrics for efficient route selection:

$$\text{RM} = C_1 * \text{TIM} + C_2 * \text{EDM}. \tag{6}$$

Here C_1 and C_2 are the normalizing factors for TIM and EDM whose values range from 0 to 1. The normalizing factors C_1 and C_2 are chosen based on the weightage of interference or delay. Initially both are assigned equal values 0.5. When the interference is higher, the value of C_1 is adaptively increased and C_2 is decreased. Similarly when delay is higher, C_1 can be decreased and C_2 can be increased. Then a path with least value of RM is selected by exchanging RREQ and RREP packets. To allocate bandwidth for this selected path, a bandwidth allocation technique is given in the next section.

3.4. Bandwidth Allocation Technique. Our bandwidth allocation is based on the fair end-to-end bandwidth allocation (FEBA) approach which supports differentiated services for traffic flows. Let us consider a node maintaining two virtual queues towards any of its neighbor nodes which are the requesting queue and the granting queue. The requesting queue is the total amount of backlogged bytes directed to its neighbor. On the other side, the total amount of data enqueued at node directed to node is the occupancy of the granting queue. The mechanism works by allocating requests and grants dynamically based on the current status of the traffic load and physical transmission rates. In IEEE 802.16 bandwidth allocation process the requests and grants are expressed in units of slots.

In Figure 1 the node X is requesting the node Y. Based on the current status of the traffic load and physical transmission

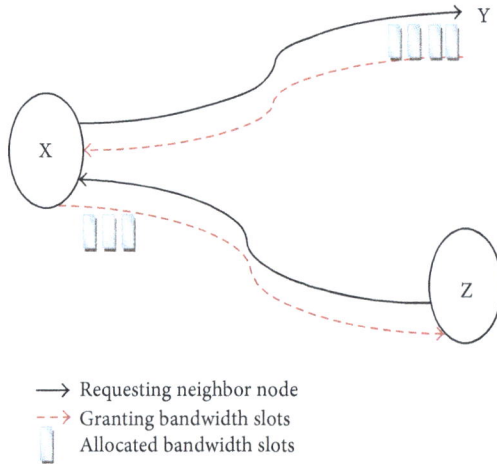

FIGURE 1: Bandwidth allocation.

$\longrightarrow$ Requesting neighbor node
$--\rightarrow$ Granting bandwidth slots
⊔ Allocated bandwidth slots

rates the node Y responds to the node X and grants the suitable bandwidth slots for the node X. In the same way the node Z requests the node X and the node X grants the suitable bandwidth slots.

3.5. Estimation of the Bandwidth. Here both requesting and granting nodes are assigned a weight (B_i) which is used by the bandwidth request/grant procedure. The allocated B_i of any queue i is calculated so that the amount of service is proportional to number of traffic flows under service, weighted based on their priorities:

$$B_i = \frac{\sum_{j \in A} P_j * I_i(j)}{\sum_{j \in A} P_j}, \tag{7}$$

where and A is the set of all active traffic flows served by this node j is an active flow with priority p_j.

$I_i(j)$ is an indicator function which equals 1 if j is under service at queue i, 0 otherwise. To provide the suitable bandwidth according to the path conditions and variations in the traffic flow during this bandwidth allocation for any queue, we also consider the traffic interference metric (TIM) in (7):

$$B_i = \frac{\sum_{j \in A} P_j * I_i(j)}{\sum_{j \in A} P_j} + \text{TIM}, \tag{8}$$

where TIM (traffic interference metric) is estimated in [12].

There are some advantages considering the TIM while allocating the bandwidth; that is, the FEBA can tackle the spatial bias problem through keeping separate queues at every node for each traversing traffic flow. And also according to the traffic flow it is possible to provide differentiated services. The FEBA can get adjusted to the short time changes in the network only by considering the variations in the traffic flow.

Algorithm 1

(1) Start

(2) Estimation of TIM

(3) If $\eta 1(D) \neq 0$

   $\text{TIM} = \text{ETT}_{ab}(D) \times L_{avg}(D)$

(4) If $\eta 1(D) = 0$

   $\text{TIM} = \text{ETT}_{ab}(D)$

(5) Estimation of the EDM

(6) $\text{EDM} = \sum_{j=1}^{h} T_k$

(7) If $\text{ELT}^2 > T_h$

   EDM is MAX

   Else

   EDM is MIN

(8) Estimation of FEBA

(9) Allocation of the Bandwidth is directly proportional to TIM

   $$B_i = \frac{\sum_{j \in A} P_j * I_i(j)}{\sum_{j \in A} P_j} + \text{TIM}$$

(10) End

In the previous algorithm, initially the TIM is found considering the set of neighbor nodes (ηl) [12], between the source and the destination nodes and difference in transmission rate and loss ratio of links. Then the end-to-end delay EDM is calculated using the time required for the transmission. This EDM is directly proportional to the ELT^2; that is, if the ELT^2 is more than the threshold value which is considered, then the EDM also increases. And the bandwidth is allocated for the path with the low delay. During this bandwidth allocation we consider TIM; depending on the TIM the bandwidth will be allocated for the particular transmission.

4. Simulation Results

4.1. Performance Metrics. We compare our traffic load and interference aware bandwidth allocation (TLIBA) technique with the FEBA [5] technique. We evaluate mainly the performance according to the following metrics, by varying the simulation time and the number of channels.

Average end-to-end delay: the end-to-end-delay is averaged over all surviving data packets from the sources to the destinations.

Received bandwidth: it is the measured at each receiver and expressed in Mb/s.

Fairness: it is the average received packets at each receiver.

4.2. Simulation Model and Parameters. We use NS-2 [18] to simulate our proposed protocol. We use the IEEE 802.16*e* simulator [19] patch for NS2 version 2.33 to simulate a WiMAX Mesh Network. It has the facility to include multiple channels and radios. It supports different types of topologies such as chain, ring, multiring, grid, binary tree, star, hexagon, and triangular. The supported traffic types are CBR, VoIP, video-on-demand (VoD), and FTP. In our simulation, mobile nodes are arranged in a ring topology of size 500 meter ∗ 500

Table 1: Simulation settings.

Number of nodes	25
Area size	500×500
Mac	802.16e
Radio range	250 m
Simulation time	100 sec
Traffic source	VoIP and VoD
VoD packet size	500 to 1500 bytes
VoD rate	100 Kb
VoIP Codec	GSM.AMR
Number of VoIP frames per packet	2
Number of traffic flows	1, 2, 3, 4, and 5
Topology type	Ring
OFDM bandwidth	10 MHz
Rate	500 Kb

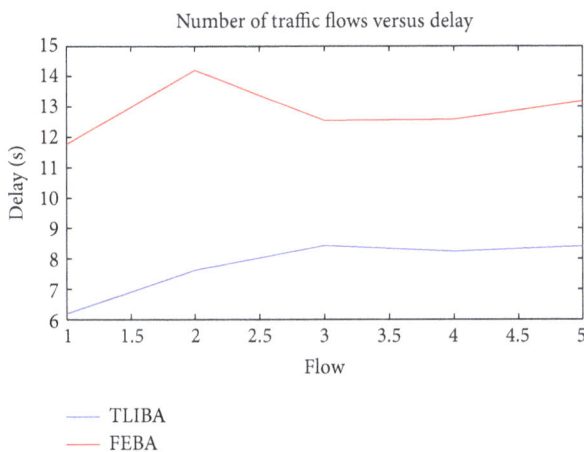

Figure 3: Flow versus received bandwidth.

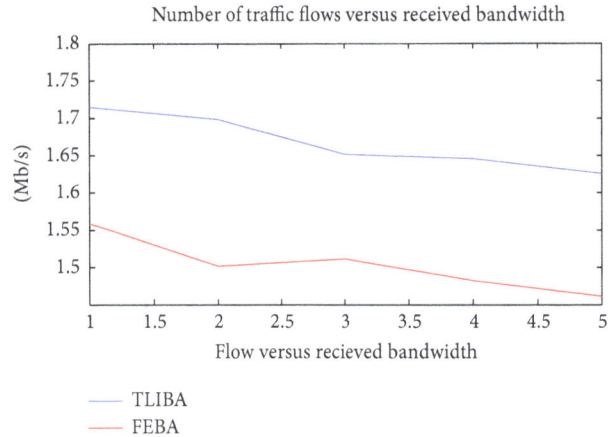

Figure 2: Flow versus delay.

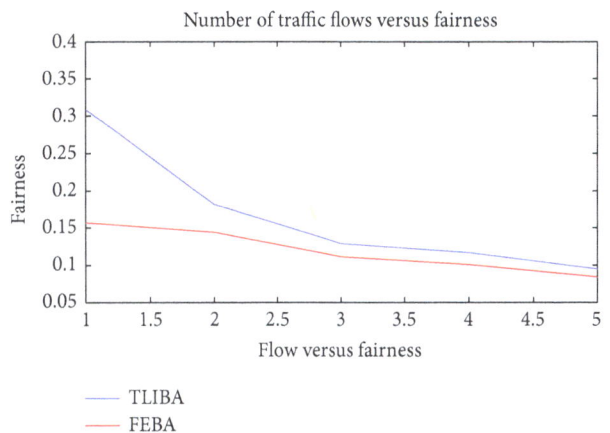

Figure 4: Flow versus fairness.

meter region. We keep the number of nodes as 25. All nodes have the same transmission range of 250 meters. A total of 4 traffic flows (one VoIP and three VoD) are used.

Our simulation settings and parameters are summarized in Table 1.

4.2.1. Performance Based on Traffic Flows. Initially we vary the number of traffic flows as 1, 2, 3, 4, and 5 with packet size as 1500 bytes.

For Figure 2, the results show that as the number of traffic flow increases, the average end-to-end delay also increases. Delay performance of TLIBA is improved significantly as compared to FEBA. In Figure 3 the results also indicate that when traffic flow increases, the bandwidth decreases but with proposed scheme more bandwidths are achieved. Figure 4 gives the fairness for both techniques when the number of traffic flows is increased. It shows that the fairness is more as compared to FEBA.

4.2.2. Performance Based on Traffic Rate. In our second experiment we vary the rate as 500 to 1500 Kb with 5 flows.

For Figure 5, the results show that when the packet size increases, the average end-to-end delay also increases and shows improvement with the proposed TLIBA scheme as compared to FEBA. In Figure 6 the results also indicate that when the packet size increases the bandwidth also increases. We can observe that TLIBA achieves more bandwidths as compared to FEBA. Figure 7 gives the fairness for both techniques when the packet size is increased. It shows that the fairness is more as compared to FEBA.

5. Conclusion and Future Work

This paper suggests a traffic load and interference based bandwidth allocation (TLIBA) technique for IEEE 802.16 mesh networks. First, we have calculated the metric of traffic interference (TIM) which considers the traffic load of interfering neighbors. End-to-end service delay (EDM) is calculated by using the expected time spent in transmitting all packets waiting for transmission through a link. Using these metrics, combined routing metric is defined for efficient route selection and the best path is selected based upon the least routing metric value. For the selected path, the bandwidth is allocated using the fair end-to-end bandwidth allocation

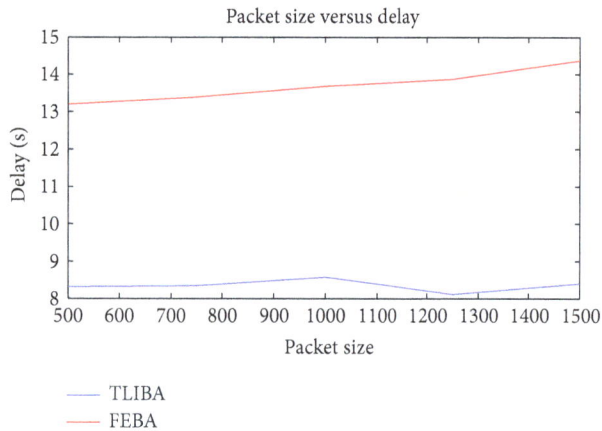

FIGURE 5: Packet size versus delay.

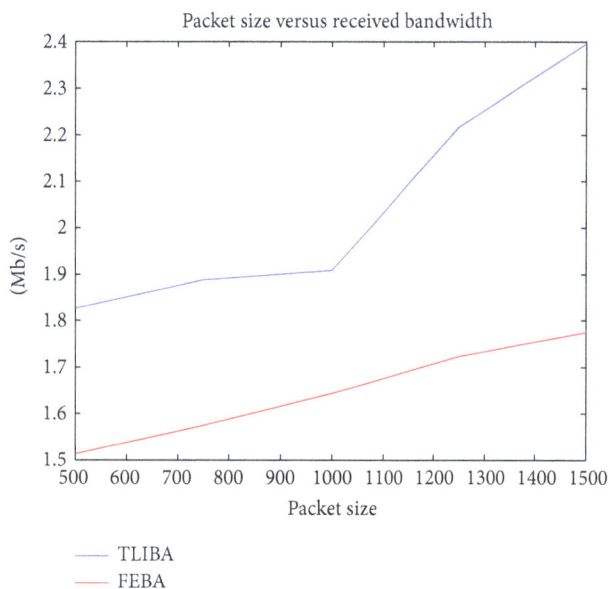

FIGURE 6: Packet size versus received bandwidth.

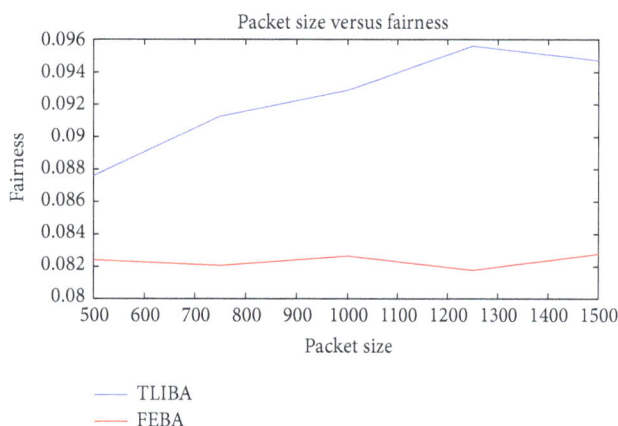

FIGURE 7: Packet size versus fairness.

(FEBA) approach. This allocation considers the load and the traffic of the link and allocates the bandwidth. The frequent changes in the path can be avoided thus increasing transmission efficiency. The main advantage of the proposed technique is that it is possible to allocate bandwidths in the wireless mesh networks according to the traffic load and interference of the network which makes it easy to achieve maximum throughput during the transmission. The simulation result shows substantial improvement in achieving better bandwidth utilization and throughput when compared with existing bandwidth allocation technique. As future work the quality of the performance evaluation can be further enhanced by considering different scenarios.

References

[1] I. F. Akyildiz and X. Wang, "A survey on wireless mesh networks," *IEEE Communications Magazine*, vol. 43, no. 9, pp. S23–S30, 2005.

[2] M. L. Sichitiu, "Wireless mesh networks: opportunities and challenges," in *Proceedings of the 6th World Wireless Congress (WWC '05)*, pp. 318–323, May 2005.

[3] H. Y. Wei, S. Ganguly, R. Izmailov, and Z. J. Haas, "Interference-aware IEEE 802.16 WiMax mesh networks," in *Proceedings of the IEEE 61st Vehicular Technology Conference (VTC '05)*, pp. 3102–3106, June 2005.

[4] S. C. Lo and L. C. Ou, "Efficient algorithms for routing and centralized scheduling for IEEE 802.16 mesh networks," in *Proceedings of the International Conference on Scalable Computing and Communications (ScalCom '09)*, pp. 212–217, September 2009.

[5] C. Cicconetti, I. F. Akyildiz, and L. Lenzini, "FEBA: a bandwidth allocation algorithm for service differentiation in IEEE 802.16 mesh networks," *IEEE/ACM Transactions on Networking*, vol. 17, no. 3, pp. 884–897, 2009.

[6] L. Peng and Z. Cao, "Fairness resource allocation and scheduling for IEEE 802.16 mesh networks," *Journal of Networks*, vol. 5, no. 6, pp. 724–731, 2010.

[7] Y. Zhang, C. Dai, and M. Song, "A novel QOS guarantee mechanism in IEEE 802.16 mesh networks," *Computing and Informatics*, vol. 29, no. 4, pp. 521–536, 2010.

[8] P. S. Mogre, N. D'Heureuse, M. Hollick, and R. Steinmetz, "CORE: centrally optimized routing extensions for efficient bandwidth management and network coding in the IEEE 802.16 MeSH mode," *Wireless Communications and Mobile Computing*, vol. 11, no. 3, pp. 338–356, 2011.

[9] F. de Rango, A. Malfitano, and S. Marano, "GCAD: a novel call admission control: algorithm in IEEE 802.16 based wireless mesh networks," *Journal of Networks*, vol. 6, no. 4, pp. 595–606, 2011.

[10] C. C. Yang, Y. T. Mai, and I. W. Lin, "Adaptive zone-based bandwidth management in the IEEE 802.16j multi-hop relay network," in *Proceedings of the International MultiConference of Engineers and Computer Scientists (IMECS '11)*, vol. 1, pp. 16–18, 2011.

[11] M. Shakeri and M. Khazaei, "Fair scheduling in the network with a mesh topology," *International Research Journal of Applied and Basic Sciences*, vol. 3, no. 5, pp. 911–918, 2012.

[12] J. Sanjeev, T. S. Vijay, and T. Sudarshan, "Delay-aware load balanced routing protocol for IEEE 802.16 wireless mesh networks," *IJCSI International Journal of Computer Science Issues*, vol. 9, no. 6, pp. 421–430, 2012.

[13] M. Liang and M. K. Denko, "A routing metric for load-balancing in wireless mesh networks," in *Proceedings of the 21st International Conference on Advanced Information Networking and Applications Workshops/Symposia (AINAW '07)*, vol. 2, pp. 409–414, May 2007.

[14] K. kelaiya, "Routing & scheduling algorithm of IEEE 802.16 mesh backhaul network for Radio Recourse Management (RRM)," in *Proceedings of the National Conference on Mobile and Pervasive Computing (CoMPC '08)*, pp. 175–179, 2008.

[15] M. J. Lee, J. Zheng, K. O. Young-Bae, and D. M. Shrestha, "Emerging standards for wireless mesh technology," *IEEE Wireless Communications*, vol. 13, no. 2, pp. 56–63, 2006.

[16] J.-S. Youn and C.-H. Kang, "Effective routing scheme through network load-aware route metric in multi-rate wireless mesh networks," in *Proceedings of the International Workshop on Mobile Services and Personalized Environments (MSPE '06)*, pp. 43–56, 2006.

[17] D. M. Shila and T. Anjali, "Load-aware traffic engineering for mesh networks through multiple gateways," in *Proceedings of the IEEE International Conference on Mobile Adhoc and Sensor Systems (MASS)*, pp. 807–812, 2006.

[18] Network Simulator, http://www.isi.edu/nsnam/ns.

[19] C. Cicconetti, I. F. Akyildiz, and L. Lenzini, "WiMsh: a simple and efficient tool for simulating IEEE 802.16 wireless mesh networks in ns-2," in *Proceedings of the 2nd International Conference on Simulation Tools and Techniques*, 2009, http://dl.acm.org/citation.cfm?id=1537624.

10

DFRFT: A Classified Review of Recent Methods with Its Application

Ashutosh Kumar Singh and Rajiv Saxena

Department of Electronics and Communication Engineering, Jaypee University of Engineering and Technology, Guna, Raghogarh, Madahya Pradesh 473226, India

Correspondence should be addressed to Ashutosh Kumar Singh; ashutoshsingh79@yahoo.com

Academic Editor: Wei-Qiang Zhang

In the literature, there are various algorithms available for computing the discrete fractional Fourier transform (DFRFT). In this paper, all the existing methods are reviewed, classified into four categories, and subsequently compared to find out the best alternative from the view point of minimal computational error, computational complexity, transform features, and additional features like security. Subsequently, the correlation theorem of FRFT has been utilized to remove significantly the Doppler shift caused due to motion of receiver in the DSB-SC AM signal. Finally, the role of DFRFT has been investigated in the area of steganography.

1. Introduction

Due to the inadequate performance of Fourier transform for the analysis of nonstationary signal, normally encountered in communication systems when either transmitter or receiver is moving, fractional Fourier transform (FRFT) can be utilized. The concept of FRFT was first introduced by N. Wiener in 1929. However, the FRFT was recognized as a *"transform method"* by mathematical bodies after the work of Victor Namias in 1980 in which the concept of FRFT had been introduced by considering fractional power of eigenfunctions of the ordinary FT. The mathematical description of FRFT was given by McBride and Keer in 1987. And finally, a general definition of FRFT for all classes of signals was given by Cariolaro et al. To compute the FRFT of any signal its discrete version was needed which initiates the work for defining discrete FRFT (DFRFT). However, due to availability of different approaches for evaluating DFRFT, a detailed study is needed in the context of suitable definition of DFRFT. This study along with comparison between the existing methods has been included in Section 3.

This paper includes five sections. The basic concept of fractional Fourier transform is described in Section 2. The detailed study of discrete FRFT (DFRFT) along with

the comparative analysis between the different methods of evaluating DFRFT is included in Section 3. Subsequently, the utility of FRFT in DSB-SC signal detection under noisy condition and the application of DFRFT in steganography are being dealt in Section 4. Finally, conclusions are made in Section 5.

2. Fractional Fourier Transform (FRFT)

The FRFT is a mathematical tool which maps a signal from one domain to other in time-frequency plane (e.g., FRFT becomes FT when it maps a signal in time domain to frequency domain), shown in Figure 1. The FRFT of a signal $x(t)$ with angle parameter α, represented by $X_\alpha(u)$ is defined as [1]:

$$F_\alpha[x(t)] = X_\alpha(u)$$

$$= \sqrt{\frac{1 - j\cot\alpha}{2\pi}} \int_{-\infty}^{\infty} x(t)\, e^{(j/2)\{(t^2 + u^2)\cot\alpha - 2tu\csc\alpha\}} dt \tag{1}$$

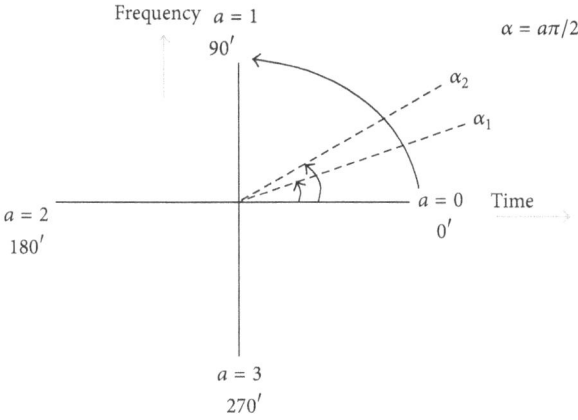

FIGURE 1: FRFT in time-frequency plane.

The angle parameter associated with the FRFT is given as:

$$\alpha = \frac{a\pi}{2}. \tag{2}$$

When $\alpha = 0$, the FRFT becomes identity operator, and for $\alpha = \pi/2$ ($a = 1$), the FRFT becomes Fourier transform (FT). Similarly, for $\alpha = \pi$ ($a = 2$) FRFT behaves as an inverse operator, and for $\alpha = 3\pi/2$ ($a = 3$), the expression converts into an inverse FT, as shown in Figure 1.

In signal processing, cross-correlation is a measure of similarity of two waveforms as a function of a time lag applied to one of them. The cross-correlation theorem for the FRFT is given as follows [2]. Let $r_{xy}(\tau)$ be defined as weighted cross-correlation of two functions $x(t)$ and $y(t)$, and $X_\alpha(u)$, $Y_\alpha(u)$, and $R_{xy,\alpha}(u)$, be defined as FRFT of $x(t)$, $y(t)$, and $r_{xy}(\tau)$, respectively. Then,

$$r_{xy}(\tau) = \int_{-\infty}^{\infty} x(t)\, y(t+\tau)\, e^{jt(\tau+t)\cot\alpha}\, dt \longleftrightarrow R_{xy,\alpha}(u)$$

$$= \sqrt{\frac{2\pi}{1-j\cot\alpha}}\, e^{-(j/2)u^2\cot\alpha} X_\alpha(-u)\, Y_\alpha(u). \tag{3}$$

This cross-correlation theorem of FRFT is utilized for detection of DSB-SC signal when the signal encounters a Doppler shift due to motion of transmitter or receiver.

3. Discrete Fractional Fourier Transform (DFRFT)

Many researchers have tried to come up with some method of evaluating DFRFT. On this journey, many methods and algorithms had come into existence. But none got acclamation due to deficiency of either not satisfying some of the prime properties that its continuous type possessses or nonexistence of a closed-form expression. Previously, many classifications have already been documented [3]. However, in this section, all the approaches available in the literature are included and divided into four classes on the basis of their methodology of evaluation. These classes are

(i) sampling-based method

(ii) linear combination method

(iii) eigenvector-based method

(iv) weighted summation-based method.

3.1. Sampling-Based Method. In this approach, by considering the samples of the kernel matrix under some predefined constraint, two different algorithms are available in the literature and they are named as sampling type DFRFT and closed-form type DFRFT. These two methods are described below:

3.1.1. Sampling Type DFRFT. The sampling theorem for the FRFT of band-limited and time-limited signals is followed from Shannon's sampling theorem [4–9]. With this approach, the simplest way to derive the DFRFT is sampling the continuous FRFT and computing it directly from the samples, but by sampling the continuous FRFT directly, the resultant discrete transform obtained will lose many important properties. The most serious problem with this type DFRFT will be noncompliance of unitary and reversibility properties. In addition, the DFRFT obtained by direct sampling of the FRFT lacks closed-form expressions and is nonadditive, which affects its application domain. In order to maintain some of the FRFT properties, a type of DFRFT was derived as a special case of the continuous FRFT as reported in [10]. Specifically, the input function was assumed as a periodic, equally spaced impulse train. Since this type of DFRFT is a special case of continuous FRFT, many properties of the FRFT exist and have the fast algorithm. However, this type of DFRFT cannot be defined for all values of α due to various imposed constraints. At the same time, Ozaktas et al. [11] had proposed two innovative approaches for obtaining the DFRFT through the sampling of the FRFT. Both methods were based on the idea of manipulating the expression of the FRFT by sampling appropriately.

In the first method, the DFRFT of a signal $x(t)$ can be obtained by multiplying the signal with a chirp function as followes:

$$x'(t) = e^{-j\pi t^2 \tan(\alpha/2)} x(t). \tag{4}$$

For this chirp multiplication, the bandwidth and time-bandwidth product of $x'(t)$ can be as large as twice of $x(t)$. Thus, the sampling of $x'(t)$ will be done at intervals of $(1/2\Delta x)$. If the samples of $x(t)$ are spaced at $(1/\Delta x)$, then this needs to be interpolated before multiplying it by the samples of the chirp function to obtain the desired samples of $x'(t)$.

The next step was to convolve the signal $x'(t)$ with another chirp function. To perform this convolution, the chirp signal was replaced by its band-limited version because $x'(t)$ is assumed as a band-limited signal.

$$x''(t) = A_\alpha \int_{-\infty}^{\infty} x(\tau)\, e^{j\pi(t-\tau)^2 \csc\alpha}\, d\tau. \tag{5}$$

The band-limited version of chirp function can be expressed in the form of Fresnel integral. Finally, again performing a chirp multiplication, the DFRFT of signal $x(t)$ can be obtained as:

$$\Im\left[x\left(t\right)\right] = X_\alpha = e^{-j\pi t^2 \tan(\alpha/2)} x''\left(t\right). \tag{6}$$

The convolution operation in the above equation can be achieved by sampling $x(t)$ and performing the convolution using the fast Fourier transform (FFT). Overall, the procedure starts with N samples spaced at $(1/\Delta x)$, which uniquely characterizes the function $x(t)$ and returns the same for X_α. For x and X_α, denote column vectors with N elements containing the samples of $x(t)$ and its DFRFT, respectively. The above procedure can be given in a matrix notation as:

$$X_\alpha = D\Lambda H\Lambda Jx. \tag{7}$$

Here, D and J are matrices representing the decimation and interpolation operations, Λ is a diagonal matrix that corresponds to chirp multiplication, and H corresponds to the convolution operation. The above method had allowed obtaining the samples of the ath transform in terms of the samples of the original function.

First method includes the analysis of one step in terms of Fresnel integral, which requires large number of computation time particularly in inverse DFRFT. To remove this discrepancy, a second method was introduced by Ozaktas et al. [11]. In this method, one assumption was taken that the Wigner distribution of function $x(t)$ could be zero outside a circle of diameter Δu centered at the origin and by limiting the order "a" to the interval $0.5 \le |a| \le 1.5$. Simultaneously, the amount of vertical shear in Wigner space resulting from the chirp modulation is bounded by $\Delta u/2$. The FRFT can be written as:

$$\Im\left[x\left(t\right)\right] = X_\alpha\left(u\right)$$
$$= A_\alpha\ e^{j\pi u^2 \cot\alpha} \int_{-\infty}^{\infty} e^{-j\pi tu \csc\alpha} \left\{e^{j\pi t^2 \cot\alpha} x\left(t\right)\right\} dt. \tag{8}$$

Thus, $e^{j\pi t^2 \cot\alpha} x(t)$ can be written by using Shannon's interpolation formula as:

$$e^{j\pi t^2 \cot\alpha} x\left(t\right)$$
$$= \sum_{n=-N}^{N} e^{j\pi(n/2\Delta u)^2 \cot\alpha} x\left(\frac{n}{2\Delta u}\right) \text{sinc}\left(2\Delta u\left(t - \frac{n}{2\Delta u}\right)\right), \tag{9}$$

where $N = (\Delta u)^2$; putting (9) in (8) and interchanging the order of integration and summation,

$$X_\alpha\left(u\right) = A_\alpha e^{j\pi u^2 \cot\alpha} \sum_{n=-N}^{N} e^{j\pi\,(n/2\Delta u)^2 \cot\alpha} x\left(\frac{n}{2\Delta u}\right)$$
$$\times \int_{-\infty}^{\infty} e^{-j\pi tu \csc\alpha} \text{sinc}\left(2\Delta u\left(t - \frac{n}{2\Delta u}\right)\right) dt. \tag{10}$$

And after some algebraic manipulation, (10) can be restructured as:

$$X_\alpha\left(\frac{m}{2\Delta u}\right) = \frac{A_\alpha}{2\Delta u}\ e^{j\pi\,(\cot\alpha - \csc\alpha)(m/2\Delta u)^2}$$
$$\times \sum_{n=-N}^{N} e^{j\pi\,(m-n/2\Delta u)^2 \csc\alpha} e^{j\pi\,(\cot\alpha - \csc\alpha)(n/2\Delta u)^2}$$
$$\times x\left(\frac{n}{2\Delta u}\right). \tag{11}$$

This summation can be interpreted as the convolution of $e^{j\pi(n/2\Delta u)^2 \csc\alpha}$ and the chirp modulated function $e^{j\pi(\cot\alpha - \csc\alpha)(n/2\Delta u)^2} x(n/2\Delta u)$. This convolution can be computed by fast Fourier transform (FFT). And the overall methodology can be written in matrix form as:

$$X_\alpha = DK_\alpha Jx, \tag{12}$$

where the K_α matrix was given as:

$$K_\alpha\left(m, n\right)$$
$$= \frac{A_\alpha}{2\Delta u} e^{j\pi\{\cot\alpha(m/2\Delta u)^2 - 2\csc\alpha(mn/(2\Delta u)^2) + \cot\alpha(n/2\Delta u)^2\}}. \tag{13}$$

In the same article, Ozaktas et al. [11] gave alternate methods to calculate the DFRFT by introducing scaling process. All these methods resulted in the same time complexity. However, in practice, these alternate methods are not preferable because they require coordinate scaling. Scaling is not an advantageous process if the original data has already been sampled. This is because scaling requires additional interpolations and so forth, which will require additional computation.

However, in both cases, the Wigner distribution of the signal was confined to a circle of diameter Δu around the origin. Thus, there might be several discrete fractional Fourier transform matrices, some of which were more elegantly expressible than the others, which gave the same result within the accuracy of this approximation. All these matrices would yield results that were in increasingly better agreement as the signal energy contained in the circle was increasingly closer to the total energy.

3.1.2. Closed-Form Type DFRFT. Pei and Ding [12] had derived a different type of DFRFT which was neat in concept and reversible but lacks the additivity property. Due to the orientation of practical usage, two types of DFRFT were derived which are different in parameterization. The parameters of the first type have directly linked to the continuous FRFT and suit the applications of computing the continuous FRFT. On the other hand, the second type has the simpler parameters set and allows more elegant expression for the operator kernels and more is suitable for other applications of DFRFT, such as the filter design, pattern recognition, and phase retrieval. Considering the expression of continuous FRFT:

$$F_\alpha(u)$$
$$= \sqrt{\frac{1 - j\cot\alpha}{2\pi}} e^{j(u^2/2)\cot\alpha} \int_{-\infty}^{\infty} e^{-jtu\csc\alpha} e^{j(t^2/2)\cot\alpha} f(t)\, dt. \tag{14}$$

And to derive the DFRFT, sampling of $f(t)$ and $F_\alpha(u)$ with the interval Δt and Δu needs to be performed as:

$$y(n) = f(n * \Delta t), \qquad Y_\alpha(m) = F_\alpha(m * \Delta u), \tag{15}$$

where $n = -N, \ldots, N$ and $m = -M, \ldots, M$. By using this sampling, FRFT can be converted [13] as:

$$Y_\alpha(m) = \sqrt{\frac{1 - j\cot\alpha}{2\pi}} \Delta t e^{j(m^2\Delta u^2/2)\cot\alpha}$$
$$\times \sum_{n=-N}^{N} e^{-jnm\Delta t\Delta u\csc\alpha} e^{j(n^2\Delta t^2/2)\cot\alpha} y(n). \tag{16}$$

The above can be written as:

$$Y_\alpha(m) = \sum_{n=-N}^{N} F_\alpha(m, n)\, y(n), \tag{17}$$

where

$$F_\alpha(m, n) = \sqrt{\frac{1 - j\cot\alpha}{2\pi}} \Delta t e^{j(m^2\Delta u^2/2)\cot\alpha}$$
$$\times e^{-jnm\Delta t\Delta u\csc\alpha} e^{j(n^2\Delta t^2/2)\cot\alpha}. \tag{18}$$

And $Y_\alpha(m)$ can be reversible only if the inverse transform of $F_\alpha(m, n)$, for all $M \geq N$, would be Hermitian (conjugate and transpose), given [12] as:

$$y(n) = \sum_{m=-M}^{M} F_\alpha^*(m, n)\, Y_\alpha(m) \quad \text{for } M \geq N. \tag{19}$$

From (17) and (19),

$$y(n) = \sum_{m=-M}^{M} \sum_{k=-N}^{N} F_\alpha^*(m, n) F_\alpha(m, k)\, y(k). \tag{20}$$

After some manipulation,

$$y(n) = \frac{\Delta t^2}{2\pi |\sin\alpha|} \sum_{m=-M}^{M} \sum_{k=-N}^{N}$$
$$\times e^{j((k^2-n^2)\Delta t^2/2)\cot\alpha} e^{jm(n-k)\Delta t\Delta u\csc\alpha}\, y(k). \tag{21}$$

The summation for m in (21) should be equal to $\delta(n - k)$, then

$$\Delta t\Delta u = \frac{S2\pi\sin\alpha}{2m + 1}, \tag{22}$$

where $|S|$ is some integer prime to $(2m + 1)$; in this case, (18) becomes

$$F_\alpha(m, n) = \sqrt{\frac{1 - j\cot\alpha}{2\pi}} \Delta t e^{j(m^2\Delta u^2/2)\cot\alpha} e^{-j(S2\pi nm/(2m+1))}$$
$$\times e^{j(n^2\Delta t^2/2)\cot\alpha},$$

$$\sum_{m=-M}^{M} \sum_{k=-N}^{N} F_\alpha^*(m, n) F_\alpha(m, k)\, y(k) = \frac{2M + 1}{2\pi |\sin\alpha|} \Delta t^2 y(n). \tag{23}$$

After normalization, the transform matrix [12] can be given as:

$$F_\alpha(m, n) = \sqrt{\frac{|\sin\alpha| - j\,\text{sgn}(\sin\alpha)\cos\alpha}{2M + 1}} e^{j(m^2\Delta u^2/2)\cot\alpha}$$
$$\times e^{-j(\text{sgn}(\sin\alpha)2\pi nm/(2M+1))} e^{j(n^2\Delta t^2/2)\cot\alpha}. \tag{24}$$

The definition of DFRFT is further divided into two parts [12]. The first part was defined for "$\sin\alpha > 0$" as:

$$Y_\alpha(m) = \sqrt{\frac{\sin\alpha - j\cos\alpha}{2M + 1}} e^{j(m^2\Delta u^2/2)\cot\alpha}$$
$$\times \sum_{n=-N}^{N} e^{-j(2\pi nm/(2M+1))} e^{j(n^2\Delta t^2/2)\cot\alpha} y(n). \tag{25}$$

And the second part was defined for "$\sin\alpha < 0$" as:

$$Y_\alpha(m) = \sqrt{\frac{-\sin\alpha + j\cos\alpha}{2M + 1}} e^{j(m^2\Delta u^2/2)\cot\alpha}$$
$$\times \sum_{n=-N}^{N} e^{j(2\pi nm/(2M+1))} e^{j(n^2\Delta t^2/2)\cot\alpha} y(n). \tag{26}$$

This definition of DFRFT possesses reversibility property and the periodicity property:

$$Y_\alpha(-m) = Y_{\alpha+\pi}(m), \qquad Y_\alpha(m) = Y_{\alpha+2\pi}(m). \tag{27}$$

The DFRFT of type 1 has a very important advantage in terms of its efficiency in calculating and implementing the DFRFT [12]. Due to two chirp multiplications and one FFT required for the implementation of this type of DFRFT, the total number of the multiplication operations required was "$2P + (P/2)\log_2 P$," for length of the output as $P = 2M + 1$.

Subsequently, another type of DFRFT, named as type 2, was introduced by modifying the definitions obtained for discrete affine Fourier transform (DAFT) and presented as:

$$Y_p(M) = \sqrt{\frac{1}{2M + 1}} e^{j(m^2 p/2)}$$
$$\times \sum_{n=-N}^{N} e^{\pm j\,(2\pi nm/(2M+1))} e^{j(n^2 p/2)} y(n), \quad \forall M \geq N. \tag{28}$$

For the DFRFT of type 2, the parameter "p" is used to control the variation of the chirp in frequency domain [13]. The DFRFT of type 2 also needed "$2P + (P/2)\log_2 P$" multiplication operations. It does not follow the additivity property but observes the convertibility operation.

These definitions of DFRFT are better to previous ones, that is, sampling type, linear combination type, and eigenfunction type, as it require less number of multiplications, having less complexity. Also it has a closed form expression being rich in properties.

3.2. Linear Combination Type DFRFT.

In [14–17], the discrete fractional Fourier transform was derived by using the linear combination of identity operation (F_0), discrete Fourier transform ($F_{\pi/2}$), time inverse operation (F_π), and inverse discrete Fourier transform ($F_{3\pi/2}$). In [15], the concept of DFRFT was given by linear combination of Lagrange interpolation polynomial of degree 3. For any function $x(v)$

$$x(v) = \sum_{k=0}^{3} \phi_k(v) g(\lambda_k), \tag{29}$$

where

$$\lambda_k = e^{jk\pi/2}, \quad \text{for } 0 \le k \le 3. \tag{30}$$

And $\phi_k(v)$ is the Lagrange interpolation polynomial of degree 3 taking value "one" at $v = \lambda_k$ and "zero" otherwise. It follows that

$$\phi_k(v) = \frac{1}{4} \sum_{i=0}^{3} u^i e^{-jki\pi/2}. \tag{31}$$

Finally, by taking the principle nth-root of eigenvalue λ_k, the fractional transform can be given as:

$$F^n = \sum_{i=0}^{3} F^i \alpha_i(n), \tag{32}$$

where

$$\alpha_i(n) = \frac{1}{4} \sum_{k=1}^{4} e^{-jk(n-i)\pi/2}, \quad \text{for } 0 \le i \le 3. \tag{33}$$

Subsequently, in [14, 15], it had also been shown that the operator defined by (17) is unitary; satisfies angle additivity and angle multiplicity properties, and the operator is periodic with a fundamental period of four. One analogous approach of this method was presented in [17], in which the DFRFT was calculated by first chirping the signal in time domain, then taking its Fourier transform, and finally chirping again in frequency domain. Depending on the character of the chirped signal, the embedded FT can be one of the two known classes in discrete case—discrete-time and periodic discrete-time.

For discrete-time DFRFT, the expression for DFRFT of signal $x(t)$ was given as:

$$X_\alpha(f) = \sum_{k=-\infty}^{\infty} Tx(kT) K_\alpha(f, kT). \tag{34}$$

And its inverse transform as:

$$x(kT) = \int_0^{F_p} X_\alpha(f) K_\alpha^*(f, kT) df, \tag{35}$$

where $F_p = 2/T \csc \alpha$ and $K_\alpha(t, u)$ is the kernel of FRFT. Similarly, for periodic discrete-time signal $x(t)$ with period T_p, the expression for DFRFT was given as:

$$X_\alpha(nf) = \sum_{k=-\infty}^{\infty} Tx(kT) K_\alpha(nF, kT). \tag{36}$$

And its inverse transform as:

$$x(kT) = \sum_{n=0}^{N-1} FX_\alpha(nF) K_\alpha^*(nF, kT), \tag{37}$$

where $T_p = NT$ and $F_p = NF = 2/T \csc \alpha$

In this type of DFRFT, the transform matrix is orthogonal and satisfies the additivity property along with the reversibility property. However, the main problem with this method is that the transform result is not matched to the continuous FRFT. Besides, it performed very similarly to the original Fourier transform or the identity operation but lost the important characteristic of fractionalization.

3.3. Eigenvector Type DFRFT.

In this classification, the basic approach of eigenvalues and eigenvectors of the kernel matrix of the FRFT is considered to define the discrete FRFT. Here, the eigenvalues of the discrete Fourier transform (DFT) are considered first and then by obtaining the powers of these eigenvectors of DFT matrix, the corresponding eigenvalues and eigenvectors of DFRFT matrix are derived. This classification includes three different approaches, named as, eigenvector decomposition method, random type method, and $\overline{V}$ method DFRFT.

3.3.1. Eigenvector Decomposition Type DFRFT.

The authors had derived another type of discrete fractional Fourier transform as reported in [13, 18–25] by searching the eigenvectors and eigenvalues of the DFT matrix followed by computing the fractional power of the DFT matrix based on Hermite function. This type of DFRFT worked very similarly to the continuous FRFT, and it fulfills the properties of orthogonality, additivity, and reversibility. The fractional power of matrix could be calculated from its eigendecomposition and the power of eigenvalues. Unfortunately, there exist two types of ambiguity in deciding the fractional power of the DFT kernel matrix.

(i) *Ambiguity in Deciding the Fractional Powers of Eigenvalues.* The square roots of unity are "1" and "−1" from elementary mathematics. This indicates that there exists root ambiguity in deciding the fractional power of eigenvalues.

(ii) *Ambiguity in Deciding the Eigenvectors of the DFT Kernel Matrix.* The DFT eigenvectors constitute four major eigensubspaces; therefore, the choices for the DFT eigenvectors to construct the DFRFT kernel are multiple and not unique.

Because of the above mentioned ambiguities, several DFRFT kernel matrices were possible which could obey the rotational properties. The idea for developing the DFRFT is to find the discrete form of the following:

$$K_\alpha(t, u) = \sum_{n=0}^{\infty} e^{-jn\alpha} H_n(t) H_n(u), \qquad (38)$$

where α indicates the rotation angle associated with FRFT of signal and $H_n(t)$ represents the nth-order normalized Hermite function with unit variance. The nth-order normalized Hermite function with variance σ is defined as:

$$H_n(t) = \frac{1}{2^n n! \sigma \sqrt{\pi}} h_n\left(\frac{t}{\sigma}\right) e^{-(t^2/2\sigma^2)}, \qquad (39)$$

where, $h_n(\cdot)$ represents the nth-order Hermite polynomial. However, the normalized Hermite functions with unitary variance $H_n(\cdot)$ indicate the eigenfunction of the FRFT. With the help of Hermite functions, the FRFT of a signal can be computed [24] as:

$$\begin{aligned} X_\alpha(u) &= \int_{-\infty}^{\infty} x(t) K_\alpha(t, u)\, dt \\ &= \sum_{n=0}^{\infty} H_n(u) e^{-jn\alpha} \left\{ \int_{-\infty}^{\infty} x(t) H_n(t)\, dt \right\}. \end{aligned} \qquad (40)$$

The FRFT can be interpreted as a weighting summation of Hermite functions, as given in (40). The weighting coefficients are obtained by multiplying the phase term $e^{-jn\alpha}$ and the inner product of the input signal with the corresponding Hermite functions [24]. The development of this DFRFT was based on the eigendecomposition of the DFT kernel. The eigenvalues of DFT kernel "F" can be given as $\{1, -j, -1, j\}$, and corresponding to each eigenvalue one eigenvector can be evaluated. The continuous FRFT has a Hermite function with unitary variance as its eigenfunction. The corresponding eigenfunction property for the DFT can be given as:

$$F^{2\alpha/\pi}[\hat{u}_n] = e^{-jn\alpha}\, \hat{u}_n, \qquad (41)$$

where $\hat{u}_n$ represents the eigenvector of DFT corresponding to the nth-order discrete Hermite functions. In order to retain the eigenfunction properties, the unit variance Hermite function is sampled with a period $T_s = \sqrt{2\pi/N}$. In the case of continuous FRFT, the terms of the Hermite functions are summed up from order zero to infinity. However, for the discrete case, only eigenvectors for the DFT Hermite eigenvectors can be added. The selection of the DFT Hermite eigenvectors is usually made from low to high orders, due to small approximation error of the low DFT Hermite eigenvectors [24].

Finally, the transform kernel of DFRFT can be defined as:

$$F^{2\alpha/\pi} = \hat{U} D^{2\alpha/\pi} \hat{U}^T$$

$$= \begin{cases} \displaystyle\sum_{k=0}^{N-1} e^{-jk\alpha} \hat{u}_k \hat{u}_k^T, & \forall N = 4m+1 \\[4mm] \left\{ \displaystyle\sum_{k=0}^{N-2} e^{-jk\alpha} \hat{u}_k \hat{u}_k^T \right\} + e^{-jN\alpha} \hat{u}_N \hat{u}_N^T, \\[4mm] & \forall N = 4m, \end{cases} \qquad (42)$$

where $\hat{U} = [\hat{u}_0|\hat{u}_1|\cdots \hat{u}_{N-1}|]$ for N as odd and $\hat{U} = [\hat{u}_0|\hat{u}_1|\cdots \hat{u}_{N-2}|\hat{u}_N]$ for N is even, $\hat{u}_k$ represents the normalized eigenvector corresponding to the kth-order discrete Hermite function, and D can be defined as follows.

For N odd,

$$D^{2\alpha/\pi} = \begin{bmatrix} e^{-j0} & & & 0 \\ & e^{-j\alpha} & & \\ & & \ddots & \\ 0 & & & e^{-j\alpha(N-1)} \end{bmatrix}. \qquad (43)$$

And for N even,

$$D^{2\alpha/\pi} = \begin{bmatrix} e^{-j0} & & & 0 \\ & \ddots & & \\ & & e^{-j\alpha(N-2)} & \\ 0 & & & e^{-jN\alpha} \end{bmatrix} \qquad (44)$$

The convergence for the eigenvectors obtained from these matrixes is not so fast for the high-order Hermite functions. To ensure orthogonality of the DFT Hermite eigenvectors $\hat{u}_k$, the Gram-Schmidt algorithm (GSA), or the orthogonal procrustes algorithm (OPA) can be used [24]. The GSA minimizes the errors between the samples of the Hermite functions and the orthogonal DFT Hermite eigenvectors. On the other hand, the OPA minimizes the total errors between those samples. The main difference between the approach proposed in [24] and similar approaches proposed in [21–23] is found in the eigenvectors obtained. The eigenvectors obtained by method reported in [24] were discrete Mathieu function, although the Mathieu functions can converge to Hermite functions as obtained by the methods reported in [21–23]. Given the number of points N and the rotation angle α, the DFT Hermite eigenvectors can be computed followed by determining the eigenvalues of DFRFT. The computation of the DFRFT can be implemented only by a transform kernel matrix multiplication. The complexity of computing the DFRFT is $O(N^2)$ same as in the DFT case. By adjusting the rotational angles, the method for implementing the DFRFT of a signal x can be given as:

$$F^{2\alpha/\pi} = \hat{U} D^{2\alpha/\pi} \hat{U}^T x = \sum_{k=0}^{N-1} a_k e^{-jk\alpha} \hat{u}_k. \qquad (45)$$

The coefficients a_k's are the inner products of signal and eigenvectors, and they can be computed in advance. If rotation angle changed then only the diagonal matrix $D^{2\alpha/\pi}$ needs to be recomputed. Additionally, the authors [24] investigated the relationship between the FRFT and the DFRFT and found that for a sampling period equal to $\sqrt{2\pi/N}$, the DFRFT performs a circular rotation of the signal in the time-frequency plane. However, the DFRFT becomes an elliptical rotation in the continuous time-frequency plane for sampling periods different from $\sqrt{2\pi/N}$ [24]. Therefore, for these elliptical rotations an angle modification and a postphase compensation in the DFRFT have been required to obtain results similar to the continuous FRFT [24]. This approach has been extended to the so-called multiple-parameter discrete fractional Fourier transform (MPDFRFT) [20, 25]. In fact, the MPDFRFT maintains all desired properties and reduces to the DFRFT when all of its order parameters are the same.

A similar approach to [24] has been proposed in [19]. However, authors in [19] believe that the discrete time counterparts of the continuous time Hermite-Gaussians maintained the same properties because these discrete time counterparts exhibited better approximations than the other proposed approaches [18]. In order to resolve this issue about the approximation of Hermite-Gaussian functions, a nearly tridiagonal commuting matrix of the DFT and a corresponding version of the DFRFT were proposed in [13]. Most of the eigenvectors of this proposed nearly tridiagonal matrix result in a good approximation of the continuous Hermite-Gaussian functions by providing a smaller approximation error in comparison to the previous approaches.

3.3.2. Random Type DFRFT. Pei and Hsue [26] had introduced a modified method to calculate the DFRFT based on the random DFT eigenvectors and eigenvalues. In this method, a new commuting matrix with random DFT eigenvectors was constructed. Then, a random discrete fractional Fourier transform (RDFRFT) kernel matrix with random DFT eigenvectors and eigenvalues was developed. The magnitude and phase of the transformed output are found to be random in RDFRFT. This can be considered as a special feature of RDFRFT. Previously, the DFT was randomized to define the discrete random Fourier transform (DRFT) in [27] by taking random powers of eigenvalues of the DFT matrix. However, the eigenvectors of the DRFT were not random and were computed using the same method as proposed in [23]. Although, the phase of the DRFT was random, its magnitude was not random, as shown in [27]. The eigen-decomposition of DFT matrix "F" was given as:

$$F = \sum_{k=0}^{N-1} \lambda_k e_k e_k^T. \tag{46}$$

The DFRFT with one order parameter "a" was defined by [25]:

$$F^a = \sum_{k=0}^{N-1} \lambda_k^a e_k e_k^T. \tag{47}$$

The definition of RDFRFT was based on the definition of multiple-parameter DFRFT (MPDFRFT) [25], given as:

$$F^{\widehat{a}} = \sum_{k=0}^{N-1} \lambda_k^{a_k} e_k e_k^T, \tag{48}$$

where the $1 \times N$ parameter vector $\widehat{a}$ was defined as:

$$\widehat{a} = [a_0, a_1, \ldots, a_{N-1}]. \tag{49}$$

This order parameter of the MPDFRFT can be taken as independent random numbers to define an MPDFRFT with the following random eigenvalues [26], $\lambda_k^{a_k}, k = 0, 1, \ldots, N-1$, and random eigenvalues were given as:

$$\lambda_k^{a_k} = \begin{cases} \left(e^{-j2\pi}\right)^{a_k}, & \text{for } \lambda_k = 1 \\ \left(e^{-j\pi/2}\right)^{a_k}, & \text{for } \lambda_k = -j \\ \left(e^{-j\pi}\right)^{a_k}, & \text{for } \lambda_k = -1 \\ \left(e^{-j3\pi/2}\right)^{a_k}, & \text{for } \lambda_k = j. \end{cases} \tag{50}$$

The generation of RDFRFT is given in the following section. In this generation of RDFRFT, the DFT-commuting matrix with orthonormal random DFT eigenvectors is used.

First defining a $N \times N$ real random Matrix "A" being "K" symmetric and generated by $A = KAK$, where K represents circular reversal matrix, which is given by:

$$K = F^2 = F^{-2} = \begin{bmatrix} 1 & 0 \\ 0 & J_{N-1} \end{bmatrix}, \tag{51}$$

where, J_{N-1} is the $(N-1) \times (N-1)$ reversal matrix, having nonzero entries "ones" placed on the antidiagonal locations. Taking the Ksymmetric part "E" of the random matrix "A" as $B = \{(A + KAK)/2\}$, with the help of this matrix, a random matrix "C" was generated as $C = (B + B^T)/2$, where "T" denotes the matrix transpose operation, and finally a DFT commuting matrix "D" will be formed [27] as $D = C+FCF^{-1}$. Subsequently, with the random DFT-commuting matrix "D" and the MPDFRFT, the random DFRFT (RDFRFT) with the $1 \times N$ parameter vector $\widehat{a} = [a_0, a_1, \ldots, a_{N-1}]$ of a signal "x" can be determined [26] as:

$$X_{(\widehat{a},D)} = F_D^{\widehat{a}} * x = \left\{ \sum_{k=0}^{N-1} \lambda_k^{a_k} r_k r_k^T \right\} * x, \tag{52}$$

where r_k represents the orthonormal random DFT eigenvectors computed from "D" and λ_k represents the DFT eigenvalue corresponding to r_k, given by (50).

The RDFRFT kernel matrix $F_D^{\widehat{a}}$ satisfied various properties of the FRFT like—additivity, FT convertibility, and angle additivity. The computational complexity of the -point RDFRFT was $O(N^2)$ same as that of the MPDFRFT, but RDFRFT has an advantage over MPDFRFT in terms of one more free parameter associated with it as parameter vector $\widehat{a}$. The magnitude and phase responses of the transformed output, with the RDFRFT, are random because the eigenvectors and eigenvalues of the RDFRFT matrix are random [26], as shown in (52). This aspect establishes its advantageous role in security areas like—cryptography, encryption, and watermarking.

3.3.3. The $\overline{V}$ Method DFRFT. In this work [28], authors focus on a different approach to derive eigenvectors of the DFT matrix in order to build the DFRFT matrix that mimics the properties of its continuous counterpart. A straightforward and refined derivation of the eigenvectors of the DFT matrix is proposed without using any commuting matrices, but using only the factored form of a basic property of the DFT matrix stating that four consecutive Fourier transforms is the identity transform

$$W_N^4 = I_N, \tag{53}$$

where W_N^4 and I_N are the centered DFT (CDFT) and the identity matrices of order N, respectively. We utilize factorization of the CDFT matrix to obtain new matrices, whose columns are eigenvectors of the CDFT matrix. Thereafter, GSA was used to find the orthonormal eigenvectors obtained by factorization of the CDFT matrix. By employing a DFT-shift matrix K, eigenvectors of the ordinary-DFT matrix are obtained by using the CDFT matrix. Later on, the S and T matrices together are also used to boost the performance.

In [24], Pei et al. compute the samples of the Hermite-Gauss functions and employ Gram-Schmidt algorithm (GSA) to orthogonalize these samples by projecting on the S matrix, which is equivalent to orthogonalizing the samples of the Hermite-Gauss functions without using the S matrix. Hanna et al. [29] claim to find the eigenvectors of the DFT matrix using a similar method. However, they use the ordinary-DFT matrix as a basis to their work, but Serbes and Durak-Ata [28] employs the CDFT matrix. Additionally, the eigenvectors are nonorthogonal and dissimilar to the samples of Hermite-Gauss functions. Serbes and Durak-Ata [28] work is different from these methods in the sense that the non-orthogonal eigenvectors of the CDFT matrix is determined without utilizing any S matrix or samples of the Hermite-Gauss functions, and then orthogonalization of eigenvectors is obtained by employing the GSA.

The DFT maps the signal from $[0, N-1]$ discrete input space to $[0, 2\pi]$ discrete frequency space, whereas the CDFT maps $[-(N-1)/2, (N-1)/2]$ to $[-\pi, \pi]$. This allows CDFT to define even and odd functions such as Hermite-Gauss functions, that is contrary to the ordinary DFT definition. In order to approximate the samples of the continuous FrFT and to imply the rotation property, Hermite-Gauss-like eigenvectors of the DFT matrix have to be obtained. Serbes and Durak-Ata [28] use the CDFT matrix instead of the ordinary DFT matrix, because the Hermite-Gauss-like vectors are eigenvectors of only the CDFT matrix, not the DFT matrix, that is, when a discrete Hermite-Gauss-like eigenvector is transformed using the ordinary DFT matrix, the output is the shifted version of the Hermite-Gauss vector multiplied by a complex sinusoidal.

The FRFT operator of order α can be defined as the αth-power of the ordinary DFT operator W_N. Hence, the DFRFT matrix can be expressed by means of its eigenvector decomposition,

$$W_N^\alpha = U_N \Lambda_N^\alpha U_N^T, \tag{54}$$

where Λ_N^α is given as—$\Lambda_N^\alpha = \text{diag}(e^{-j0}, e^{-j\pi/2a}, \ldots, e^{-j\pi/2\,a(N-2)}, e^{-j\pi/2a(N-1)})$ for the centered-DFRFT.

The centered-DFRFT matrix can be obtained by calculating the Hermite-Gaussian-like eigenvectors of DFRFT as followes:

$$W_N^\alpha = \overline{V_1 \Lambda_{k_1}^\alpha} \, \overline{V}_1^T + \overline{V_2 \Lambda_{k_2}^\alpha} \, \overline{V}_2^T + \overline{V_3 \Lambda_{k_3}^\alpha} \, \overline{V}_3^T + \overline{V_4 \Lambda_{k_4}^\alpha} \, \overline{V}_4^T, \tag{55}$$

where

$$
\begin{aligned}
\overline{\Lambda}_{k_1}^\alpha &= \text{diag}\left(e^{-j2\pi(k_1-1)\,a}, \ldots, e^{-j2\pi a}, e^{-j0}\right), \\[4pt]
\overline{\Lambda}_{k_2}^\alpha &= \text{diag}\left(e^{-j(\pi+(k_2-1)2\pi)a}, \ldots, e^{-j3\pi a}, e^{-j\pi a}\right), \\[4pt]
\overline{\Lambda}_{k_3}^\alpha &= \text{diag}\left(e^{-j(3\pi/2+(k_3-1)2\pi)a}, \ldots, e^{-j7\pi a/2}, e^{-j3\pi a/2}\right), \\[4pt]
\overline{\Lambda}_{k_4}^\alpha &= \text{diag}\left(e^{-j(\pi/2+(k_4-1)2\pi)a}, \ldots, e^{-j5\pi a/2}, e^{-j\pi a/2}\right).
\end{aligned} \tag{56}
$$

The above mentioned method satisfies unitary and angle additivity properties. Also, it reduces to the ordinary DFT when $a = 1$ and approximates to the samples of the continuous FRFT for fractional values of a.

Performance of the Serbes and Durak-Ata [28] algorithm can be improved by using it in combination with S and T matrices. Pei et al. [13] uses the S method together with the Grunbaum's T matrix [30] as a linear combination, in which it has been observed that Hermite-Gauss-like eigenvectors of $(S + 15T)$ are more accurate than both S and T alone. Serbes and Durak-Ata [28] proposed that eigenvectors of linear combinations of S, T, and V_T matrices as $k_1 S + k_1 T + k_1 V_T$ produce more accurate eigenvectors. Pei and Hsue [25] showed that using the linear combination of S and T as $(S + 15T)$ produces better results than the S or T alone. Whereas, Serbes and Durak-Ata [28] showed that using $(S + 30T - 7V_T)$ produces a more accurate approximation to the continuous FRFT. Although, better commuting matrix generation method is available in the literature [31–33], but all these methods are explained in the context of DFT and some other transform only, excluding DFRFT. Therefore, the betterment offered by these methods can be utilized in devising the new algorithm of DFRFT based on these commuting matrixes.

3.4. Weighted Summation Type DFRFT. Yeh and Pei [34] had developed a new method for DFRFT computation. The idea of the developed method was to compute the DFRFT at any angle by a weighted summation of the DFRFT's with the special angles. In this method DFRFT computation with odd point of length can be realized by the weighted summation of the DFRFTs in special angles. The special angles were multiples of $2\pi/N$ for the odd case and multiples of $2\pi/(N+1)$ for even length. Regardless of even- or odd-length cases, the weighting coefficients were obtained from an IDFT operation. The N-point IDFT was needed for odd length, and the $(N+1)$-point was computed for even length.

TABLE 1: Comparison of Different Types of DFRFT Algorithms.

Properties/parameters	Sampling-based method		Linear combination-based method	Eigenvector-based method			Weighted summation-based method
	Sampling type	Closed-form type	Linear combination type	Eigenvector decomposition type	Random type	$\overline{V}$ Method type	Weighted summation Type
Reversibility	NO	YES	YES	YES	YES	YES	YES
Additivity	NO	NO	YES	YES	YES	YES	YES
Periodicity	YES	YES	YES	YES	YES	YES	YES
Symmetry	YES	YES	NO	YES	YES	YES	YES
Angle additivity	NO	YES	NO	YES	YES	YES	YES
Closed-form expression	YES	YES	YES	YES	YES	YES	YES
Similarity with FRFT	YES	YES	NO	YES	YES	YES	NO
Complexity	$O(N\log_2 N)$	$O(N\log_2 N)$	$O(N\log_2 N)$	$O(N^2)$	$O(N^2)$	$O(N\log_2 N)$	$O(N^2)$
Constraints	Many	Fewer	Many	Many	Very less	Many	Fewer

For discrete signal "x" having odd length "N", the DFRFT of "x" for rotation angle "α" can be computed as:

$$X_\alpha = \sum_{n=0}^{N-1} B_{n,\alpha} X_{n,\beta} \quad \forall \beta = \frac{2\pi}{N}, \qquad (57)$$

where the weighting coefficients $B_{n,\alpha}$ was given as:

$$B_{n,\alpha} = \text{IDFT}\left[e^{-jk\alpha}\right]$$

$$\forall k = 0, 1, \ldots, N-1 = \frac{1}{N}\sum_{n=0}^{N-1} e^{-jk\alpha} e^{j\,(2\pi/N)n\,k}. \qquad (58)$$

The weighted summation $B_{n,\alpha}$ has a closed-form solution:

$$B_{n,\alpha} = \begin{cases} \dfrac{1}{N}\dfrac{1 - e^{-j(N-1)(\alpha-n\beta)}}{1 - e^{-j(\alpha-n\beta)}}, & \forall \alpha \neq k\beta \\ \delta(n-k), & \forall \alpha = k\beta. \end{cases} \qquad (59)$$

Similarly, for discrete signal "x" having even length N, the DFRFT of "x" for rotation angle "α" can be computed by the following expression:

$$X_\alpha = \sum_{n=0}^{N} B_{n,\alpha} X_{n,\beta} \quad \forall \beta = \frac{2\pi}{N+1}. \qquad (60)$$

The weighting coefficients, $B_{n,\alpha}$ can be computed as:

$$B_{n,\alpha} = \text{IDFT}\left[e^{-jk\alpha}\right]$$

$$\forall k = 0, 1, \ldots, N = \frac{1}{N+1}\sum_{n=0}^{N} e^{-jk\alpha} e^{j\,(2\pi/(N+1))nk}. \qquad (61)$$

Similar to the odd length, the weighted summation $B_{n,\alpha}$ for even case has also a closed-form solution:

$$B_{n,\alpha} = \begin{cases} \dfrac{1}{N+1}\dfrac{1 - e^{-jN(\alpha-n\beta)}}{1 - e^{-j(\alpha-n\beta)}} & \forall \alpha \neq k\beta \\ \delta(n-k) & \forall \alpha = k\beta \end{cases} \qquad (62)$$

The eigenvector decomposition methods require N^2 storages to store the transform kernel. If the computing angle is changed, the transform kernel is also changed and needs to be recomputed. In this algorithm, the DFRFT of the specified angle was stored. The DFRFT at any angle can be obtained by weighted summation of these specified DFRFTs. The weighted computation in this algorithm still takes $O(N^2)$ multiplications; therefore, its computation load is still $O(N^2)$. In [34], two implementation methods for the DFRFT computation algorithm were introduced. One was called the parallel method and the other cascade method. Similar to the parallel method, the special angle and numbers of terms of cascade forms are also different for the even and odd cases. Both of the implementation methods can have the advantages over each other for different application. The parallel method was suitable for the signal, whose DFRFTs with special angles were already known. Hence, the computation of the DFRFT will become only a linear combination of the DFRFT's in special angles. And, the cascade method means that the computation of the DFRFT can be realized from the DFRFT with only one specified angle. If the DFRFT with the specified angle can be computed efficiently, the computation of the DFRFT will become efficient. Therefore, it has been found suitable for VLSI implementation [34].

Although, the existence of one well-accepted closed-form definition of DFRFT is not available; however many definitions with their pros and cons are accessible. A comparison of all the proposed classes of DFRFT based on some properties and complexity is shown in Table 1. In this Table,

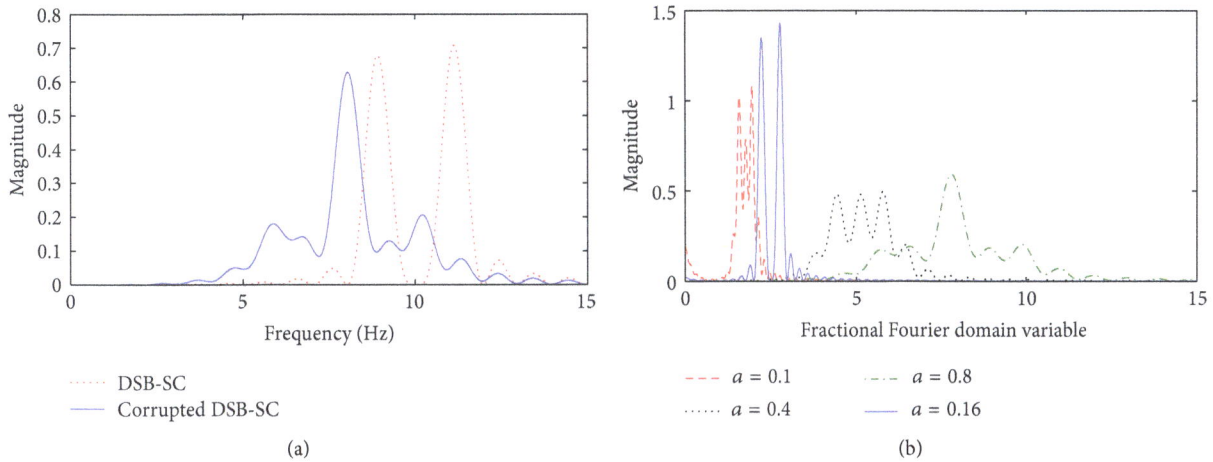

FIGURE 2: Removal of Doppler component introduced in DSB-SC signal. (a) FRFT of weighted autocorrelation of DSB-SC signal and corrupted DSB-SC signal at the FRFT order of $a = 1$ (i.e., Fourier transform domain) and (b) FRFT of weighted auto-correlation of corrupted DSB-SC signal at different FRFT order.

yes is included against the method which possesses given property/parameter, otherwise NO is mentioned.

From the above mentioned comparative analysis, it can be recapitulated that RDFRFT provides a better solution for computation of the DFRFT. The computation of RDFRFT requires more time than sampling type, linear combination type, and closed form type; however, it satisfies more properties than any other method with very less number of imposed constraints. Previously, IEEE has conferred to use the Eigenvectors decomposition type DFRFT as the most appropriate definition of DFRFT. In the year 2006, Pei et al. [25] had established the superiority of their method (MPDFRFT) over the earlier reported methods. Subsequently, in the year 2009, the MPDFRFT was modified by Pei et al. [26] and given as RDFRFT. It was also established that RDFRFT has one more degree of freedom, while satisfying the same set of properties as satisfied by MPDFRFT with equal computational complexity. Although, Serbes and Durak-Ata [28] proposed a modified version of eigenvalue decomposition technique with slight betterment but this does not redeem the advantage offered by RDFRFT, that is, security of the data. This makes the RDFRFT as a better choice in security applications.

4. Application of DFRFT

In this section, the application of DFRFT has been investigated in two areas. First, the detection of Doppler shifted DSB-SC signal will be performed than DFRFT has been applied in the image processing application, particularly in steganography.

4.1. Removal of Doppler Shift in DSB-SC AM Signal. The weighted autocorrelation theorem defined for FRFT [2] is utilized to remove the unwanted frequency components introduced in double sideband suppressed-carrier amplitude

modulated (DSB-SC-AM) signal. It is assumed that a Doppler component is introduced in the carrier frequency due to motion of transmitter and/or receiver. The transmitted DSB-SC amplitude-modulated signal represented by "$e(t)$" with modulating signal frequency as f_m and carrier frequency as f_c is given as:

$$e(t) = \cos\{2\pi f_m t\} \cos\{2\pi f_c t\}. \tag{63}$$

Similarly, the received signal, which is corrupted by noise (Doppler frequency component—f_d), represented as "$r(t)$" is given by:

$$r(t) = \cos\{2\pi f_m t\} \cos\{2\pi (f_c + f_d) t\}. \tag{64}$$

The values of "f_m, f_c, and f_d" are taken as 1 Hz, 10 Hz and as 10% of carrier frequency, respectively. The simulation is performed on the platform of MATLAB software (version 7.1) on a system having configuration: processor as Pentium-4, Intel (R) CPU-1.8 GHz processor having 1 GB RAM. The effect of Doppler shift in the carrier is illustrated in Figure 2(a), where Fourier transform of original signal and Doppler corrupted signal is plotted.

It is evident from Figure 2(a) that for a received Doppler corrupted DSB-SC signal, the magnitude spectrum of the signal in Fourier domain is not giving the exact information about original frequency component present in the transmitted signal due to only one peak present in the spectrum in Fourier domain. Whereas, when the same corrupted DSB-SC signal is analyzed in fractional Fourier domain, the magnitude spectrum of FRFT of weighted auto-correlation of corrupted DSB-SC signal gives two clear peak at an angle corresponding to the value of $a = 0.16$, as shown in Figure 2(b). Subsequently, the actual frequency components present in transmitted DSB-SC can be estimated by observing the location of two peaks in this FRFT domain. From the observed value of FRFT domain variable "u_{peak}" at which the peak is obtained,

(a) Cover image

(b) Secret image

(c) Steganography Image

(d) Retrieved secret image at $a_1 = -0.2$

(e) Retrieved secret image at $a_1 = -0.4$

(f) Retrieved secret image at $a_1 = -0.22$

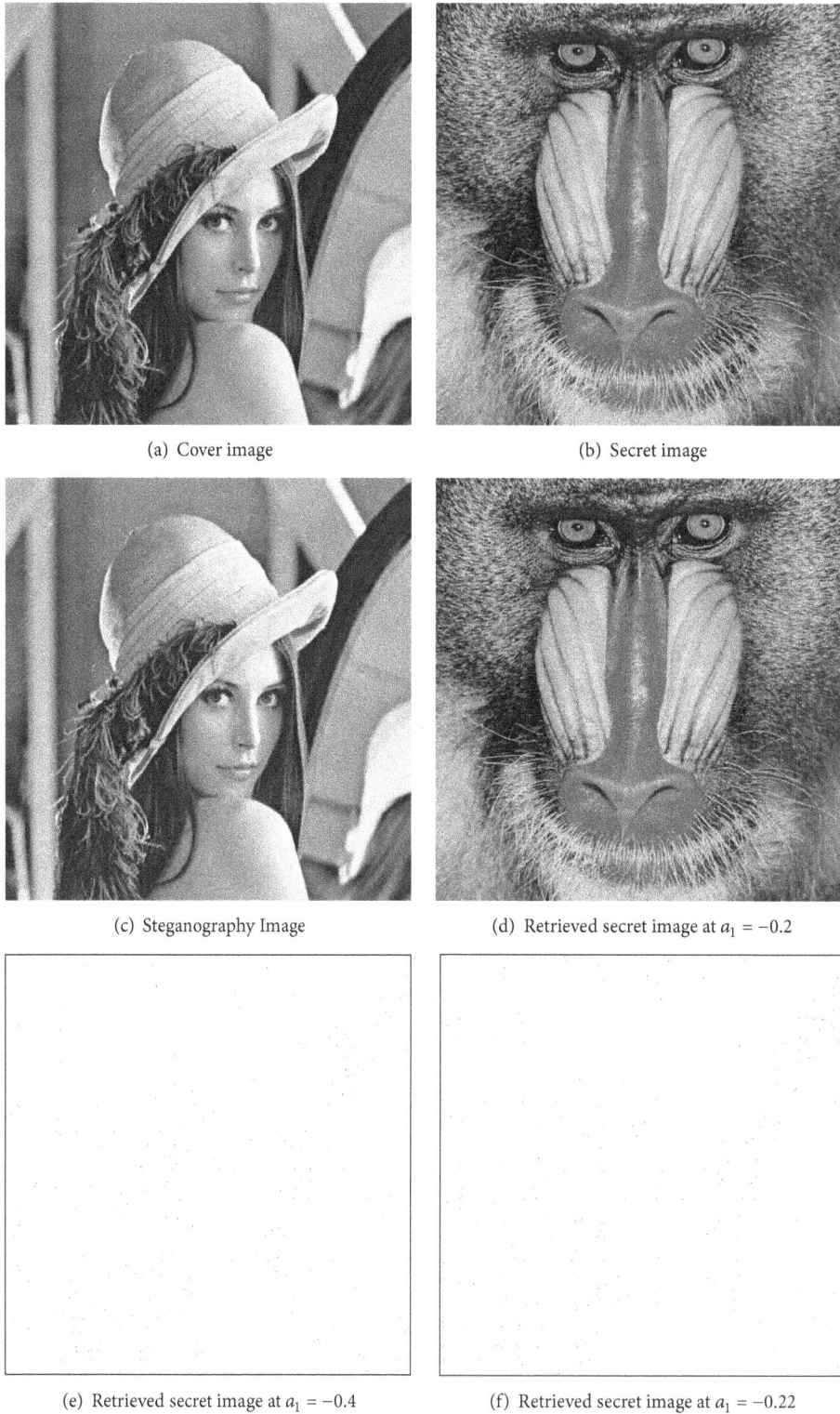

FIGURE 3: Steganography using DFRFT.

the frequency component present in the signal can be estimated by:

$$\text{Estimated value} = \frac{u_{\text{peak}}}{\cos\{(1-a)(\pi/2)\}}. \quad (65)$$

In Figure 2(b), two simulated values of u_{peak} is obtained as $u_{\text{peak,low}}$ and $u_{\text{peak,high}}$ for $a = 0.16$. Then these values of u_{peak} is used in (65) to obtain the desired frequency components. The assumed frequency component value and the estimated ones are shown in Table 2.

TABLE 2: Frequency estimation in DSB-SC signal.

Assume value		Calculated value		Simulated value		Estimated value	
f_m	f_c	$f_c + f_m$	$f_c - f_m$	$u_{\text{peak,high}}$	$u_{\text{peak,low}}$	$f_c + f_m$	$f_c - f_m$
1	10	11	9	2.769	2.22	11.13	8.927

From Table 2, it can be clearly observed that the calculated value of higher and lower frequency components present in DSB-SC AM signal is matched nicely with the estimated value of higher and lower frequency components even when the signal is corrupted with the presence of a Doppler frequency component in the signal itself.

4.2. Data Hiding by Steganography in Fractional Fourier Domain. Steganography is the technique of hiding confidential information within any other media. The objective of steganography is to hide a secret message within a cover-media in such a way that others cannot determine the presence of the hidden message. Hiding information into a media requires the following elements:

(i) the cover media (C) that will hold the hidden data,

(ii) the secret message (M), may be text, image, or any type of data,

(iii) the stego function and its inverse,

(iv) stego-key (K) may be used to hide and unhide the message.

In this exercise, the cover and secret messages are assumed as Lena image and Mandrill image, respectively. The stego function is defined as

Stego Image

$$= \mathscr{F}_a^{-1} \left[\mathscr{F}_a \{\text{Cover Image}\} + b * \mathscr{F}_a \{\text{Secret Image}\} \right]. \tag{66}$$

Here, first, the DFRFT ($\mathscr{F}_a$) of cover image and secret image is obtained, then the steganographed image can be achieved by using the stego function defined in (66), where $\mathscr{F}_a^{-1}$ is representing the inverse DFRFT and b is used as stego weight. Similarly, the inverse stego function for obtaining the secret information can be written as:

Secret image

$$= \mathscr{F}_{a_1}^{-1} \left[\frac{\mathscr{F}_{a_1} \{\text{Stego Image}\} - \mathscr{F}_{a_1} \{\text{Secret Image}\}}{b} \right]. \tag{67}$$

In this process, the values of stego weight "b" and DFRFT order "a" have been assumed as 0.05 and 0.2, respectively. For retrieval of secret image, the DFRFT order "a_1" in (67) should be equal to the negative of the DFRFT order "a" in (66). The results obtained are shown in Figure 3. In this simulation exercise, the detection of hidden image is also performed for two other DFRFT angle parameters, both are in other side of actual, that is, at $a_1 = -0.4$ and $a_1 = -0.22$ in Figures 3(e) and

3(f), respectively. It can be easily verified from these results that for obtaining the hidden image correctly, the correct value of angular parameter of DFRFT should be known; hence, this correct value can also be used as a key other than stego weight. This extra key is only possible with the DFRFT and not with DFT, as there no such variation of angle is possible with DFT. This again confirms the superiority of DFRFT over DFT in the security applications.

5. Conclusion

In this paper, the different approaches of obtaining discrete fractional Fourier transform (DFRFT) have been analyzed. These different approaches are classified in four categories based on their methodology of evaluation. Further, with the help of a comparative analysis, it is established that the random type DFRFT emerged out as a better choice for calculating DFRFT. Thereafter, the random type DFRFT along with weighted auto-correlation theorem defined for the FRFT has been employed for the successful removal of the Doppler frequency component introduced in the DSB-SC AM signal due to time-variant motion of transmitter or receiver. Subsequently, the role of DFRFT in steganography has also been successfully performed with the availability of one extra key, in the form of DFRFT angular parameter. The results obtained are in the conformity of the superiority of the fractional Fourier transform over Fourier transform.

References

[1] L. B. Almeida, "The fractional Fourier transform and time-frequency representations," *IEEE Transaction on Signal Processing*, vol. 42, no. 11, pp. 3084–3091, 1994.

[2] A. K. Singh and R. Saxena, "Correlation theorem for fractional Fourier transform," *International Journal of Signal Processing, Image Processing and Pattern Recognition*, vol. 4, no. 2, pp. 31–40, 2011.

[3] A. K. Singh and R. Saxena, "Recent developments in FRFT, DFRFT with their applications in signal and image processing," *Recent Patents on Engineering*, vol. 5, no. 2, pp. 113–138, 2011.

[4] A. I. Zayed and A. G. Garcia, "New sampling formulae for the fractional Fourier transform," *Signal Processing*, vol. 77, no. 1, pp. 111–114, 1999.

[5] Ç. Candan and H. M. Ozaktas, "Sampling and series expansion theorems for fractional Fourier and other transforms," *Signal Processing*, vol. 83, no. 11, pp. 2455–2457, 2003.

[6] K. K. Sharma and S. D. Joshi, "Fractional Fourier transform of band-limited periodic signals and its sampling theorems," *Optics Communications*, vol. 256, no. 4–6, pp. 272–278, 2005.

[7] R. T. Tao, B. Deng, W. Q. Zhang, and Y. Wang, "Sampling and sampling rate conversion of band limited signals in the fractional Fourier transform domain," *IEEE Transactions on Signal Processing*, vol. 56, no. 1, pp. 158–171, 2008.

[8] T. Erseghe, P. Kraniauskas, and G. Cariolaro, "Unified fractional Fourier transform and sampling theorem," *IEEE Transactions on Signal Processing*, vol. 47, no. 12, pp. 3419–3423, 1999.

[9] X. G. Xia, "On bandlimited signals with fractional fourier transform," *IEEE Signal Processing Letters*, vol. 3, no. 3, pp. 72–74, 1996.

[10] O. Arıkan, M. A. Kutay, H. M. Ozaktas, and O. K. Akdemir, "The discrete fractional Fourier transformation," in *Proceeding IEEE-SP International Symposium on Time-Frequency and Time-Scale Analysis*, pp. 205–207, Atlanta, Ga, USA, June 1996.

[11] H. M. Ozaktas, O. Ankan, M. Alper Kutay, and G. Bozdagi, "Digital computation of the fractional fourier transform," *IEEE Transactions on Signal Processing*, vol. 44, no. 9, pp. 2141–2150, 1996.

[12] S. C. Pei and J. J. Ding, "Closed-form discrete fractional and affine Fourier transforms," *IEEE Transaction on Signal Processing*, vol. 48, no. 5, pp. 1338–1353, 2000.

[13] S. C. Pei, W. L. Hsue, and J. J. Ding, "Discrete fractional Fourier transform based on new nearly tridiagonal commuting matrices," *IEEE Transactions on Signal Processing*, vol. 54, no. 10, pp. 3815–3828, 2006.

[14] B. Santhanam and J. H. McClellan, "DRFT—a rotation in time-frequency space," in *Proceedings of the 20th International Conference on Acoustics, Speech, and Signal Processing*, pp. 921–924, May 1995.

[15] B. Santhanam and J. H. McClellan, "Discrete rotational Fourier transform," *IEEE Transactions on Signal Processing*, vol. 44, no. 4, pp. 994–998, 1996.

[16] B. W. Dickinson and K. Steiglitz, "Eigenvectors and functions of the discrete Fourier transform," *IEEE Transactions on Acoustics, Speech, and Signal Processing*, vol. 30, no. 1, pp. 25–31, 1982.

[17] G. Cariolaro, T. Erseghe, P. Kraniauskas, and N. Laurent, "A unified framework for the fractional fourier transform," *IEEE Transactions on Signal Processing*, vol. 46, no. 12, pp. 3206–3219, 1998.

[18] C. Candan, "On higher order approximations for Hermite-Gaussian functions and discrete fractional fourier transforms," *IEEE Signal Processing Letters*, vol. 14, no. 10, pp. 699–702, 2007.

[19] C. Candan, M. Alper Kutay, and H. M. Ozaktas, "The discrete fractional fourier transform," *IEEE Transactions on Signal Processing*, vol. 48, no. 5, pp. 1329–1337, 2000.

[20] J. G. Vargas-Rubio and B. Santhanam, "On the multiangle centered discrete fractional Fourier transform," *IEEE Signal Processing Letters*, vol. 12, no. 4, pp. 273–276, 2005.

[21] S. C. Pei, C. C. Tseng, M. H. Yeh, and J. J. Shyu, "Discrete fractional hartley and fourier transforms," *IEEE Transactions on Circuits and Systems II*, vol. 45, no. 6, pp. 665–675, 1998.

[22] S. C. Pel, C. C. Tseng, and M. H. Yeh, "A new discrete fractional fourier transform based on constrained eigendecomposition of dft matrix by largrange multiplier method," *IEEE Transactions on Circuits and Systems II*, vol. 46, no. 9, pp. 1240–1245, 1999.

[23] S. C. Pei and M. H. Yeh, "Improved discrete fractional Fourier transform," *Optical Letters*, vol. 22, no. 14, pp. 1047–1049, 1997.

[24] S. C. Pei, M. H. Yeh, and C. C. Tseng, "Discrete fractional Fourier transform based on orthogonal projections," *IEEE Transactions on Signal Processing*, vol. 47, no. 5, pp. 1335–1348, 1999.

[25] S. C. Pei and W. L. Hsue, "The multiple-parameter discrete fractional Fourier transform," *IEEE Signal Processing Letters*, vol. 13, no. 6, pp. 329–332, 2006.

[26] S. C. Pei and W. L. Hsue, "Random discrete fractional fourier transform," *IEEE Signal Processing Letters*, vol. 16, no. 12, pp. 1015–1018, 2009.

[27] Z. Liu and S. Liu, "Randomization of the Fourier transform," *Optical Letters*, vol. 32, no. 5, pp. 478–480, 2007.

[28] A. Serbes and L. Durak-Ata, "The discrete fractional Fourier transform based on the DFT matrix," *Signal Processing*, vol. 91, no. 3, pp. 571–581, 2011.

[29] M. T. Hanna, N. P. A. Seif, and W. A. E. M. Ahmed, "Hermite-Gaussian-like eigenvectors of the discrete Fourier transform matrix based on the singular-value decomposition of its orthogonal projection matrices," *IEEE Transactions on Circuits and Systems I*, vol. 51, no. 11, pp. 2245–2254, 2004.

[30] F. A. Grünbaum, "The eigenvectors of the discrete Fourier transform: a version of the Hermite functions," *Journal of Mathematical Analysis and Applications*, vol. 88, no. 2, pp. 355–363, 1982.

[31] A. Serbes and L. Durak-Ata, "Efficient computation of DFT commuting matrices by a closed-form infinite order approximation to the second differentiation matrix," *Signal Processing*, vol. 91, no. 3, pp. 582–589, 2011.

[32] S. C. Pei, W. L. Hsue, and J. J. Ding, "DFT-commuting matrix with arbitrary or infinite order second derivative approximation," *IEEE Transactions on Signal Processing*, vol. 57, no. 1, pp. 390–394, 2009.

[33] S. C. Pei, J. J. Ding, W. L. Hsue, and K. W. Chang, "Generalized commuting matrices and their Eigenvectors for DFTs, offset DFTs, and other periodic operations," *IEEE Transactions on Signal Processing*, vol. 56, no. 8, pp. 3891–3904, 2008.

[34] M. H. Yeh and S. C. Pei, "A method for the discrete fractional Fourier transform computation," *IEEE Transactions on Signal Processing*, vol. 51, no. 3, pp. 889–891, 2003.

11

Optical Properties of One-Dimensional Structured GaN:Mn Fabricated by a Chemical Vapor Deposition Method

Sang-Wook Ui, In-Seok Choi, and Sung-Churl Choi

Division of Materials Science and Engineering, College of Engineering, Hanyang University, 17 Haengdang-dong, Seongdong-gu, Seoul 133-791, Republic of Korea

Correspondence should be addressed to Sung-Churl Choi; choi0505@hanyang.ac.kr

Academic Editor: Georgiy B. Shul'pin

Group III nitride semiconductors with direct band gaps have recently become increasingly important in optoelectronics and microelectronics applications due to their direct band gaps, which cover the whole visible spectrum and a large part of the UV range. Major developments in wide band gap III–V nitride semiconductors have recently led to the commercial production of high-temperature, high-power electronic devices, light-emitting diodes (LEDs), and laser diodes (LDs). In this study, GaN nanowires were grown on horizontal reactors by chemical vapor deposition (CVD) employing a vapor-solid mechanism. Many studies have described how to control the diameters of wires in the liquid phase catalytic process, but one-dimensional nanostructures, which are grown using a noncatalytic process, are relatively unexplored due to the challenge of producing high-quality synthetic materials of controlled size. However, vapor-solid mechanisms to make synthesized nanowires are simple to implement. We obtained results from GaN nanostructures that were a preferential c-axis orientation from the substrate. The morphology and crystallinity of the GaN nanowires were characterized by field-emission scanning electron microscopy and X-ray diffraction. The chemical compositions of GaN with Mn were analyzed by energy dispersive X-ray spectroscopy. Optical properties were investigated using photo luminescence and cathode-luminescence measurements.

1. Introduction

As a wide band gap semiconductor having a direct gap-type band gap, GaN is a material that has been studied since the 1970s for the purpose of applying various photoelectric elements and protection films, including blue luminous elements [1]. GaN has been actively researched all over the world. Since GaN has consecutive solid solubility with III–V series nitride semiconductors, such as InN (Eg = 1.92 eV) and AlN (Eg = 6.2 eV), it forms ternary series nitride homogenous solid solutions such as InxGa1-xN and GaxAl1-xN. Because the composition of a band gap can change to a linear function depending on the composition of these ternary-series nitrides, it is possible to manufacture luminous elements and light-receiving elements that include all of the red visible ray fields of ultraviolet rays by controlling the composition of III–V nitrides [2, 3].

Gallium nitride (GaN) is one of the most common semiconductor materials since it has a wide band gap of 3.39 eV

and very good chemical stability. It is a very ecofriendly material that does not contain substances such as As and Hg [4, 5]. With its strong electron affinity, GaN has especially superior voltage characteristics such as electron mobility, saturated electron velocity, and electric field breakdown, as well as superior optical characteristics. Thus, it has numerous applications related to optical and electronic elements [6]. Sapphire, SiC, Si, and GaAs substrates are used to grow GaN semiconductors. The homoepitaxy growth method, which grows GaN on a GaN substrate, has recently attracted attention to reduce lattice defects [7].

In addition to green and red optical elements produced using GaAs and InP compound semiconductors, blue wavelength optical elements using GaN make it possible to fabricate displays that have natural colors. These displays have high potential for application in fields related to graphics and visual display terminals such as instrument panels of electronic devices [8–11]. Due to the development of high-speed broadband information and communication networks

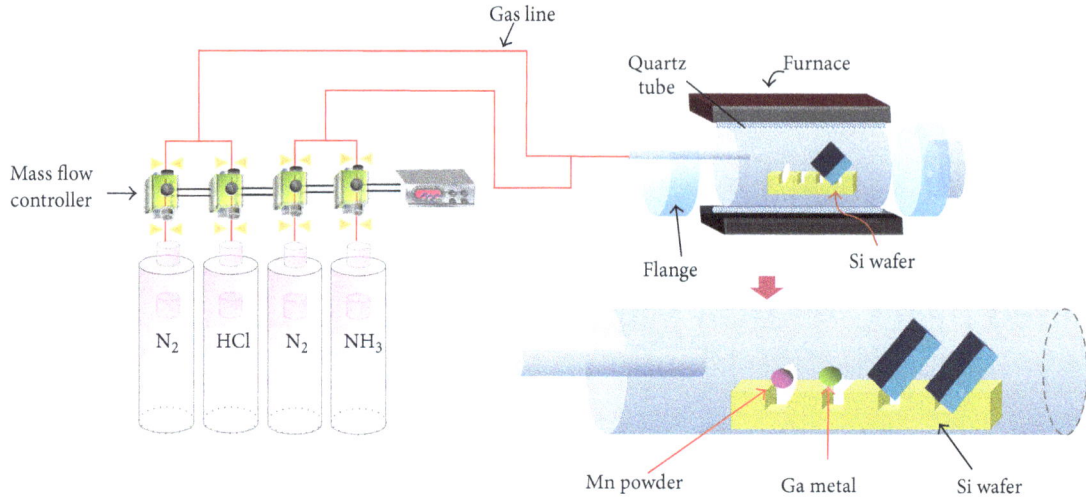

FIGURE 1: Schematic experimental setup for the growth of GaN and doped Mn.

and the expansion of fiber to the home (FTTH), there is also increased demand to transmit and process high-capacity broadband information. In this field, the capacity of an optical information processing system is in inverse proportion $(1/\lambda^2)$ to the square of luminous source wavelength used for the system. Thus, if a short-wavelength semiconductor laser using GaN in the blue and ultraviolet ray domain is developed, it will be possible to effect a large increase in the rate of information processing [12–14].

In this paper, we have performed a systematic study of the properties of Mn-doped GaN nanowires.

GaN doped with Mn is a candidate for spintronics devices since a Curie temperature above room temperature has been reported. Additional interest in wide-gap nitride semiconductors arises when they are doped with transition metal Mn, due to their potential use in spintronics devices. The aim of this study was to improve the understanding of how the manganese is incorporated into the GaN layers, and how this process affects the defect structure of these wires.

2. Experimental

2.1. Materials and Reagents. For the growth of Mn-doped GaN wires, we used Ga metal (purity of 99.99%, density of 5.904 g/mL at 25°C, mp 29.8°C) and Mn powder ($\geq$99%, −325 mesh, density of 7.3 g/mL at 25°C, mp 1244°C), with high-purity N_2 (99.999%) as a carrier gas and NH_3 (99.9995%) as a reaction gas. A mass flow controller (MFC) was used to regulate the partial pressures of N_2 and NH_3 gases. A silicon wafer substrate (100) was used for the growth of the GaN wires.

2.2. Experiments and Analysis. To grow GaN nanowires, we employed the chemical vapor deposition (CVD) method, and the configuration used for experiments is shown in Figure 1.

After a certain amount of gallium metal ($\fallingdotseq$ 100 mg) and manganese powder ($\fallingdotseq$ 5 mg) was put into the alumina boat, the substrate was inserted into the frame in the back of the

TABLE 1: Experimental operating parameters.

No.	Temperature (°C)	N_2 (SCCM)	NH_3 (SCCM)
a	1000	175	150
b	1050	175	150
c	1100	175	150
d	1200	175	150
e	1300	175	150
f	1400	175	150

boat, and then it was placed in the center of a tube furnace. While the temperature of the tube furnace was increased to achieve the reaction temperature according to each set of conditions (Table 1), N_2 (99.999%), NH_3 (99.9995%), and HCl (N_2 mixture gas, 10 mol%) gases were injected. The heating rate was 5°C/min, and the holding time was 1 hour. During this time, the temperature and the total flow were regulated in the ranges of 950–1400°C and 490–625 sccm, respectively. The location of the gallium metal was occasionally changed with that of the manganese powder. Images of synthesized GaN nanowires were analyzed through field-emission scanning electron microscopy (FE-SEM, JEOL JSM-7000F) to determine the growth conditions, and the wires were also qualitatively analyzed through energy dispersion X-ray spectroscopy (EDS). The crystal structure was analyzed through X-ray diffraction (XRD, Rigaku, D/max-2500/PC), and, for an analysis of the optical characteristics, photo luminescence (PL) and cathode luminescence (CL) measurements were taken.

3. Results and Discussion

3.1. Growth of GaN Nanowires. Figure 2 shows GaN nanowires grown on the surface of a Si substrate observed through FE-SEM at various reaction temperatures. The experiment was conducted by increasing the flux of N_2 gas and NH_3 gas by 150 sccm, and the furnace temperature was increased

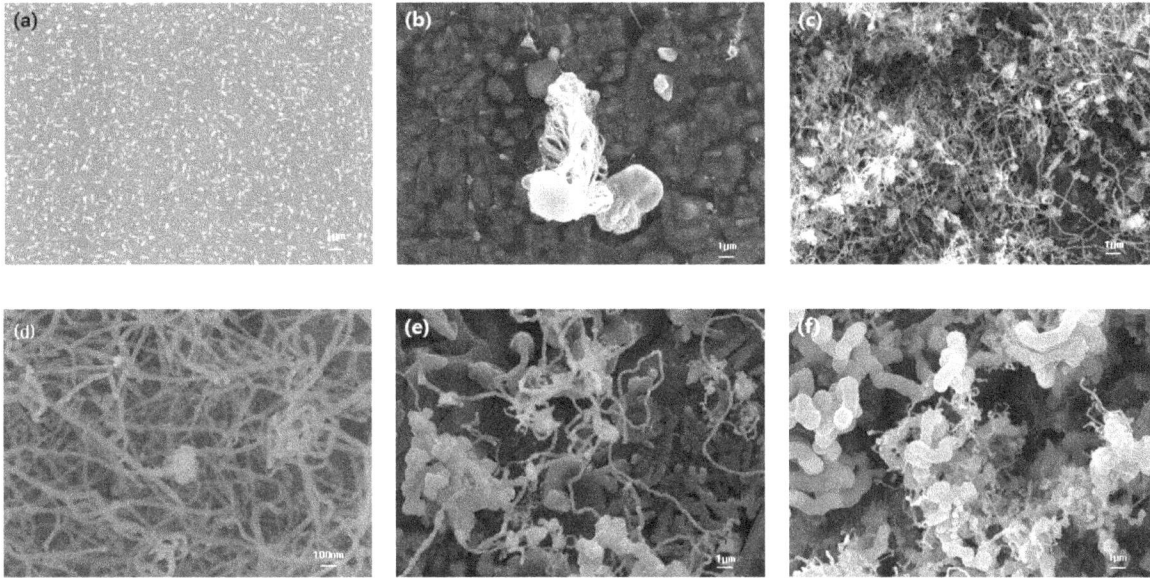

FIGURE 2: Structural characteristics of GaN:Mn nanowires grown on Si substrates at different temperatures: (a) 1000°C, (b) 1050°C, (c) 1100°C, (d) 1200°C, (e) 1300°C and (f) 1400°C.

5°C/min, and the reaction time was maintained for 1 hour. As shown in the FE-SEM picture of Figure 2, the experiment was carried out for 1 hour in the following conditions: (a) 1000°C, (b) 1050°C, (c) 1100°C, (d) 1200°C, (e) 1300°C, and (f) 1400°C.

As shown in Figure 2(a), GaN was not deposited on the surface of the Si substrate for a reaction temperature of 1000°C. However, in Figures 2(b), 2(c), 2(d), 2(e), and 2(f), GaN was deposited on the surface of the Si substrate for higher temperatures. As shown in the FE-SEM image, Figure 2(b) shows that GaN started growing at 1050°C, but only a fraction of the Si substrate showed GaN growth. On the other hand, Figures 2(c) and 2(d) show that GaN nanowires grew throughout the temperature range of 1100°C to 1200°C. In Figures 2(d), 2(e), and 2(f), GaN nanowires were produced, but they were thicker than those shown in Figure 2(c). It was also observed that lumps of wires were created in places. In particular, it was observed in Figure 2(f) that wires were created on the surface of a large-sized lump, but when the temperature was 1400°C, there was a high probability that the Si substrate could melt away. Thus, no additional experiments were conducted at temperatures above 1400°C. Overall, when other variables were fixed, and the reaction temperatures varied, the GaN started being deposited at over 1050°C, and its growth was best at temperatures between 1100°C and 1200°C.

3.2. XRD Analysis. Figure 3 shows XRD measurements of GaN nanowires. Based on the XRD diffraction pattern, we found peaks in the (0002) direction and (0004) direction in the GaN Wurtzite structure. Thus, the crystal structure of GaN grown is a Wurtzite structure having lattice constants where $a = 3.189$ Å and $c = 5.185$ Å.

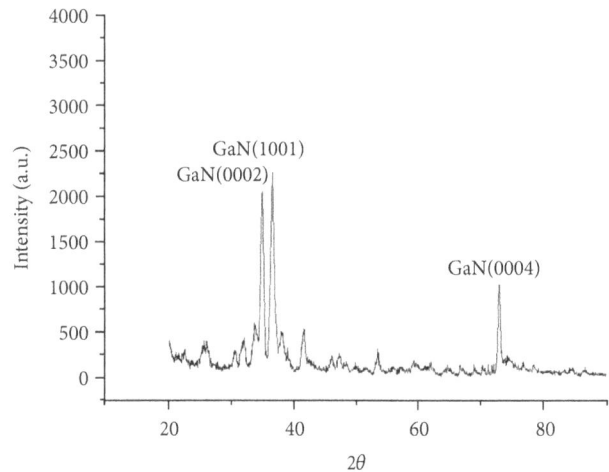

FIGURE 3: XRD pattern of GaN:Mn nanowires.

3.3. EDS Analysis. Figure 4 shows EDS measurements of GaN nanowires to determine the chemical composition of the GaN nanowires. Analyses of regions where GaN nanowires were deposited showed that they were chemically composed of Ga, N, and Mn. As a result, nanowires grown with those components deposited were found to be GaN. Interestingly, too much Mn was detected, although it was used for doping, and thus another experiment was carried out, regulating the amount of Mn to find the correct amount to use for proper doping based on the EDS results.

3.4. PL Analysis. Photoluminescence (PL) was measured to investigate the light-emitting characteristics of GaN doped

(a)   (b)

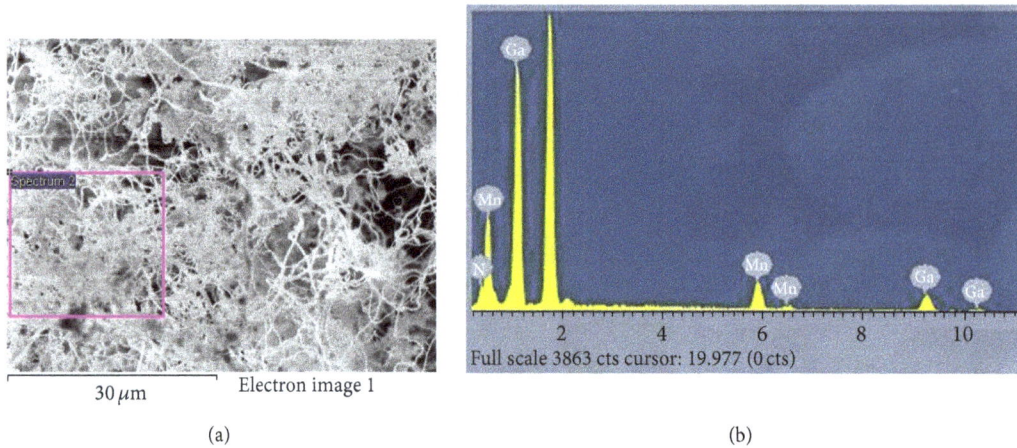

FIGURE 4: Typical EDS spectrum data of the GaN:Mn nanowires.

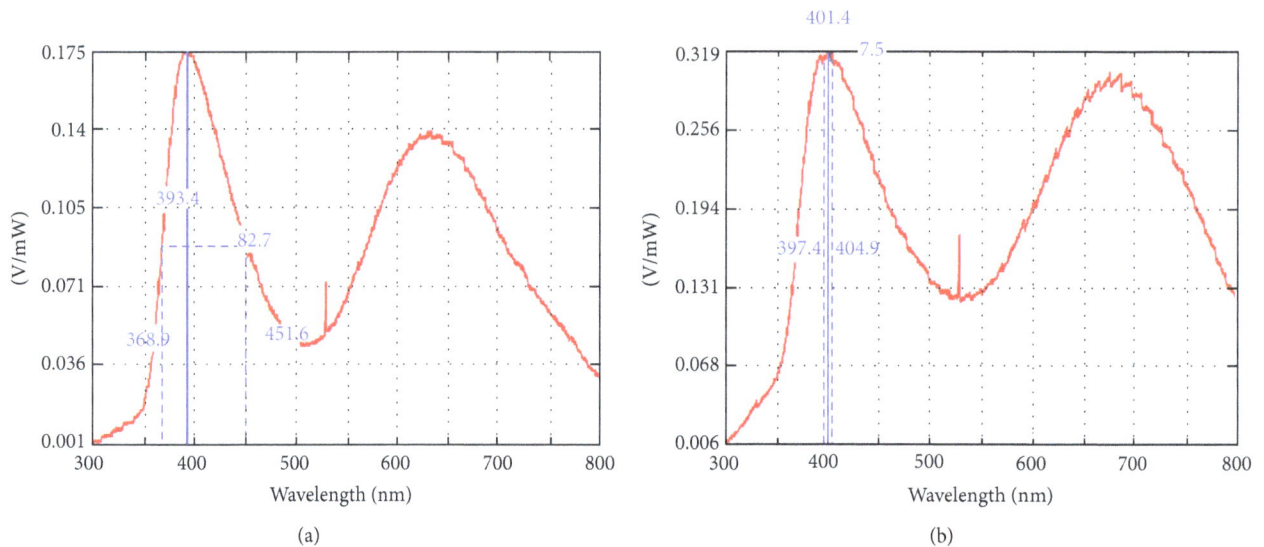

(a)   (b)

FIGURE 5: PL spectrum of GaN:Mn nanowires (a) PL data of GaN nanowires and (b) PL data of GaN:Mn nanowires.

with Mn. Figure 5 shows the PL analysis of the GaN nanowires (Figure 5(a)) and the PL analysis of the nanowires where GaN was doped with Mn (Figure 5(b)). The value of the GaN light-emitting wavelength was 393.4 nm. When the GaN was doped with Mn, the light-emitting wavelength was 401.4 nm. We found that as the amount of Mn doping increased, the light-emitting wavelength increased up to 700 nm (red), the maximum. In addition, even when the experiment was carried out with the same ratio of Ga metal to Mn powder, the value of the light-emitting wavelength was found to be 644.4 nm while keeping the temperature maintained at 1300°C.

3.5. CL Analysis. Cathode-luminescence(CL) indicates luminescence caused by the recombination of positively charged holes made by incident electrons inside a crystal phosphor with electrons mainly floating on the conduction band. Figure 6 shows CL images of GaN nanowires grown on the surface of the Si substrate [14].

4. Conclusion

We succeeded in depositing nanostructured GaN on the surface of a Si substrate using the CVD method. The optimal conditions for the growth of GaN were determined by regulating the reaction temperature, the flux of reaction gases, the mixture ratio of gallium metal and manganese powder, and whether to use HCl gas and control of the reaction time [15]. The optimal conditions through the CVD method were 1200°C in temperature, 165 sccm in the flux of NH_3 gas, 1 hour of deposition time, and 20 : 1 in terms of the quantity (g) ratio of Ga metal to Mn powder [16].

GaN nanowires were deposited on the surface of a Si substrate using the CVD method, as observed through field-emission scanning electron microscopy (FE-SEM). As a result, we found that as-grown GaN nanowires were between 90 and 200 nm in diameter [17, 18].

XRD and EDS analysis of the crystal structure and chemical composition of GaN nanowires deposited on the surface

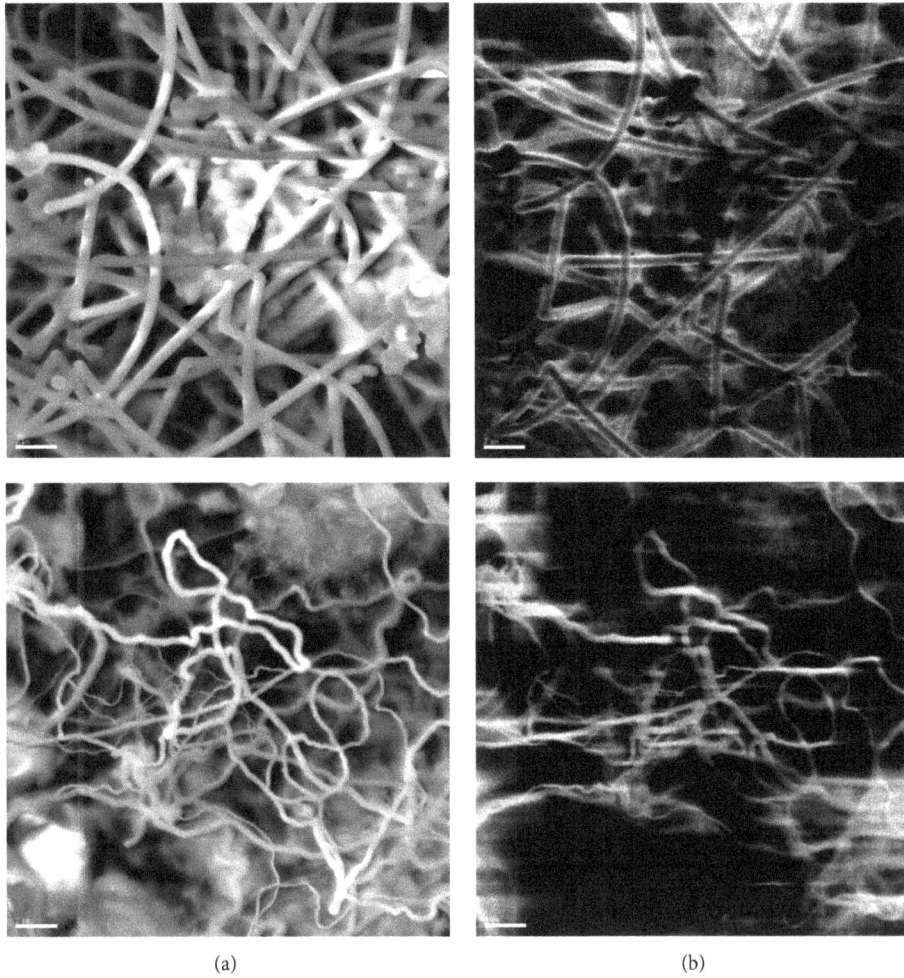

(a)   (b)

FIGURE 6: CL data of GaN:Mn nanowires (a) SEM image and (b) CL image.

of a Si substrate using the CVD method and showed that the wires had wurtzite structure.

Measurements of the light-emitting characteristics of Mn-doped GaN nanowire PL showed blue light emission. This study demonstrates the potential for applying GaN nanowires in blue light-emitting diodes (LEDs) [19, 20].

References

[1] K. Ando, H. Saito, V. Zayets, and M. C. Debnath, "Optical properties and functions of dilute magnetic semiconductors," *Journal of Physics*, vol. 16, no. 48, pp. S5541–S5548.

[2] J. M. Baik, H. K. Jo, T. W. Kang, and J. L. Lee, "Effects of implanted nitrogen on the microstructural, optical, and magnetic properties of Mn-implanted GaN," *Journal of the Electrochemical Society*, vol. 152, no. 8, pp. G608–G612, 2005.

[3] Y. Baik, J. M. Shon, T. W. Kang, and J. L. Lee, "Fabrication of (Ga, Mn)N nanowires with room temperature ferromagnetism using nitrogen plasma," *Applied Physics Letters*, vol. 87, Article ID 042105, 2005.

[4] F. Virot, R. Hayn, and A. Boukortt, "Electronic structure and Jahn-Teller effect in GaN:Mn and ZnS:Cr," *Journal of Physics*, vol. 23, no. 2, Article ID 025503, 2011.

[5] W. Burdett, O. Lopatiuk, L. Chernyak, M. Hermann, M. Stutzmann, and M. Eickhoff, "Electron injection-induced effects in Mn-doped GaN," *Journal of Applied Physics*, vol. 96, no. 6, pp. 3556–3558, 2004.

[6] I. A. Buyanova, M. Izadifard, L. Storasta et al., "Optical and electrical characterization of (Ga,Mn)N/InGaN multiquantum well light-emitting diodes," *Journal of Electronic Materials*, vol. 33, no. 5, pp. 467–471, 2004.

[7] X. H. Chen, S. J. Lee, and M. Moskovits, "Modification of the electronic properties of GaN nanowires by Mn doping," *Applied Physics Letters*, vol. 91, Article ID 082109, 2007.

[8] H. J. Choi, H. K. Seong, and U. Kim, "Diluted magnetic semiconductor nanowires," *Nano*, vol. 3, no. 1, pp. 1–19, 2008.

[9] X. G.) Cui, Z. K. Tao, R. Zhang et al., "Structural and magnetic properties in Mn-doped GaN grown by metal organic chemical vapor deposition," *Applied Physics Letters*, vol. 92, Article ID 152116, 2008.

[10] X. G. Cui, R. Zhang, Z. K. Tao et al., *Chinese PhysicsLetters*, vol. 26, Article ID 038103, 2009.

[11] X. Y. Cui, B. Delley, A. J. Freeman, and C. Stampfl, "Magnetic metastability in tetrahedrally bonded magnetic III-nitride semiconductors," *Physical Review Letters*, vol. 97, Article ID 016402, 2006.

[12] G. P. Das, B. K. Rao, P. Jena, and Y. Kawazoe, "Dilute magnetic III-V semiconductor spintronics materials: a first-principles approach," *Computational Materials Science*, vol. 36, no. 1-2, pp. 84–90, 2006.

[13] A. Debernardi, "Mn-doped GaN/AlN heterojunction for spintronic devices," *Superlattices and Microstructures*, vol. 40, no. 4–6, pp. 530–532, 2006.

[14] A. Boukortt, R. Hayn, and F. Virot, "Optical properties of Mn-doped GaN," *Physical Review B*, vol. 85, Article ID 033302, 2012.

[15] H. Ham and J. M. Myoung, "Doping concentration dependence of ferromagnetic ordering in (Ga, Mn)N nanowires," *Applied Physics Letters*, vol. 89, no. 17, Article ID 173117, 2006.

[16] S. Han, J. Park, K. W. Rhie, S. Kim, and J. Chang, "Ferromagnetic Mn-doped GaN nanowires," *Applied Physics Letters*, vol. 86, no. 3, Article ID 032506, 2005.

[17] E. Han, H. Oh, J. J. Kim, H. K. Seong, and H. J. Choi, "Observation of hysteretic magnetoresistance in Mn-doped GaN nanowires with the mesoscopic Co and Ti/Au contacts," *Applied Physics Letters*, vol. 87, Article ID 062102, 2005.

[18] N. Smolentsev, G. Smolentsev, S. Wei, and A. V. Soldatov, "Local atomic structure around Mn ions in GaN:Mn thin films: quantitative XANES analysis," *Physica B*, vol. 406, no. 14, pp. 2843–2846, 2011.

[19] X. L. Hu, J. Q. Li, Y. F. Zhang, H. H. Li, and T. Li, "Density functional theory study on the optical properties of Mn-doped GaN," *Jiegou Huaxue*, vol. 29, no. 3, pp. 476–482, 2010.

[20] M. F. Huang and T. H. Lu, "Optimization of the active-layer structure for the deep-UV AlGaN light-emitting diodes," *IEEE Journal of Quantum Electronics*, vol. 42, no. 8, pp. 820–826, 2006.

12

Optimization of Aqueous Two-Phase Systems for the Recovery of Soluble Proteins from Tannery Wastewater Using Response Surface Methodology

Selvaraj Raja and Vytla Ramachandra Murty

Department of Biotechnology, Manipal Institute of Technology, Manipal, Karnataka 576104, India

Correspondence should be addressed to Selvaraj Raja; rajaselvaraj@gmail.com

Academic Editor: Run-Cang Sun

Aqueous two-phase system (ATPS) composed of polyethylene glycol 6000 (PEG 6000) and sodium citrate (SC) has been proposed to recover the valuable soluble proteins from tannery wastewater. A sequential optimization strategy which included fractional factorial design (fFD) and central composite design (CCD) was employed to enhance the recovery. From this strategy, a second-order polynomial model was obtained for the protein recovery and it was validated. The optimum recovery was found as 93.46% when pH, NaCl concentration, and temperature were kept at 7.5, 0.1 M, and 33°C, respectively, for a phase system composed of 20% (w/w) PEG 6000-15% (w/w) SC. Thus the proposed ATPS can serve as an alternative to the conventional precipitation method to recover the soluble proteins from tannery wastewater.

1. Introduction

The leather industry is one of the major foreign exchange earners in India nearly over last thirty years [1]. It is reported that the Indian market has been fragmented with about 2200 tanneries [2]. During the traditional leather processing, the skin and hides are subjected to various operations such as soaking, dehairing, liming, deliming, bating, degreasing, and pickling [3]. When the skin is subjected to the alkali treatment, the soluble proteins present on the surface of the skin are discharged as waste. Yearly, nine million tons of skins/hides are being processed worldwide [4]. Literature reveals that, for every 100 kg of raw hides, 15 kg of solubilized protein is discharged as waste in the early stages of the process of transforming hides into leather [5]. The value of these solubilized proteins is enormous and they find applications in food and pharmaceutical industries [6]. The presence of these proteins in the tannery effluents increases the biological oxygen demand and chemical oxygen demand and leads to pollution. By removing these proteins from the waste, a reduction of the tax on wastewater can be achieved and the recovered proteins can be used in food and pharmaceutical industries.

There are a few reports available in the literature for the recovery of soluble proteins from industry effluents using membrane separation processes, but the major drawback of these processes is membrane fouling [7, 8]. The conventional method of protein recovery from tannery wastewater is "precipitation method" which has been well addressed by Kabdasli et al. [9] and Marsal et al. [10]. However the precipitation method has the limitation of protein denaturation and low recovery. For example, only a recovery of 50–70% and 68–78% of soluble proteins from the effluents of tannery beam-house operations was possible in the precipitation methods developed by Kabdasli et al. [9] and Marsal et al. [10], respectively.

In the context of clean environment and pollution prevention, nanotechnology could play a key role [11]. However, aqueous two-phase system (ATPS) attracts more attention because of simple process, low cost, and easy scale-up procedures. For example, very recently, ATPS composed of PEG and sodium citrate was successfully used for textile effluent dye removal [12]. It has been proven by many researchers that ATPS is an efficient and economical process when compared to other separation processes like precipitation, chromatography, and so forth [13–15].

Therefore, a benign technique for the proteins, ATPS, has been proposed in the present study in order to recover the soluble proteins from the tannery wastewater. ATPS is a downstream processing method which uses the principles of liquid-liquid extraction. It can be formulated by mixing two hydrophilic polymers (Poly Ethylene Glycol (PEG)-Dextran) with water or one hydrophilic polymer (PEG) and inorganic salts (phosphates, sulfates, and citrates) with water [16]. It has been reported that polymer-salt-based ATPS has many advantages over polymer-polymer ATPS and few of them includes easy scale-up, low cost, low interfacial tension, possibility of process integration, and less viscous phases [17].

For the first time, Saravanan et al. [18] research group developed an ATPS made up of PEG/sulfate to recover the soluble proteins from tannery wastewater. Since then, ATPS has received much attention to recover the biomolecules from various industrial effluents such as fish industry [19], dairy industry [20], and prawn industry [21]. Recently, our research group has addressed the partitioning of tannery wastewater proteins in ATPS composed of PEG 10000 plus different citrate salts [22]. The proposed method of ATPS in this paper is environment-benign since the phase components used are nontoxic and biodegradable.

For the maximization of recovery of proteins from tannery wastewater, a sequential method of optimization using response surface methodology (RSM) has been employed which includes the following steps:

(i) screening of significant process variables which affect the protein partitioning in ATPS by a fractional factorial design (fFD),

(ii) crude optimization of the most significant variables by a full factorial design (FFD) with center points,

(iii) final optimization of the most significant variables by central composite design (CCD) using response surface methodology (RSM),

(iv) development and verification of mathematical model and expressing the relationship between the protein partitioning and significant process variables.

2. Materials and Methods

2.1. Materials. PEG 6000 was purchased from Merck and used without further purification. Tri-sodium citrate, citric acid, and sodium chloride were also purchased from Merck and Millipore-Milli-Q water was used in all the experiments.

2.2. Preparation of Tannery Wastewater. The tannery wastewater sample was prepared as discussed elsewhere [23]. In this method, known weight of raw skin/hides was treated with alkali solution. This sectional stream wastewater was used as a protein source in partition experiments to recover the soluble proteins.

2.3. Preparation of Two-Phase Systems. Calculated amounts of tri-sodium citrate and citric acid were taken, and pH

of the system was adjusted. ATPS was prepared by mixing appropriate amounts of PEG and citrate solutions, with tannery waste sample as described in the previous section in 15 mL graduated tubes. By the addition of water, the weight of the system was maintained at 10 g. The systems were well mixed in a vortex mixer and left in a water bath at various temperatures for overnight.

2.4. Quantification of Tannery Wastewater Soluble Proteins. The soluble protein from the tannery wastewater was quantified by Bradford method [24]. For the determination of protein concentration, samples were withdrawn from each phase and diluted if necessary with distilled water, and its absorbance was measured using Shimadzu spectrophotometer at 595 nm.

2.5. Partition Coefficient and Recovery. The partitioning of soluble proteins in ATPS is characterized by two factors, namely, partition coefficient K_p and the percentage bottom phase protein recovery $R_{p,B}$:

$$K_p = (\text{Concentration of soluble protein in top phase})$$

$$/ (\text{Concentration of soluble protein in bottom phase})$$

$$R_{p,B}\,(\%) = \frac{\text{amount of soluble protein in bottom phase}}{\text{total amount of soluble protein in the system}}$$

$$= \frac{100}{1 + (V_{\text{PEG}}/V_{\text{SC}})\,K_p},$$

$$(1)$$

where V_{PEG} and V_{SC} are the volume of PEG rich and sodium citrate phases, respectively.

2.6. Screening of Significant Process Variables. Partitioning of proteins in ATPS is a complex phenomenon. It depends on many factors like type and concentration of phase-forming components, pH, temperature, presence of neutral salts, and so forth. Based on prior experiments (data not shown), the following five factors namely, concentration of PEG 6000, concentration of SC, pH of the system, concentration of NaCl, and temperature were chosen as the factors that affect the protein partitioning.

Consequently, a 1/2 fraction, 2-level factorial design for five factors ($2^{5-1} = 16$ experiments) was employed to investigate the significant factors. Table 1 gives both coded and uncoded values of these factors in fFD with the percentage recovery. The table shows a wide variation in percentage recovery ranging from 34% to 83% which reflects the importance to attain higher percentage recovery.

2.7. Crude Optimization. It has been observed from the analysis that the three factors, namely, pH, NaCl, and temperature are the significant factors which enhance the protein recovery. Since PEG and SC do not play a significant role in partitioning, PEG was fixed at 20% (positive effect) and SC was fixed at 15% (negative effect) for all the upcoming

TABLE 1: Coded and uncoded values of factors of 2^{5-1} fractional factorial design (fFD).

Experiment no.	Coded[a] and uncoded[b] values of variables					Recovery, $R_{p,B}$ (%)
	PEG (% w/w)	SC (% w/w)	pH	NaCl (M)	Temperature (°C)	
1	12 (−1)	23(+1)	8 (+1)	0.3 (+1)	20 (−1)	67.24
2	20 (+1)	15 (−1)	8 (+1)	0.1 (−1)	40 (+1)	68.24
3	12 (−1)	15 (−1)	6 (−1)	0.1 (−1)	40 (+1)	33.95
4	20 (+1)	23 (+1)	8 (+1)	0.3 (+1)	40 (+1)	83.42
5	12 (−1)	23 (+1)	6 (−1)	0.3 (+1)	40 (+1)	52.71
6	12 (−1)	23 (+1)	8 (+1)	0.1 (−1)	40 (+1)	50.49
7	20 (+1)	23 (+1)	8 (+1)	0.1 (−1)	20 (−1)	52.21
8	12 (−1)	23 (+1)	6 (−1)	0.1 (−1)	20 (−1)	37.29
9	20 (+1)	15 (−1)	6 (−1)	0.1 (−1)	20 (−1)	29.89
10	20 (+1)	15 (−1)	8 (+1)	0.3 (+1)	20 (−1)	71.77
11	20 (+1)	23 (+1)	6 (−1)	0.3 (+1)	20 (−1)	54.59
12	20 (+1)	15 (−1)	6 (−1)	0.3 (+1)	40 (+1)	64.49
13	12 (−1)	15 (−1)	8 (+1)	0.3 (+1)	40 (+1)	75.33
14	12 (−1)	15 (−1)	8 (+1)	0.1 (−1)	20 (−1)	61.46
15	20 (+1)	23 (+1)	6 (−1)	0.1 (−1)	40 (+1)	48.93
16	12 (−1)	15 (−1)	6 (−1)	0.3 (+1)	20 (−1)	58.49

[a] Data in brackets; [b] Data without brackets.

TABLE 2: Central composite design (CCD) of variables with % recovery as response.

Experiment no.		Coded[a] and uncoded[b] values of variables			Recovery, $R_{p,B}$ (%)	
		pH	NaCl (M)	Temperature (°C)	Experimental	Predicted
1		6.4 (−1)	0.14 (−1)	24 (−1)	76.78	78.30
2		7.6 (+1)	0.14 (−1)	24 (−1)	87.69	86.51
3		6.4 (−1)	0.26 (+1)	24 (−1)	79.68	78.88
4	2^3 Factorial points	7.6 (+1)	0.26 (+1)	24 (−1)	84.35	84.82
5		6.4 (−1)	0.14 (−1)	35 (+1)	88.90	88.45
6		7.6 (+1)	0.14 (−1)	36 (+1)	89.56	90.39
7		6.4 (+1)	0.26 (+1)	36 (+1)	76.85	78.06
8		7.6 (−1)	0.26 (+1)	36 (+1)	79.21	77.72
9		7 (0)	0.2 (0)	30 (0)	90.66	89.81
10	I set of center points	7 (0)	0.2 (0)	30 (0)	90.50	89.81
11		7 (0)	0.2 (0)	30 (0)	89.65	89.81
12		6 (−1.68)	0.2 (0)	30 (0)	83.22	82.36
13		8 (+1.68)	0.2 (0)	30 (0)	88.15	88.98
14	Star points	7 (0)	0.1 (−1.68)	30 (0)	92.46	92.05
15		7 (0)	0.3 (+1.68)	30 (0)	81.51	81.88
16		7 (0)	0.2 (0)	20 (−1.68)	88.62	83.64
17		7 (0)	0.2 (0)	40 (+1.68)	78.55	78.51
18		7 (0)	0.2 (0)	30 (0)	89.56	89.81
19	II set of center points	7 (0)	0.2 (0)	30 (0)	88.62	89.81
20		7 (0)	0.2 (0)	30 (0)	89.85	89.81

experiments. The three significant variables were further optimized using a 2^3 FFD (Table 2, Experiment nos. 1–8) with three center points (Table 2, Experiment nos. 9–11) to determine the optimum operating conditions. These experiments were done to make sure that the proposed optimization process was in the appropriate region [25].

2.8. Final Optimization. In order to include the curvature, few more experiments were done by adding 6 axial (star) points (Table 2, Experiment nos. 12–17) and 3 more center points (Table 2, Experiment nos. 18–20) to the previous FFD set-up. This entire set of 20 experiments is a central composite design (CCD) for three factors, an RSM technique [26].

TABLE 3: ANOVA table for fFD.

Source	Degrees of freedom	Sum of squares (SS)	Mean square	F value	P value
PEG	1	83.63	83.63	5.93	0.055
pH	1	1402.88	1402.88	99.50	0.000*
NaCl	1	1324.60	1324.60	93.95	0.000*
Temperature	1	124.43	124.43	8.83	0.014*
PEG*Temperature	1	294.29	294.29	20.87	0.001*
Error	10	140.99	14.10		
Total	15	3370.82			

$S = 3.75490$; R-Sq = 95.82%; R-Sq(adj) = 93.73%.
*Significant at 95% confidence level.

In this methodology, the effects of the variables on the protein recovery were fit to the second-order polynomial model according to the following equation:

$$Y = a_o + \sum a_i F_i + \sum a_{ii} F_i^2 + \sum a_j F_i F_j, \quad (2)$$

where Y is the response variable (percentage recovery), F_i and F_j are the independent variables in coded units, a_o is the average response, and a_i, a_{ii}, and a_j are the measures of the F_i, F_j, F_i^2, and $F_i F_j$ of linear, quadratic, and interaction effects, respectively.

For the statistical calculations, the variables were coded according to the following equation:

$$F_i = \frac{f_i - f_o}{\Delta f_i}, \quad (3)$$

where F_i is the independent variable in the coded unit, f_i is the real value of independent variable, f_o is center point the real value of the independent variable, and Δf_i is the step change value. By analyzing the contour plots, the optimum values of the significant variables were obtained. The statistical analysis of the model was represented in the form of analysis of variance (ANOVA).

3. Results and Discussion

3.1. Screening of Significant Process Variables by fFD and Crude Optimization. The fFD showing the recovery of protein from each experiment combination shown in Table 1 was used for statistical analysis. The results were analysed by MINITAB-15.0 (CA, USA). Figure 1 represents the normal probability plot of the effect estimates. This plot is used to analyze the significant factors based on the α (=0.05) value. The significant factors do not conform to the normal plot and lie away from the normal line. From the figure, it is clear that the factors pH, NaCl, temperature, and the interaction between PEG concentration and temperature are significant.

The plot of the mean percentage recovery and experiment levels (Figure 2) illustrates the main effects of the operating conditions on the recovery. PEG, pH, NaCl, and temperature increased the protein recovery at high level. A decrease in mean percentage recovery was observed for SC.

Table 3 summarizes the analysis of variance for fFD. The model sum of squares is $SS_{Model} = SS_{PEG} + SS_{pH} + SS_{NaCl} + SS_{Temperature} + SS_{PEG\ Temperature} = 3229.83$, and this accounts

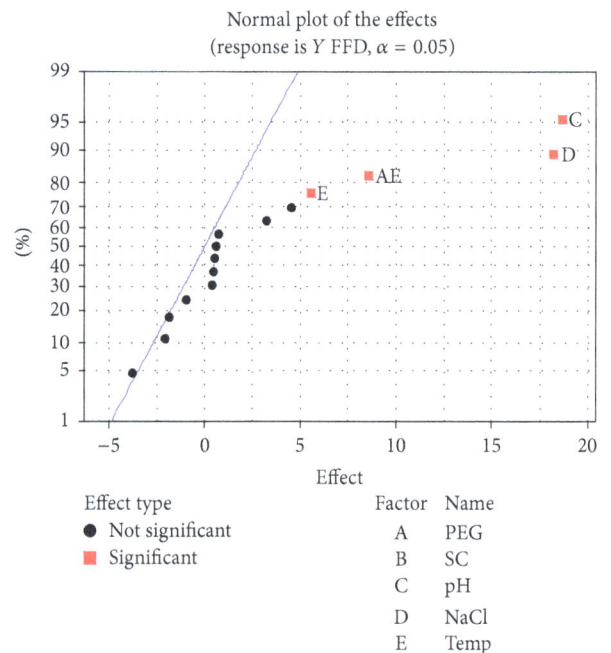

FIGURE 1: Normal probability plot of the effects.

for 95.82% (high R^2) of the total variability in recovery. Moreover, the variables are statistically significant when the P value (defined as the smallest level of significance which leads to the rejection of null hypothesis) is less than 0.05 (95% confidence level). Based on this, pH, NaCl, temperature, and the interaction between PEG and temperature are considered as the significant factors.

The positive effect of PEG at high concentration may be because of the volume occupied by the PEG molecules with the increase in concentration decreases the free space available for the molecules in the top phase. Therefore, because of "volume exclusion effect" all the biomolecules tend to partition towards the bottom phase and thus percentage recovery in the bottom phase increases [27].

In contrast to this, SC had a negative effect on percentage recovery which can be explained based on the "salting-out effect". At high salt concentrations of salt, the ions decrease the solubility of biomolecules which makes them to move to the PEG rich top phase and therefore the percentage recovery in the bottom phase decreases [28].

TABLE 4: Estimated regression coefficients for percentage recovery.

Term	Coefficient	Standard error of coefficient	T	P
Constant	89.8802	1.1292	79.597	0.000
pH	1.9683	0.7492	2.627	0.025*
NaCl	−3.0216	0.7492	−4.033	0.002*
Temperature	−0.7985	0.7492	−1.066	0.312
pH*pH	−1.9375	0.7492	−2.657	0.024*
NaCl*NaCl	−1.4779	0.7492	−2.026	0.070
Temperature*Temperature	−2.6800	0.7492	−3.675	0.004*
pH*NaCl	−0.5688	0.7492	−0.581	0.574
pH*Temperature	−1.5687	0.7492	−1.603	0.140
NaCl*Temperature	−2.7438	0.7492	−2.803	0.019*

$R^2 = 84.86\%$; R^2 (adj) $= 71.22\%$.
*Significant at 95% confidence level.

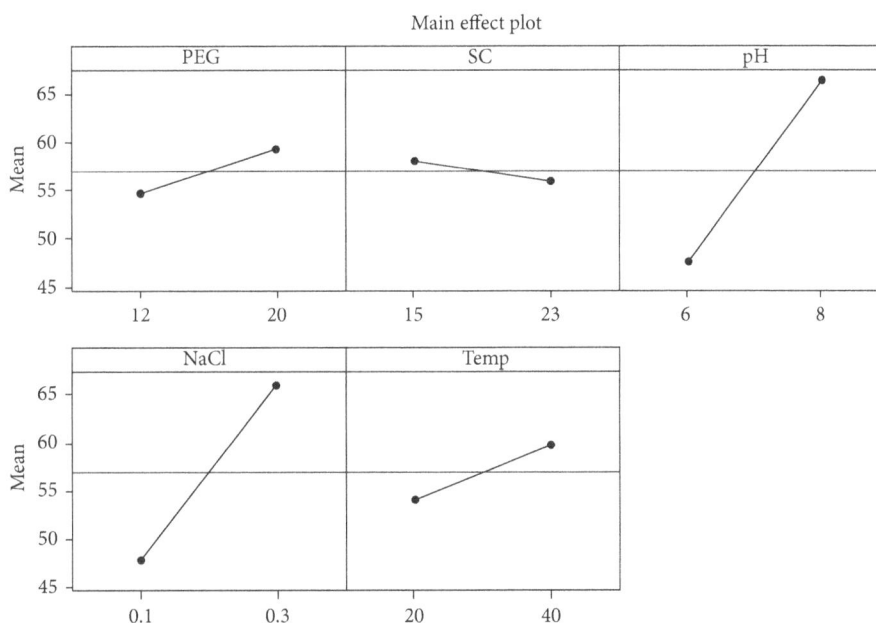

FIGURE 2: Main effect plot for recovery (%).

The pH presented a statistically significant positive effect for the partitioning of proteins to the bottom phase. It can be explicated with respect to the isoelectric point of the proteins. The wastewater proteins present in tannery wastewater are soluble and globular proteins [29] and therefore they have net negative charge at pH > 5. Hence, the negatively charged protein molecules partition to the bottom phase at high pH. Similar results were obtained by many researchers [30–32].

NaCl presence in the ATPS showed a significant positive effect which may be due to the alterations of hydrophobic interactions or changes in the electrostatic potential difference. For the NaCl concentrations studied in this study (0.1 M to 0.3 M), the interaction of biomolecules with the salt rich bottom phase increases because of the changes in the electrostatic potential difference [33].

The temperature also indicated a significant positive effect on the percentage recovery. The increase in temperature not only alters the structure of biomolecules but also changes the phase composition of the ATPS. Therefore the increase in temperature increases the protein recovery in the bottom phase [34].

As a conclusion from fFD, the factors pH, NaCl, and temperature are confirmed as significant factors and therefore selected for further optimization to maximize the percentage recovery. From the Table 2, it is evident that the average recovery in the center of the experimental region is 90.27%, while the average recovery at the corners is 82.88%. Since this difference is significant, the recovery will be a curved function of all three factors. Moreover, because of the presence of curvature, the response could not be explained by a linear

TABLE 5: ANOVA for % recovery.

Source	DF	Seq SS	Adj SS	Adj MS	F	P
Regression	9	429.487	429.487	47.721	6.23	0.004[*]
Linear	3	186.308	186.308	62.103	8.10	0.005[*]
Square	3	160.677	160.677	53.559	6.99	0.008[*]
Interaction	3	82.501	82.501	27.500	3.59	0.054
Residual Error	10	76.654	76.654	7.665		
Lack-of-Fit	5	73.949	73.949	14.790	27.34	0.001[*]
Pure Error	5	2.704	2.704	0.541		
Total	19	506.140				

[*]Significant at 95% confidence level.

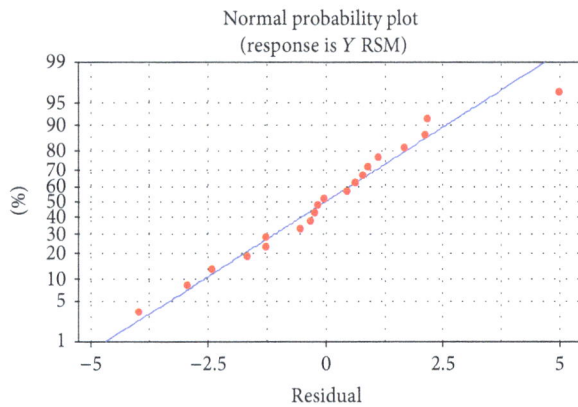

FIGURE 3: Residual plot with outlier.

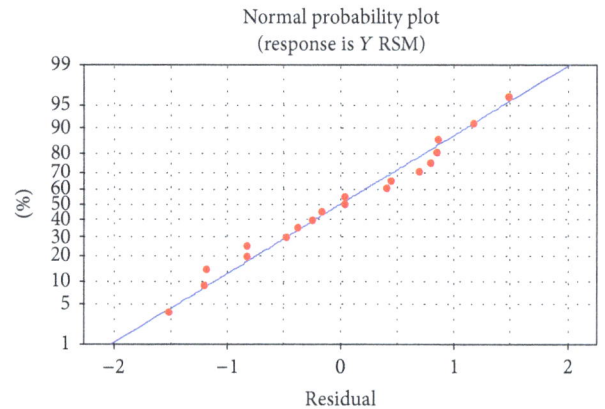

FIGURE 4: Residual plot without outlier.

model and there is a need for a quadratic *model* which is discussed in the following section.

3.2. Final Optimization.

After crude optimization of pH, NaCl, and temperature by 2^3 FFD with three center points and ascertaining of optimal region, additional experiments were performed with 6 axial points and 3 more center points to frame a complete CCD (Table 2). The CCD of 20 experimental runs was used to analyze and optimize the significant factors. Table 4 lists the Minitab output of estimated regression coefficients, standard errors, t-values, and P values.

As discussed earlier, at 95% confidence level, the terms having P values <0.05 are considered to be statistically significant. By substituting these statistically significant parameters' regression coefficients in (2), the following model was obtained in coded units:

$$R_{p,B} = 89.88 + 1.96\,\text{pH} - 3.02\,\text{NaCl} - 1.94\,\text{pH} * \text{pH}$$
$$- 2.68\,\text{Temperature} * \text{Temperature} \qquad (4)$$
$$- 2.74\,\text{NaCl} * \text{Temperature}.$$

In addition to the linear effects, RSM helps to evaluate the interaction and quadratic effects. It is clear from the table that interaction effects are also significant in this process. The regression coefficients R^2 and R^2_{adj} values were determined as 84.86% and 71.22%, respectively.

Table 5 represents the ANOVA for the quadratic model developed. Higher F values indicate that the term is statistically significant. Another convenient measure to test the significance of the terms is P value. It is evident that linear, interaction, and quadratic effects were statistically significant ($P < 0.05$) for the developed model. Nevertheless, a low value for lack of fit indicated that it is also statistically significant, and therefore it is necessary to identify possible outliers. It is done by examination of residual plot as shown in Figure 3 which suggested that a data point corresponding to experimental run 16 could be a possible outlier.

Consequently, this data point was omitted and the regression was repeated for the remaining data. The regression coefficients and the ANOVA values after the omission of outlier are given in Tables 6 and 7, respectively.

Substituting the new regression coefficients into (2) gives the following new modified model:

$$Y_B = 89.81 + 1.96\,\text{pH} - 3.02\,\text{NaCl} - 1.46\,\text{pH} * \text{pH}$$
$$- 1.00\,\text{NaCl} * \text{NaCl} - 4.45\,\text{Temperature}$$
$$* \text{Temperature} - 1.57\,\text{pH} * \text{Temperature} \qquad (5)$$
$$- 2.74\,\text{NaCl} * \text{Temperature}.$$

The new regression coefficients R^2 and R^2_{adj} values were determined as 97.27% and 94.54%, respectively, which were higher than the previous values. Thus, 97.27% of variation in

TABLE 6: Estimated regression coefficients for recovery (%) without outlier.

Term	Coefficient	Standard Error of coefficient	T	P
Constant	89.8078	0.5011	179.231	0.000*
pH	1.9683	0.3324	5.922	0.000*
NaCl	−3.0216	0.3324	−9.091	0.000*
Temperature	0.7622	0.4108	1.855	0.097
pH*pH	−1.4647	0.3317	−4.416	0.002*
NaCl*NaCl	−1.0051	0.3317	−3.030	0.014*
Temperature*Temperature	−4.4476	0.4236	−10.500	0.000*
pH*NaCl	−0.5688	0.4343	−1.310	0.223
pH*Temperature	−1.5687	0.4343	−3.612	0.006*
NaCl*Temperature	−2.7438	0.4343	−6.318	0.000*

$R^2 = 97.27\%$; R^2 (adj) = 94.54%.
*Significant at 95% confidence level.

TABLE 7: ANOVA for recovery (%) without outlier.

Source	DF	Seq SS	Adj SS	Adj MS	F	P
Regression	9	483.698	483.698	53.744	35.62	0.000*
Linear	3	180.715	182.794	60.931	40.39	0.000*
Square	3	220.481	220.481	73.494	48.71	0.000*
Interaction	3	82.501	82.501	27.500	18.23	0.000*
Residual Error	9	13.578	13.578	1.509		
Lack-of-Fit	4	10.874	10.874	2.718	5.03	0.053
Pure Error	5	2.704	2.704	0.541		
Total	18	497.275				

*Significant at 95% confidence level.

TABLE 8: Experimental verification of the model.

Optimized input process parameters			Modified value of input process parameters			Predicted recovery	Experimental recovery
pH	NaCl (M)	Temperature (°C)	pH	NaCl (M)	Temperature (°C)		
7.45	0.1	32.72	7.5	0.1	33	94.40%	93.46 ± 0.7%

yield was explained while only 2.73% was left to the residuals. Apart from this, the P value for lack of fit increased from 0.001 to 0.053 (Table 7) after the omission of the outlier. From Table 7, it is obvious that the new model is valid and linear, and quadratic and interaction terms should be included in the model. A good normal distribution of the model (Figure 4) with a linear line for the percentage recovery confirmed that the model was well fitted with the experimental results and all the major assumptions of the model [35] have been validated.

The three-dimensional response surface plots (Figures 5(a) and 5(c)) show the effects of operating parameters on the percentage recovery while the contour plots (Figures 5(b) and 5(d)) reflect the nature and degree of these effects. As seen from the figures, the response surface plots are concave, indicating that it is possible to obtain a maximum value within the range of the levels investigated. The curved lines in the contour plots confirmed that interaction between the factors ($P < 0.05$) was present and these interaction terms were included in the new model (5). The new regression model was solved for the maximum recovery using the response optimizer tool in MINTAB 15.0, and the optimum values of pH, 7.45; NaCl, 0.1 M; and temperature, 32.72°C

were obtained with a maximum predicted response of 94.40% recovery.

3.3. Verification of the Model by Experiment. In order to validate these results, experiments were done in triplicates (Table 8) by using the modified optimized values (pH: 7.5; NaCl: 0.1 M; Temperature: 33°C). A minimum partition coefficient of 0.056 was obtained with a recovery of 93.46 ± 7%. The partition coefficient obtained is in good agreement with the literature for the soluble protein BSA [36]. The good correlation between the observed and predicted recoveries confirmed that the validity of the new model was adequate. This recovery is comparatively high when compared to the precipitation method as discussed earlier [9, 10]. Therefore, this ATPS can be an alternative to the conventional method of protein recovery from tannery wastewater. Moreover, in the current system studied, most of the protein is partitioned to the salt rich bottom phase and it is possible to recycle the PEG from the top phase by ultrafiltration [37, 38]. Therefore recycling of the phase components decreases the overall cost of the process.

(a)

(b)

(c)

(d)

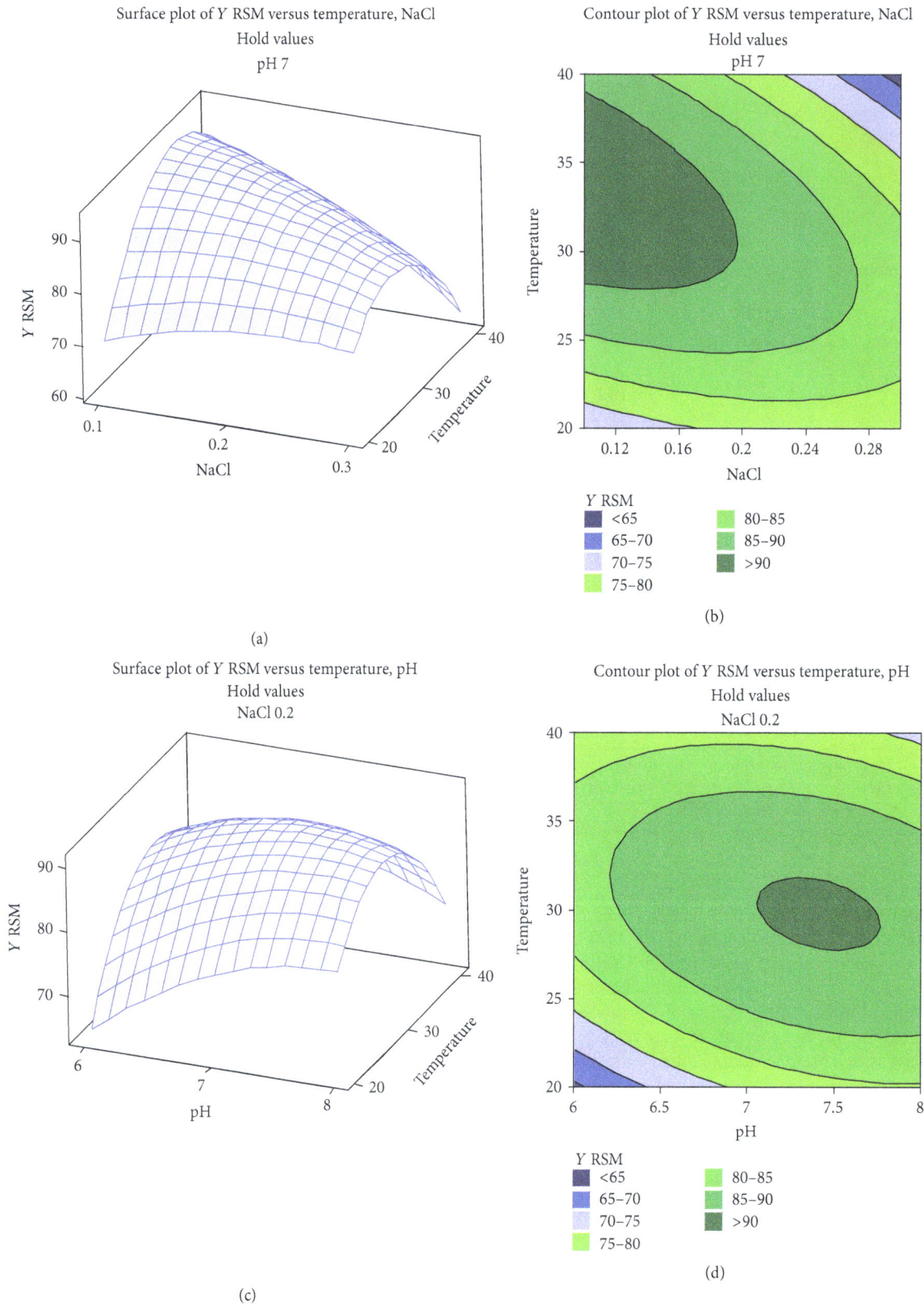

FIGURE 5: Response surface and contour plots showing the effect of process variables in uncoded values on % recovery. ((a) and (b)) NaCl and temperature; ((c) and (d)) pH and temperature.

4. Conclusions

A sequential optimization method which consisted of fFD and CCD was used to obtain the optimum values of significant factors for the recovery of soluble proteins from tannery wastewater in PEG 6000-SC ATPS. The fFD revealed that only pH, concentration of NaCl, and temperature were the significant factors. From the CCD studies, the optimized values of these significant factors were determined: pH 7.5, NaCl 0.1 M, and temperature 33°C for a phase system composed of 20% (w/w) PEG 6000-15% (w/w) SC. The predicted and observed recoveries were 94.40% and 93.46%, respectively, which confirmed that the proposed quadratic model was valid. Thus, it is concluded that ATPS can be used as an alternative method to recover the valuable soluble proteins from tannery wastewater.

Conflict of Interests

The authors declare that they have no conflict of interests.

Acknowledgment

The authors gratefully acknowledge the Department of Biotechnology, MIT, Manipal University, for providing the facilities to carry out the research work.

References

[1] S. Sanyal, S. Banerjee, and S. Majumder, "India's leather in the world market: exploration of recent trends," *Trade and Development Review*, vol. 3, no. 1, pp. 22–58, 2010.

[2] http://sinetinfo.com/pdf/chapters/leather1.pdf.

[3] N. R. Kamini, C. Hemachander, J. G. S. Mala, and R. Puvanakrishnan, "Microbial enzyme technology as an alternative to conventional chemicals in leather industry," *Current Science*, vol. 77, no. 1, pp. 80–96, 1999.

[4] P. Thanikaivelan, J. R. Rao, B. U. Nair, and T. Ramasami, "Recent trends in leather making: processes, problems, and pathways," *Critical Reviews in Environmental Science and Technology*, vol. 35, no. 1, pp. 37–79, 2005.

[5] A. Marsal, E. Bautista1, S. Cuadros, F. Maldonado, A. M. Manich, and J. Font, "Diminution of toxicity of beamhouse operations by precipitation of solubilized proteins," in *Proceedings of the 30th Congreso IULTCS*, Beijing, China, 2009.

[6] J. Benavides and M. Rito-Palomares, "Practical experiences from the development of aqueous two-phase processes for the recovery of high value biological products," *Journal of Chemical Technology and Biotechnology*, vol. 83, no. 2, pp. 133–142, 2008.

[7] H. Paredes and R. B. Bórquez, "Development of an alternative treatment system for fishing industry waste waters using ultrafiltration," *Latin American Applied Research*, vol. 31, pp. 359–365, 2001.

[8] M. D. Afonso and R. B. Bórquez, "Nanofiltration of wastewaters from the fish meal industry," *Desalination*, vol. 151, pp. 131–138, 2002.

[9] I. Kabdasli, T. Olmez, and O. Tunay, "Nitrogen removal from tannery wastewater by protein recovery," *Water Science Technology*, vol. 48, pp. 215–223, 2003.

[10] A. Marsal, E. Hernández, S. Cuadros, R. Puig, E. Bautista, and J. Font, "Recovery of proteins from wastewater of tannery beamhouse operations: influence on the main pollution parameters," *Water Science Technology*, vol. 62, no. 3, pp. 658–666, 2010.

[11] C. Darnault, K. Rockne, A. Stevens, G. A. Mansoori, and N. Sturchio, "Fate of environmental pollutants," *Water Environmental Research*, vol. l77, pp. 2576–2658, 2005.

[12] D. Z. Ivetic, M. B. Sciban, V. M. Vasic, D. V. Kukic, J. M. Prodanovic, and M. G. Antov, "Evaluation of possibility of textile dye removal from wastewater by aqueous two-phase extraction," *Desalination and Water Treatment*, vol. 51, pp. 1–6, 2012.

[13] O. Aguilar, V. Albiter, L. Serrano-Carreón, and M. Rito-Palomares, "Direct comparison between ion-exchange chromatography and aqueous two-phase processes for the partial purification of penicillin acylase produced by *E. coli*," *Journal of Chromatography B*, vol. 835, no. 1-2, pp. 77–83, 2006.

[14] K. Naganagouda and V. H. Mulimani, "Aqueous two-phase extraction (ATPE): an attractive and economically viable technology for downstream processing of *Aspergillus oryzae* α-galactosidase," *Process Biochemistry*, vol. 43, no. 11, pp. 1293–1299, 2008.

[15] V. Yazbik and M. Ansorge-Schumacher, "Fast and efficient purification of chloroperoxidase from *C. fumago*," *Process Biochemistry*, vol. 45, no. 2, pp. 279–283, 2010.

[16] P. A. Albertsson, *Partitioning of Cell Particles and Macromolecules*, John Wiley and Sons, New York, NY, USA, 3rd edition, 1987.

[17] R. Hatti-Kaul, *Aqueous Two Phase Systems: Methods and Protocols*, Humana Press, Totowa, NJ, USA, 2000.

[18] S. Saravanan, J. R. Rao, T. Murugesan, B. U. Nair, and T. Ramasami, "Recovery of value-added globular proteins from tannery wastewaters using PEG-salt aqueous two-phase systems," *Journal of Chemical Technology and Biotechnology*, vol. 81, no. 11, pp. 1814–1819, 2006.

[19] R. Iyyaswami, P. D. Belur, B. Girish, and V. H. Nagaraj, "Development and evaluation of PEG-lithium citrate salt based aqueous two phase system and its application in partitioning of proteins from fish industry Effluent," *Separation Science and Technology*, vol. 47, pp. 591–598, 2012.

[20] M. Perumalsamy and T. Murugesan, "Extraction of Cheese whey proteins (alpha-lactalbumin and beta-lactoglobulin) from dairy effluents using environmentally benign aqueous biphasic system," *International Journal of Chemical and Environmental Engineering*, vol. 3, pp. 55–59, 2012.

[21] D. Ramyadevi, A. Subathira, and S. Saravanan, "Central composite design application for optimization of aqueous two-phase extraction of protein from shrimp waste," *Journal of Chemical and Pharmaceutical Research*, vol. 4, pp. 2087–2095, 2012.

[22] S. Raja and V. R. Murty, "Development and evaluation of environmentally benign aqueous two phase systems for the recovery of proteins from tannery waste water," *ISRN Chemical Engineering*, vol. 2012, Article ID 290471, 9 pages, 2012.

[23] S. Saravanan, J. R. Rao, T. Murugesan, B. U. Nair, and T. Ramasami, "Recovery of value-added globular proteins from tannery wastewaters using PEG-salt aqueous two-phase systems," *Journal of Chemical Technology and Biotechnology*, vol. 81, no. 11, pp. 1814–1819, 2006.

[24] M. M. Bradford, "A rapid and sensitive method for the quantitation of microgram quantities of protein utilizing the principle

of protein dye binding," *Analytical Biochemistry*, vol. 72, no. 1-2, pp. 248–254, 1976.

[25] J. Lawson and J. Erjavec, *Modern Statistics for Engineering and Quality Improvement*, Thomson Publihsers, Australlia, 1st edition, 2002.

[26] D. C. Montgomery, *Design and Analysis of Experiments*, John Wiley and Sons, Hoboken, NJ, USA, 1st edition, 2005.

[27] T. S. Porto, G. M. Medeiros e Silva, C. S. Porto et al., "Liquid-liquid extraction of proteases from fermented broth by PEG/citrate aqueous two-phase system," *Chemical Engineering and Processing: Process Intensification*, vol. 47, no. 4, pp. 716–721, 2008.

[28] F. Luechau, T. C. Ling, and A. Lyddiatt, "Selective partition of plasmid DNA and RNA in aqueous two-phase systems by the addition of neutral salt," *Separation and Purification Technology*, vol. 68, no. 1, pp. 114–118, 2009.

[29] S. Saravanan, J. R. Rao, T. Murugesan, B. U. Nair, and T. Ramasami, "Partition of tannery wastewater proteins in aqueous two-phase poly (ethylene glycol)-magnesium sulfate systems: effects of molecular weights and pH," *Chemical Engineering Science*, vol. 62, no. 4, pp. 969–978, 2007.

[30] L. P. Malpiedi, G. Picó, and B. Nerli, "Features of partitioning pattern of two pancreatic enzymatic precursors: trypsinogen and chymotrypsinogen in polyethyleneglycol-sodium citrate aqueous biphasic systems," *Journal of Chromatography B*, vol. 870, no. 1, pp. 1–7, 2008.

[31] A. Boaglio, G. Bassani, G. Picó, and B. Nerli, "Features of the milk whey protein partitioning in polyethyleneglycol-sodium citrate aqueous two-phase systems with the goal of isolating human alpha-1 antitrypsin expressed in bovine milk," *Journal of Chromatography B*, vol. 837, no. 1-2, pp. 18–23, 2006.

[32] G. Tubío, L. Pellegrini, B. B. Nerli, and G. A. Picó, "Liquid-liquid equilibria of aqueous two-phase systems containing poly(ethylene glycols) of different molecular weight and sodium citrate," *Journal of Chemical and Engineering Data*, vol. 51, no. 1, pp. 209–212, 2006.

[33] M. Perumalsamy and T. Murugesan, "Partition behavior of bovine serum albumin in peg2000-sodium citrate-water based aqueous two-phase system," *Separation Science and Technology*, vol. 42, no. 9, pp. 2049–2065, 2007.

[34] C. A. S. da Silva, J. S. R. Coimbra, E. E. G. Rojas, L. A. Minim, and L. H. M. da Silva, "Partitioning of caseinomacropeptide in aqueous two-phase systems," *Journal of Chromatography B*, vol. 858, no. 1-2, pp. 205–210, 2007.

[35] D.C. Montomery and G.C. Runger, *Applied Statistics and Probability For Engineers*, John Wiley and Sons, New York, NY, USA, 5th edition, 2011.

[36] M. Perumalsamy and T. Murugesan, "Partition behavior of bovine serum albumin in peg2000-sodium citrate-water based aqueous two-phase system," *Separation Science and Technology*, vol. 42, no. 9, pp. 2049–2065, 2007.

[37] Y. M. Lu, Y. Z. Yang, X. D. Zhao, and C. B. Xia, "Bovine serum albumin partitioning in polyethylene glycol (PEG)/potassium citrate aqueous two-phase systems," *Food and Bioproducts Processing*, vol. 88, no. 1, pp. 40–46, 2010.

[38] G. Johansson, *Methods in Enzymology, Aqueous Two-Phase Systems*, vol. 228, Academic Press, San Diego, Calif, USA, 1994.

13

Dynamic Characters of Stiffened Composite Conoidal Shell Roofs with Cutouts: Design Aids and Selection Guidelines

Sarmila Sahoo

Department of Civil Engineering, Heritage Institute of Technology, Kolkata 700107, India

Correspondence should be addressed to Sarmila Sahoo; sarmila_ju@yahoo.com

Academic Editor: Jun Li

Dynamic characteristics of stiffened composite conoidal shells with cutout are analyzed in terms of the natural frequency and mode shapes. A finite element code is developed for the purpose by combining an eight-noded curved shell element with a three-noded curved beam element. The code is validated by solving benchmark problems available in the literature and comparing the results. The size of the cutouts and their positions with respect to the shell centre are varied for different edge constraints of cross-ply and angle-ply laminated composite conoids. The effects of these parametric variations on the fundamental frequencies and mode shapes are considered in details. The results furnished here may be readily used by practicing engineers dealing with stiffened composite conoids with cutouts central or eccentric.

1. Introduction

Laminated composite structures are gaining wide importance in various fields of aerospace and civil engineering. Shell roof structures can be conveniently built with composite materials that have many attributes, besides high specific strength and stiffness. Among the different shell panels which are commonly used as roofing units in civil engineering practice, the conoidal shell has a special position due to a number of advantages it offers. Conoidal shells are often used to cover large column-free areas. Being ruled surfaces, they provide ease of casting and also allow north light in. Hence, this shell is preferred in many places, particularly in medical, chemical, and food processing industries where entry of north light is desirable. Application of conoids in these industries often necessitates cutouts for the passage of light, service lines, and also sometimes for alteration of resonant frequency. In practice, the margin of the cutouts must be stiffened to take account of stress concentration effects. An in-depth study including bending, buckling, vibration, and impact is required to exploit the possibilities of these curved forms. The present investigation is, however, restricted only to the free vibration behaviour. A generalized formulation for the doubly curved laminated composite shell has been presented using the eight-noded curved quadratic isoparametric finite element including three radii of curvature. Some of the important contributions on the investigation of conoidal shells are briefly reviewed here.

The research on conoidal shell started about four decades ago. In 1964, Hadid [1] analysed static characteristics of conoidal shells using the variational method. The research was carried forward and improved by researchers like Brebbia and Hadid [2], Choi [3], Ghosh and Bandyopadhyay [4, 5], Dey et al. [6], and Das and Bandyopadhyay [7]. Dey et al. [6] provided a significant contribution on static analysis of conoidal shell. Chakravorty et al. [8] applied the finite element technique to explore the free vibration characteristics of shallow isotropic conoids and also observed the effects of excluding some of the inertia terms from the mass matrix on the first four natural frequencies. Chakravorty et al. [9–11] published a series of papers where they reported on free and forced vibration characteristics of graphite-epoxy composite conoidal shells with regular boundary conditions. Later, Nayak and Bandyopadhyay [12–15] reported free vibration of stiffened isotropic and composite conoidal shells. Das and Chakravorty [16, 17] considered bending and free vibration characteristics of unpunctured and unstiffened composite conoids. Hota and Chakravorty [18] studied isotropic punctured conoidal shells with complicated boundary conditions along the four edges, but no such study about composite conoidal shells is available in the literature. Also, they did not furnish any information on vibration mode shapes. It is also seen from the recent reviews [19, 20] that dynamic characteristics of stiffened conoidal shells with cutout are still missing in the literature. The present study thus focuses on the free vibrations of graphite-epoxy laminated composite stiffened conoids with cutout both in terms of the natural frequencies and mode shapes. The results so obtained may be readily used by practicing engineers dealing with stiffened composite conoids with cutouts. The novelty of the present study lies in the consideration of vibration mode shapes of stiffened composite conoids in presence of cutouts.

2. Mathematical Formulation

A laminated composite conoidal shell of uniform thickness h (Figure 1), and radius of curvature R_y, and radius of cross curvature R_{xy} is considered. Keeping the total thickness the same, the thickness may consist of any number of thin laminae each of which may be arbitrarily oriented at an angle θ with reference to the x-axis of the coordinate system. The constitutive equations for the shell are given by (a list of notations is separately given)

$$\{F\} = [E]\{\varepsilon\}, \tag{1}$$

where

$$\{F\} = \left\{N_x, N_y, N_{xy}, M_x, M_y, M_{xy}, Q_x, Q_y\right\}^T,$$

$$[E] = \begin{bmatrix} [A] & [B] & [0] \\ [B] & [D] & [0] \\ [0] & [0] & [S] \end{bmatrix}, \tag{2}$$

$$\{\varepsilon\} = \left\{\varepsilon_x^0, \varepsilon_y^0, \gamma_{xy}^0, k_x, k_{xy}, \gamma_{xz}^0, \gamma_{yz}^0\right\}^T.$$

The force and moment resultants are expressed as

$$\left\{N_x, N_y, N_{xy}, M_x, M_y, M_{xy}, Q_x, Q_y\right\}^T$$

$$= \int_{-h/2}^{h/2} \left\{\sigma_x, \sigma_y, \tau_{xy}, \sigma_z \cdot z, \sigma_y \cdot z, \tau_{xy} \cdot z, \tau_{xz}, \tau_{yz}\right\}^T dz. \tag{3}$$

The submatrices $[A]$, $[B]$, $[D]$, and $[S]$ of the elasticity matrix $[E]$ are functions of Young's moduli, shear moduli, and Poisson's ratio of the laminates. They also depend on the angle which the individual lamina of a laminate makes with the global x-axis. The detailed expressions of the elements of the elasticity matrix are available in several references including Vasiliev et al. [21] and Qatu [22].

The strain-displacement relations on the basis of improved first-order approximation theory for thin shell (Dey et al. [6]) are established as

$$\left\{\varepsilon_x, \varepsilon_y, \gamma_{xy}, \gamma_{xz}, \gamma_{yz}\right\}^T$$

$$= \left\{\varepsilon_x^0, \varepsilon_y^0, \gamma_{xy}^0, \gamma_{xz}^0, \gamma_{yz}^0\right\}^T + z\left\{k_x, k_y, k_{xy}, k_{xz}, k_{yz}\right\}^T, \tag{4}$$

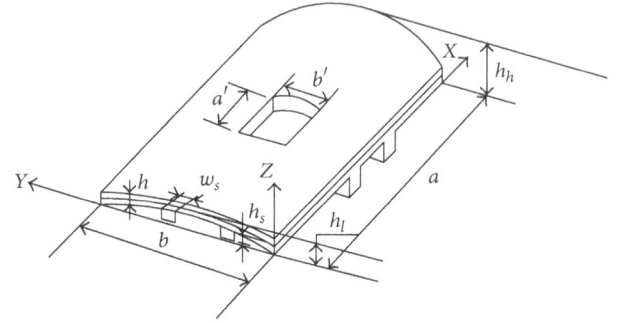

FIGURE 1: Conoidal shell with a concentric cutout stiffened along the margins.

where the first vector is the midsurface strain for a conoidal shell and the second vector is the curvature.

3. Finite Element Formulation

3.1. Finite Element Formulation for Shell. An eight-noded curved quadratic isoparametric finite element is used for conoidal shell analysis. The five degrees of freedom taken into consideration at each node are u, v, w, α, β. The following expressions establish the relations between the displacement at any point with respect to the coordinates ξ and η and the nodal degrees of freedom

$$u = \sum_{i=1}^{8} N_i u_i, \qquad v = \sum_{i=1}^{8} N_i v_i, \qquad w = \sum_{i=1}^{8} N_i w_i,$$

$$\alpha = \sum_{i=1}^{8} N_i \alpha_i, \qquad \beta = \sum_{i=1}^{8} N_i \beta_i, \tag{5}$$

where the shape functions derived from a cubic interpolation polynomial [6] are

$$N_i = \frac{(1 + \xi\xi_i)(1 + \eta\eta_i)(\xi\xi_i + \eta\eta_i - 1)}{4}, \quad \text{for } i = 1, 2, 3, 4,$$

$$N_i = \frac{(1 + \xi\xi_i)(1 - \eta^2)}{2}, \quad \text{for } i = 5, 7,$$

$$N_i = \frac{(1 + \eta\eta_i)(1 - \xi^2)}{2}, \quad \text{for } i = 6, 8. \tag{6}$$

The generalized displacement vector of an element is expressed in terms of the shape functions and nodal degrees of freedom as

$$[u] = [N]\{d_e\}, \tag{7}$$

that is,

$$\{u\} = \begin{Bmatrix} u \\ v \\ w \\ \alpha \\ \beta \end{Bmatrix} = \sum_{i=1}^{8} \begin{bmatrix} N_i & & & & \\ & N_i & & & \\ & & N_i & & \\ & & & N_i & \\ & & & & N_i \end{bmatrix} \begin{Bmatrix} u_i \\ v_i \\ w_i \\ \alpha_i \\ \beta_i \end{Bmatrix}. \tag{8}$$

3.1.1. Element Stiffness Matrix. The strain-displacement relation is given by

$$\{\varepsilon\} = [B]\{d_e\}, \tag{9}$$

where

$$[B] = \sum_{i=1}^{8} \begin{bmatrix} N_{i,x} & 0 & 0 & 0 & 0 \\ 0 & N_{i,y} & -\dfrac{N_i}{R_y} & 0 & 0 \\ N_{i,y} & N_{i,x} & -\dfrac{2N_i}{R_{xy}} & 0 & 0 \\ 0 & 0 & 0 & N_{i,x} & 0 \\ 0 & 0 & 0 & 0 & N_{i,y} \\ 0 & 0 & 0 & N_{i,y} & N_{i,x} \\ 0 & 0 & N_{i,x} & N_i & 0 \\ 0 & 0 & N_{i,y} & 0 & N_i \end{bmatrix}. \tag{10}$$

The element stiffness matrix is

$$[K_e] = \iint [B]^T [E] [B] \, dx \, dy. \tag{11}$$

3.1.2. Element Mass Matrix. The element mass matrix is obtained from the integral

$$[M_e] = \iint [N]^T [P] [N] \, dx \, dy, \tag{12}$$

where

$$[N] = \sum_{i=1}^{8} \begin{bmatrix} N_i & 0 & 0 & 0 & 0 \\ 0 & N_i & 0 & 0 & 0 \\ 0 & 0 & N_i & 0 & 0 \\ 0 & 0 & 0 & N_i & 0 \\ 0 & 0 & 0 & 0 & N_i \end{bmatrix},$$

$$[P] = \sum_{i=1}^{8} \begin{bmatrix} P & 0 & 0 & 0 & 0 \\ 0 & P & 0 & 0 & 0 \\ 0 & 0 & P & 0 & 0 \\ 0 & 0 & 0 & I & 0 \\ 0 & 0 & 0 & 0 & I \end{bmatrix}, \tag{13}$$

in which

$$P = \sum_{k=1}^{np} \int_{z_{k-1}}^{z_k} \rho \, dz, \qquad I = \sum_{k=1}^{np} \int_{z_{k-1}}^{z_k} z\rho \, dz. \tag{14}$$

3.2. Finite Element Formulation for Stiffener of the Shell. Three-noded curved isoparametric beam element (Figure 2) are used to model the stiffeners, which are taken to run only along the boundaries of the shell elements. In the stiffener element, each node has four degrees of freedom, that is, u_{sx}, w_{sx}, α_{sx}, and β_{sx} for X-stiffener and v_{sy}, w_{sy}, α_{sy}, and β_{sy} for Y-stiffener. The generalized force-displacement relation of stiffeners can be expressed as

X-stiffener: $\{F_{sx}\} = [D_{sx}]\{\varepsilon_{sx}\} = [D_{sx}][B_{sx}]\{\delta_{sxi}\}$,

Y-stiffener: $\{F_{sy}\} = [D_{sy}]\{\varepsilon_{sy}\} = [D_{sy}][B_{sy}]\{\delta_{syi}\}$, \hfill (15)

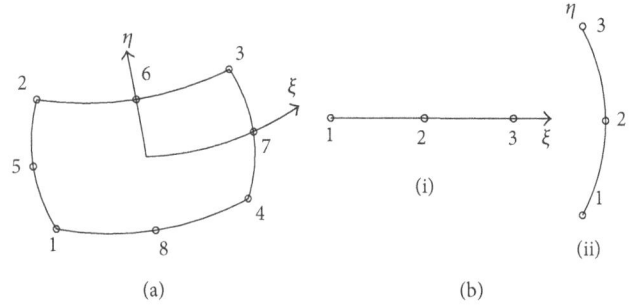

FIGURE 2: (a) Eight-noded shell element with isoparametric coordinates. (b) Three-noded stiffener element: (i) X-stiffener (ii) Y-stiffener.

where

$$\{F_{sx}\} = \begin{bmatrix} N_{sxx} & M_{sxx} & T_{sxx} & Q_{sxxz} \end{bmatrix}^T,$$

$$\{\varepsilon_{sx}\} = \begin{bmatrix} u_{sx\cdot x} & \alpha_{sx\cdot x} & \beta_{sx\cdot x} & (\alpha_{sx} + w_{sx\cdot x}) \end{bmatrix}^T,$$

$$\{F_{sy}\} = \begin{bmatrix} N_{syy} & M_{syy} & T_{syy} & Q_{syyz} \end{bmatrix}^T, \tag{16}$$

$$\{\varepsilon_{sy}\} = \begin{bmatrix} v_{sy\cdot y} & \beta_{sy\cdot y} & \alpha_{sy\cdot y} & (\beta_{sy} + w_{sy\cdot y}) \end{bmatrix}^T.$$

The generalized displacements of the Y-stiffener and the shell are related by the transformation matrix $\{\delta_{syi}\} = [T]\{\delta\}$, where

$$[T] = \begin{bmatrix} 1 + \dfrac{e}{R_y} & \text{symmetric} & & \\ 0 & 1 & & \\ 0 & 0 & 1 & \\ 0 & 0 & 0 & 1 \end{bmatrix}. \tag{17}$$

This transformation is required due to curvature of Y-stiffener, and $\{\delta\}$ is the appropriate portion of the displacement vector of the shell excluding the displacement component along the x-axis.

Elasticity matrices are as follows:

$$[D_{sx}] = \begin{bmatrix} A_{11}b_{sx} & B'_{11}b_{sx} & B'_{12}b_{sx} & 0 \\ B'_{11}b_{sx} & D'_{11}b_{sx} & D'_{12}b_{sx} & 0 \\ B'_{12}b_{sx} & D'_{12}b_{sx} & \dfrac{1}{6}(Q_{44}+Q_{66})d_{sx}b_{sx}^3 & 0 \\ 0 & 0 & 0 & b_{sx}S_{11} \end{bmatrix},$$

$$[D_{sy}] = \begin{bmatrix} A_{22}b_{sy} & B'_{22}b_{sy} & B'_{12}b_{sy} & 0 \\ B'_{22}b_{sy} & \dfrac{1}{6}(Q_{44}+Q_{66})b_{sy} & D'_{12}b_{sy} & 0 \\ B'_{12}b_{sy} & D'_{12}b_{sy} & D'_{11}d_{sy}b_{sy}^3 & 0 \\ 0 & 0 & 0 & b_{sy}S_{22} \end{bmatrix}, \tag{18}$$

where

$$D'_{ij} = D_{ij} + 2eB_{ij} + e^2 A_{ij},$$

$$B'_{ij} = B_{ij} + eA_{ij}, \tag{19}$$

and A_{ij}, B_{ij}, D_{ij}, and S_{ij} are explained in an earlier paper by Sahoo and Chakravorty [23].

Here, the shear correction factor is taken as 5/6. The sectional parameters are calculated with respect to the mid-surface of the shell by which the effect of eccentricities of stiffeners is automatically included. The element stiffness matrices are of the following forms:

$$\text{for } X\text{-stiffener: } \left[K_{xe}\right] = \int \left[B_{sx}\right]^T \left[D_{sx}\right] \left[B_{sx}\right] dx,$$

$$\text{for } Y\text{-stiffener: } \left[K_{ye}\right] = \int \left[B_{sy}\right]^T \left[D_{sy}\right] \left[B_{sy}\right] dy. \tag{20}$$

The integrals are converted to isoparametric coordinates and are carried out by 2-point Gauss quadrature. Finally, the element stiffness matrix of the stiffened shell is obtained by appropriate matching of the nodes of the stiffener and shell elements through the connectivity matrix and is given as

$$\left[K_e\right] = \left[K_{\text{she}}\right] + \left[K_{xe}\right] + \left[K_{ye}\right]. \tag{21}$$

The element stiffness matrices are assembled to get the global matrices.

3.2.1. Element Mass Matrix.

The element mass matrix for shell is obtained from the integral

$$\left[M_e\right] = \iint [N]^T [P] [N] \, dx \, dy, \tag{22}$$

where

$$[N] = \sum_{i=1}^{8} \begin{bmatrix} N_i & 0 & 0 & 0 & 0 \\ 0 & N_i & 0 & 0 & 0 \\ 0 & 0 & N_i & 0 & 0 \\ 0 & 0 & 0 & N_i & 0 \\ 0 & 0 & 0 & 0 & N_i \end{bmatrix}, \tag{23}$$

$$[P] = \sum_{i=1}^{8} \begin{bmatrix} P & 0 & 0 & 0 & 0 \\ 0 & P & 0 & 0 & 0 \\ 0 & 0 & P & 0 & 0 \\ 0 & 0 & 0 & I & 0 \\ 0 & 0 & 0 & 0 & I \end{bmatrix},$$

in which

$$P = \sum_{k=1}^{np} \int_{z_{k-1}}^{z_k} \rho \, dz, \qquad I = \sum_{k=1}^{np} \int_{z_{k-1}}^{z_k} z\rho \, dz. \tag{24}$$

Element mass matrix for stiffener element

$$\left[M_{sx}\right] = \iint [N]^T [P] [N] \, dx \quad \text{for } X\text{-stiffener,}$$

$$\left[M_{sy}\right] = \iint [N]^T [P] [N] \, dy \quad \text{for } Y\text{-stiffener.} \tag{25}$$

Here, $[N]$ is a 3×3 diagonal matrix.

Consider

$$[P] = \sum_{i=1}^{3} \begin{bmatrix} \rho \cdot b_{sx} d_{sx} & 0 & 0 & 0 \\ 0 & \rho \cdot b_{sx} d_{sx} & 0 & 0 \\ 0 & 0 & \dfrac{\rho \cdot b_{sx} d_{sx}^2}{12} & 0 \\ 0 & 0 & 0 & \dfrac{\rho \left(b_{sx} \cdot d_{sx}^3 + b_{sx}^3 \cdot d_{sx}\right)}{12} \end{bmatrix}$$

for X-stiffener,

$$[P] = \sum_{i=1}^{3} \begin{bmatrix} \rho \cdot b_{sy} d_{sy} & 0 & 0 & 0 \\ 0 & \rho \cdot b_{sy} d_{sy} & 0 & 0 \\ 0 & 0 & \dfrac{\rho \cdot b_{sy} d_{sy}^2}{12} & 0 \\ 0 & 0 & 0 & \dfrac{\rho \left(b_{sy} \cdot d_{sy}^3 + b_{sy}^3 \cdot d_{sy}\right)}{12} \end{bmatrix}$$

for Y-stiffener. (26)

The mass matrix of the stiffened shell element is the sum of the matrices of the shell and the stiffeners matched at the appropriate nodes

$$\left[M_e\right] = \left[M_{\text{she}}\right] + \left[M_{xe}\right] + \left[M_{ye}\right]. \tag{27}$$

The element mass matrices are assembled to get the global matrices.

3.3. Modeling the Cutout.

The code developed can take the position and size of cutout as input. The program is capable of generating nonuniform finite element mesh all over the shell surface. So, the element size is gradually decreased near the cutout margins. One such typical mesh arrangement is shown in Figure 3. Such finite element mesh is redefined in steps, and a particular grid is chosen to obtain the fundamental frequency when the result does not improve by more than one percent on further refining. Convergence of results is ensured in all the problems taken up here.

3.4. Solution Procedure for Free Vibration Analysis.

The free vibration analysis involves determination of natural frequencies from the condition

$$\left|[K] - \omega^2 [M]\right| = 0. \tag{28}$$

This is a generalized eigen value problem and is solved by the subspace iteration algorithm.

4. Numerical Examples

The validity of the present approach is checked through solution of benchmark problems. The first problem, free vibration of stiffened clamped conoid, was solved earlier by Nayak and Bandyopadhyay [13]. The second is the free vibration of composite conoid with cutouts solved by Chakravorty et al. [11]. The results obtained by the present method, along with the published results, are presented in Tables 1 and 2, respectively.

Additional problems for conoids with cutouts are solved, varying the size and position of cutout along both of the plan

TABLE 1: Fundamental frequencies (rad/sec) of clamped conoidal shell with central stiffeners.

| Stiffener position | Stiffener along x-direction | | | | Stiffener along y-direction | | | | Stiffener along both x- and y-directions | | | |
| | Nayak and Bandyopadhyay [13] | Present model | | | Nayak and Bandyopadhyay [13] | Present model | | | Nayak and Bandyopadhyay [13] | Present model | | |
		8×8	10×10	12×12		8×8	10×10	12×12		8×8	10×10	12×12
Concentric	17.28	17.50	17.37	17.31	20.83	21.15	20.85	20.76	20.90	21.24	21.01	20.84
Eccentric at top	17.83	17.92	17.78	17.73	21.57	21.76	21.53	21.45	22.39	22.86	22.51	22.32
Eccentric at bottom	17.55	17.80	17.61	17.52	22.31	22.66	22.40	22.24	22.92	23.25	23.00	22.87

$a = 50\,\text{m}$, $b = 50\,\text{m}$, $h = 0.2\,\text{m}$, $h_h = 10\,\text{m}$, $h_l = 2.5\,\text{m}$, $E = 25.4910 \times 10^9$, $\nu = 0.15$, $\rho = 2500\,\text{kg/m}^3$, $w_s = 0.3\,\text{m}$, and $h_s = 1\,\text{m}$.

TABLE 2: Nondimensional fundamental frequencies ($\overline{\omega}$) for laminated composite conoidal shell with cutout.

| a'/a | Corner point supported | | Simply supported | | Clamped | |
	Chakravorty et al. [11]	Present model	Chakravorty et al. [11]	Present model	Chakravorty et al. [11]	Present model
0.0	23.863	23.494	75.450	74.892	124.736	123.306
0.1	23.554	23.872	75.098	75.278	123.811	123.987
0.2	23.746	23.485	73.668	73.324	122.074	120.588
0.3	23.510	23.768	69.979	69.763	120.515	119.101
0.4	23.205	23.101	61.824	61.524	116.924	115.924

$a/b = 1$, $a/h = 100$, $a'/b' = 1$, $a/h_h = 2.5$, and $h_l/h_h = 0.25$.

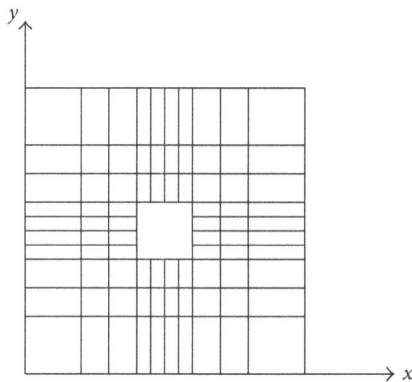

FIGURE 3: Typical 10×10 nonuniform mesh arrangements drawn to scale.

directions of the shell for different practical boundary conditions. In order to study the effect of cutout size on the free vibration response, results for unpunctured conoids are also included in the study.

5. Results and Discussion

It is found from Table 1 that the fundamental frequencies of stiffened conoids obtained by the present method agree well with those reported by Nayak and Bandyopadhyay [13]. Here, monotonic convergence is noted as the mesh is made progressively finer. Thus, the correctness of the stiffened shell element used here is established. It is evident from Table 2 that the present results agree with those of Chakravorty et al. [11], and the fact that the cutouts are properly modeled in the present formulation is thus established.

5.1. Free Vibration Behaviour of Shells with Concentric Cutouts. Tables 3 and 4 show the results for the non-dimensional fundamental frequency $\overline{\omega}$ of composite cross-ply and angle-ply stiffened conoidal shells for different cutout sizes and various combinations of boundary conditions along the four edges. The shells considered are of square planform ($a = b$), and the cutouts are also taken to be square in plan ($a' = b'$). The cutout sizes (i.e., a'/a) are varied from 0 to 0.4, and boundary conditions are varied along the four edges. Cutouts are concentric on shell surface. The stiffeners are placed along the cutout periphery and extended up to the edge of the shell. The boundary conditions are designated by describing the support clamped or simply supported as C or S taken in an anticlockwise order from the edge $x = 0$. This means that a shell with CSCS boundary is clamped along $x = 0$, simply supported along $y = 0$, clamped along $x = a$, and simply supported along $y = b$. The material and geometric properties of shells and cutouts are mentioned along with the figures.

5.1.1. Effect of Cutout Size on Fundamental Frequency. From Tables 3 and 4, it is seen that when a cutout is introduced to a stiffened shell, the fundamental frequencies increase. This increasing trend continues up to $a'/a = 0.4$ for both cross- and angle ply shells except some angle ply shells with $a'/a > 0.2$. The initial increase in frequency may be explained by the fact that when a cutout is introduced to an unpunctured surface, the number of stiffeners increases from two to four in the present study. When the cutout size is further increased, the number and dimensions of the stiffeners do not change, but the shell surface undergoes loss of both mass and stiffness. As the cutout grows in size, the loss of mass is more significant than that of stiffness, and hence

Boundary condition ↓	$\frac{a'}{a} \rightarrow$	0	0.1	0.2	0.3	0.4
CCCC						
CSCC						
CCSC						
CCCS						
CSSC						
CCSS						
CSCS						
SCSC						
CSSS						
SSSC						
SSCS						
SSSS						
Point supported						

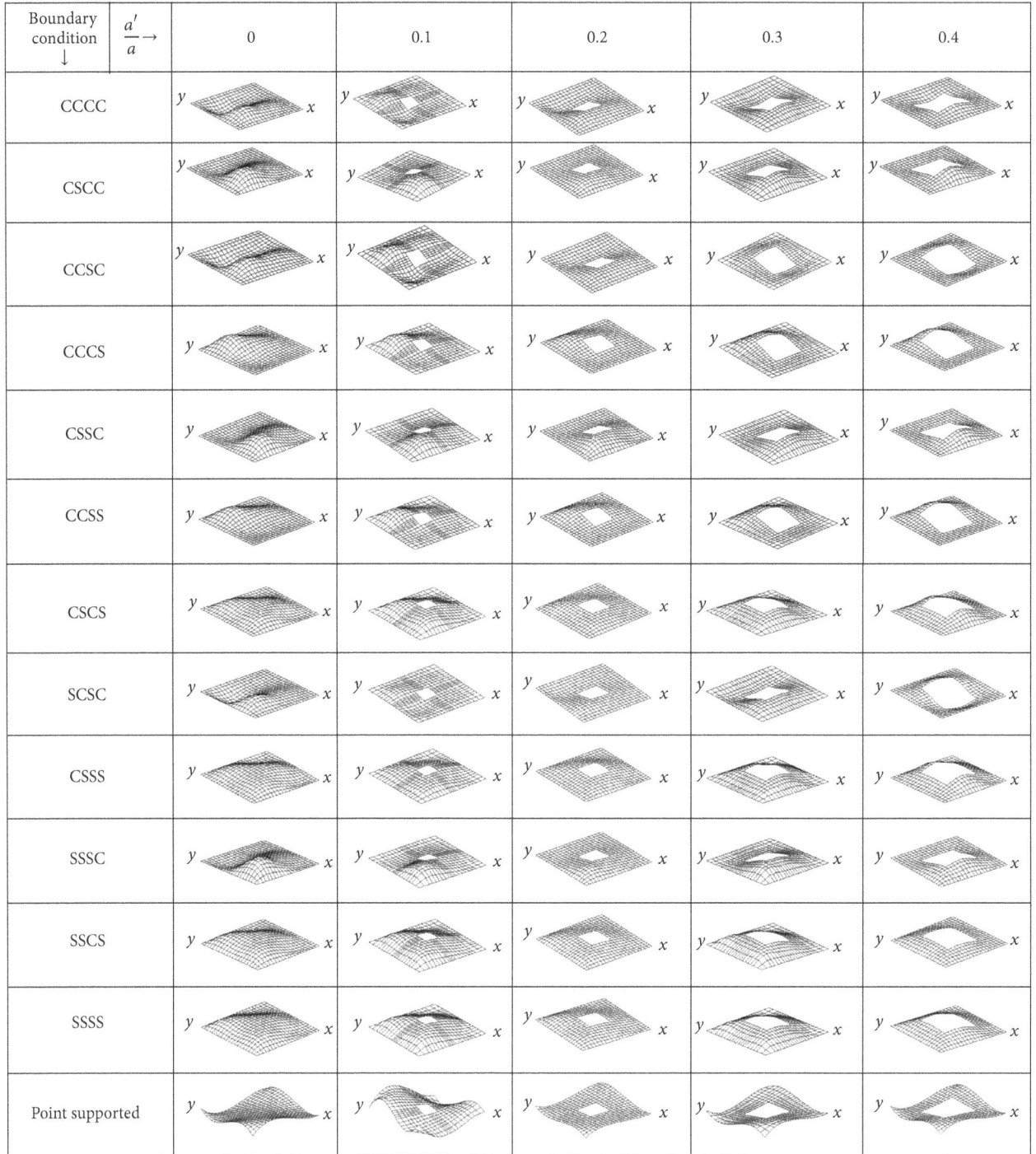

FIGURE 4: First mode shapes of laminated composite (0/90/0/90) stiffened conoidal shell for different sizes of the central square cutout and boundary conditions.

the frequency increases. But for some angle ply shells with further increase in the size of the cutout, the loss of stiffness gradually becomes more important than that of mass, resulting in decrease in fundamental frequency. This leads to the engineering conclusion that cutouts with stiffened margins may always safely be provided on shell surfaces for functional requirements.

5.1.2. Effect of Boundary Conditions on Fundamental Frequency. The boundary conditions may be divided into six groups, considering number of boundary constraints. The combinations in a particular group have equal number of boundary reactions. These groups are

Group I: CCCC shells,

Group II: CSCC, CCSC, and SCCC shells,

Boundary condition ↓ / $\frac{a'}{a} \rightarrow$	0	0.1	0.2	0.3	0.4
CCCC					
CSCC					
CCSC					
CCCS					
CSSC					
CCSS					
CSCS					
SCSC					
CSSS					
SSSC					
SSCS					
SSSS					
Point supported					

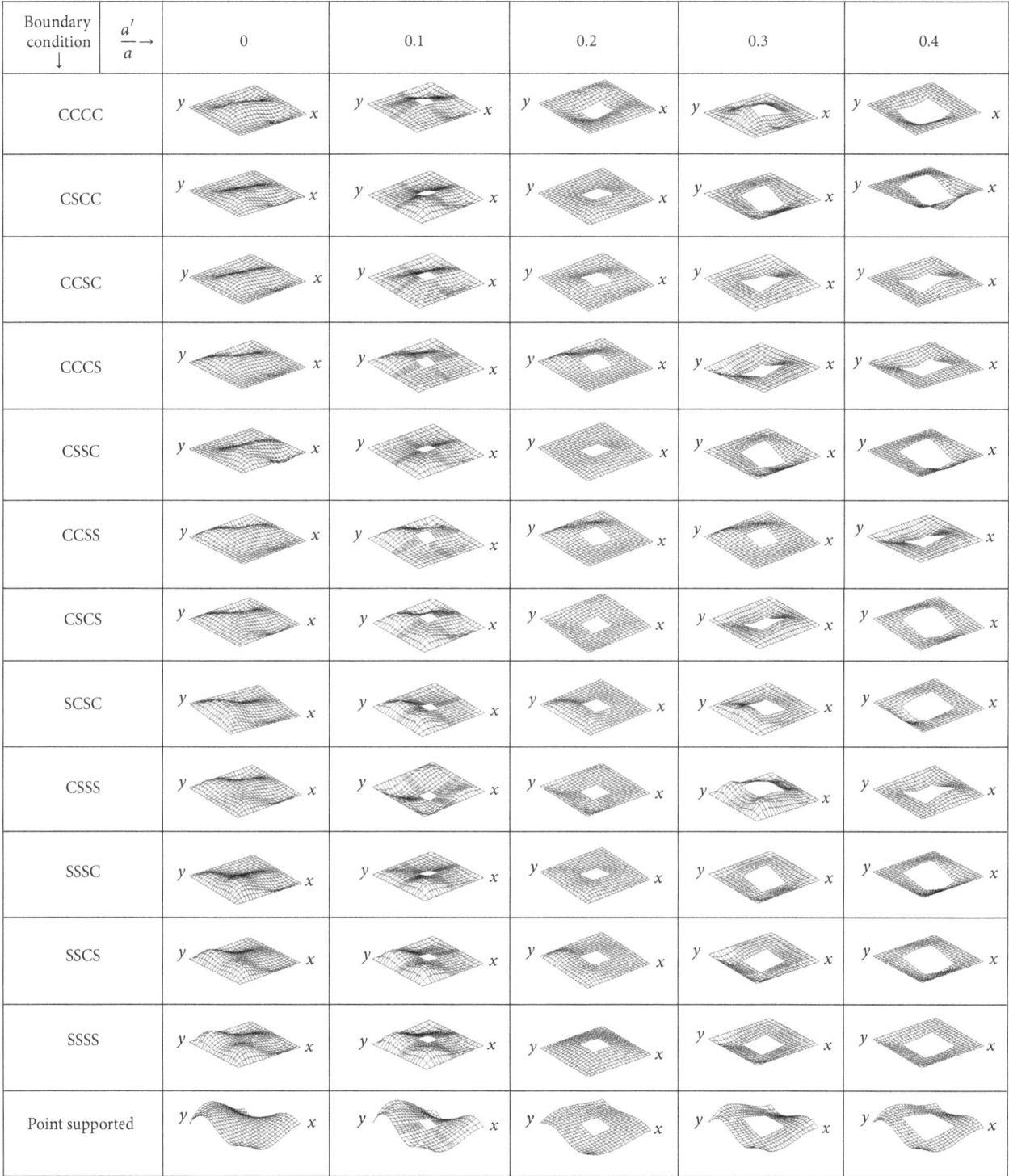

FIGURE 5: First mode shapes of laminated composite ($+45/-45/+45/-45$) stiffened conoidal shell for different sizes of the central square cutout and boundary conditions.

Group III: CSSC, SSCC, CSCS, and SCSC shells,

Group IV: CSSS, SSSC, and SSCS shells,

Group V: SSSS shells,

Group VI: Corner point supported shell.

It is seen from Tables 3 and 4 that fundamental frequencies of members belonging to the same groups of boundary combinations may not have close values. So, the different boundary conditions may be regrouped according to performance. According to the values of $\bar{\omega}$, the following groups may be identified.

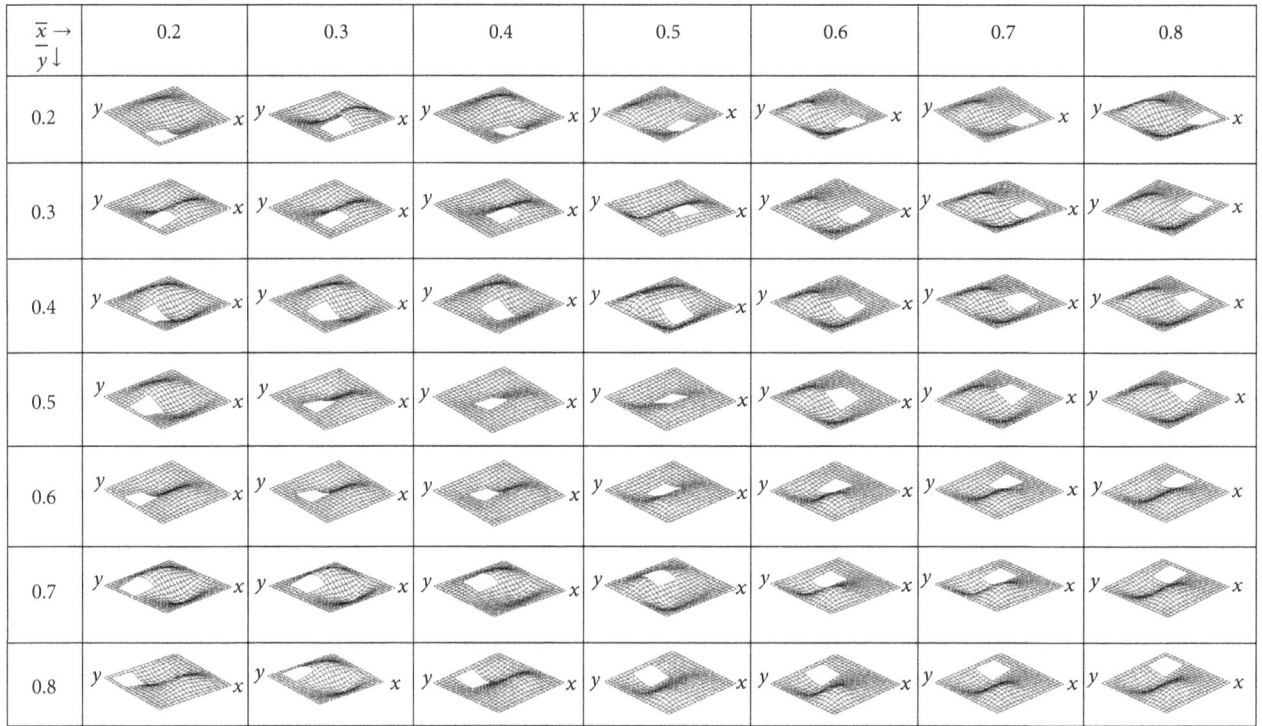

FIGURE 6: First mode shapes of laminated composite (0/90/0/90) stiffened conoidal shell for different positions of the square cutout with CCCC boundary condition.

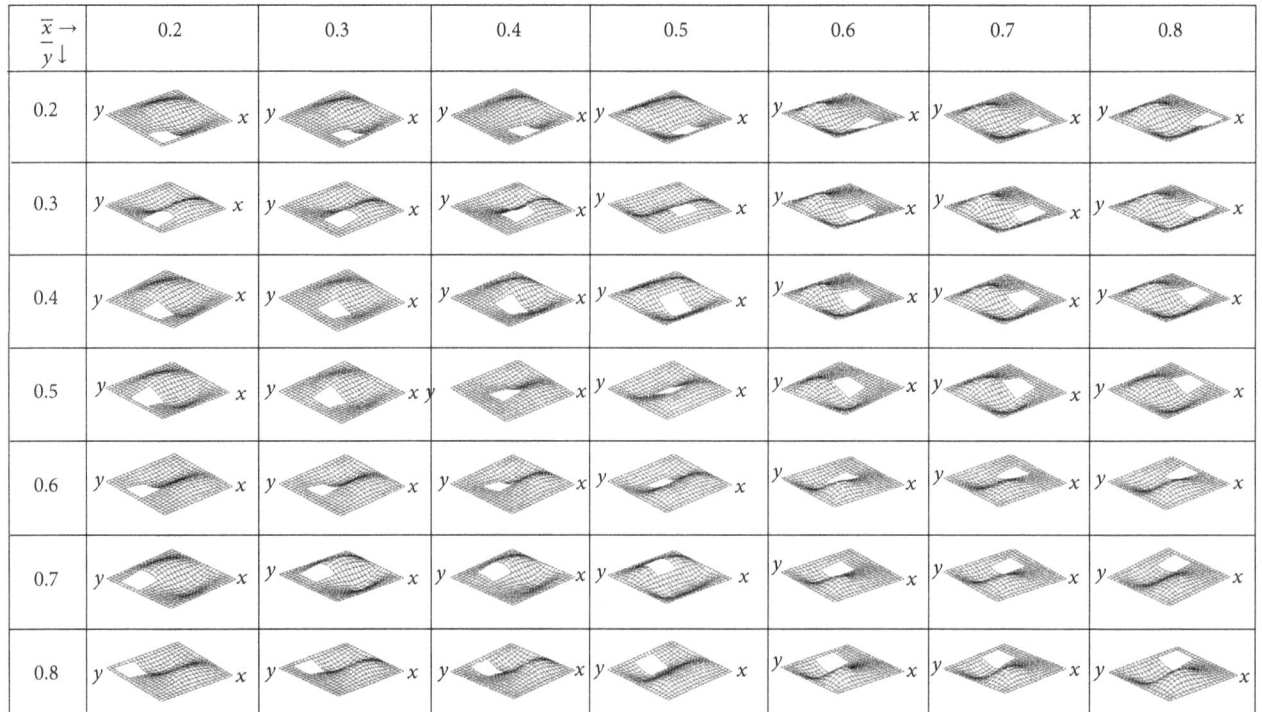

FIGURE 7: First mode shapes of laminated composite (0/90/0/90) stiffened conoidal shell for different positions of the square cutout with CCSC boundary condition.

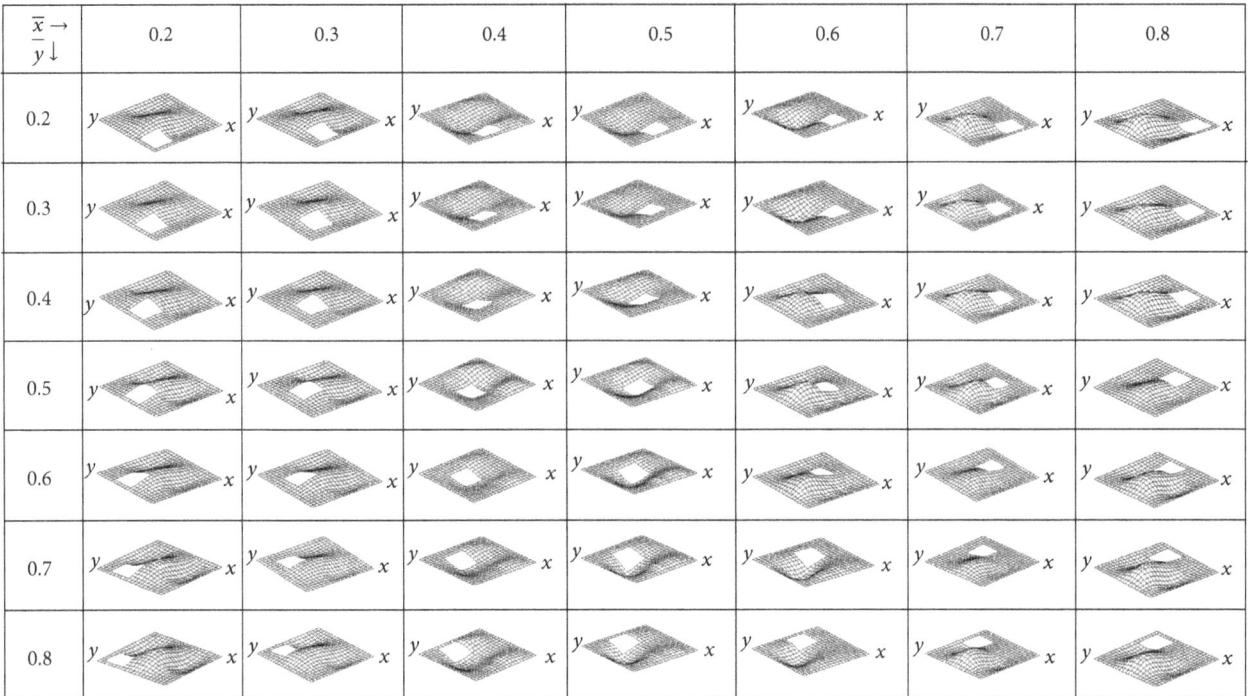

FIGURE 8: First mode shapes of laminated composite (+45/−45/+45/−45) stiffened conoidal shell for different positions of the square cutout with CCCC boundary condition.

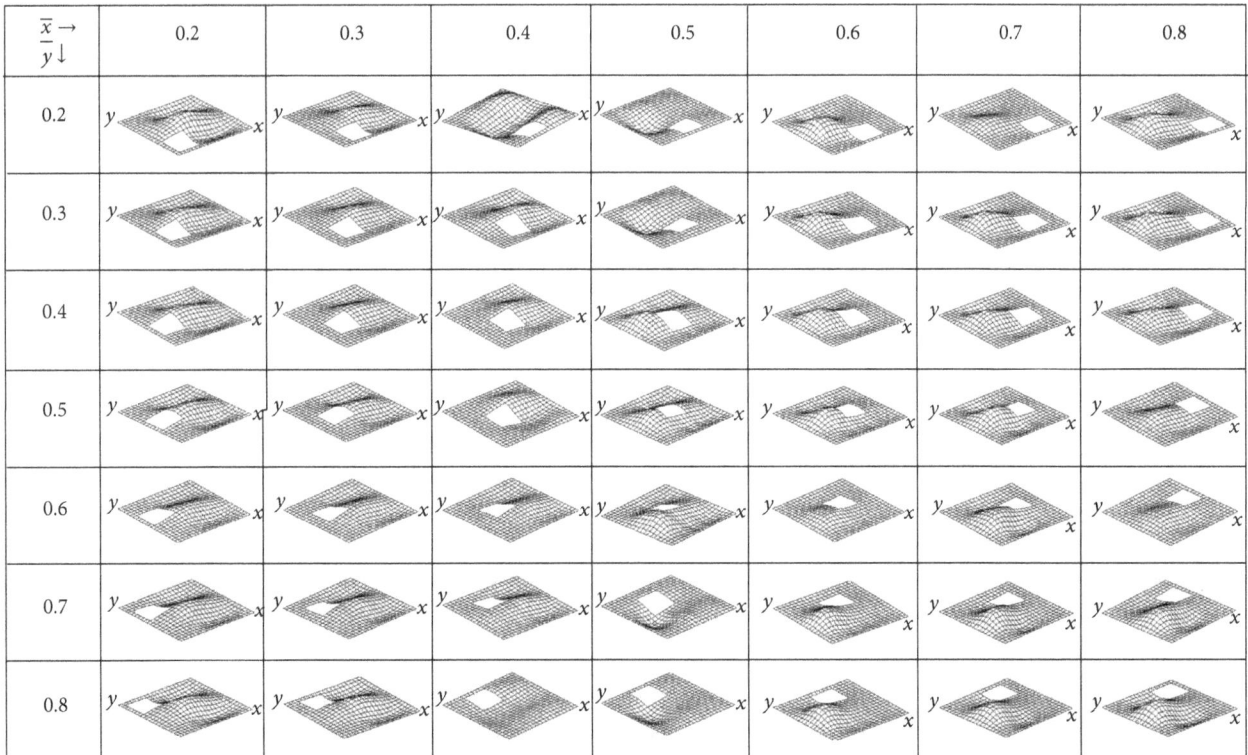

FIGURE 9: First mode shapes of laminated composite (+45/−45/+45/−45) stiffened conoidal shell for different positions of the square cutout with CCSC boundary condition.

TABLE 3: Non-dimensional fundamental frequencies ($\bar{\omega}$) for laminated composite (0/90/0/90) stiffened conoidal shell for different sizes of the central square cutout and different boundary conditions.

Boundary conditions	Cutout size (a'/a)				
	0	0.1	0.2	0.3	0.4
CCCC	105.7679	118.9136	124.386	127.4786	124.4929
CSCC	79.3544	87.8992	91.4718	96.4778	97.0499
CCSC	104.0259	117.3182	120.8486	122.5918	119.586
CCCS	79.0009	86.457	91.1477	95.8592	97.0069
CSSC	76.7919	84.417	87.445	91.1546	90.359
CCSS	76.4645	83.126	87.1684	90.6361	90.3194
CSCS	70.8733	76.0322	80.215	86.5974	91.9972
SCSC	96.4607	106.1886	112.2644	116.5409	114.8718
CSSS	65.7562	69.5302	72.8659	77.5964	81.1403
SSSC	65.7398	70.72	75.0802	79.9808	82.2204
SSCS	62.3308	65.2487	69.3811	74.7187	80.4246
SSSS	54.7734	56.6793	59.276	62.3683	65.2097
Point supported	21.1813	21.686	22.6166	24.2309	25.5361

$a/b = 1, a/h = 100, a'/b' = 1, a/h_h = 5, h_l/h_h = 0.25, E_{11}/E_{22} = 25, G_{23} = 0.2E_{22}, G_{13} = G_{12} = 0.5E_{22}$, and $\nu_{12} = \nu_{21} = 0.25$.

TABLE 4: Non-dimensional fundamental frequencies ($\bar{\omega}$) for laminated composite (+45/−45/+45/−45) stiffened conoidal shell for different sizes of the central square cutout and different boundary conditions.

Boundary conditions	Cutout size (a'/a)				
	0	0.1	0.2	0.3	0.4
CCCC	137.5363	149.748	156.932	154.6134	147.0345
CSCC	127.6366	139.9845	142.6213	142.2069	133.7437
CCSC	133.1847	157.494	165.9908	157.6214	149.8912
CCCS	123.9139	136.05	139.2721	140.9553	134.9193
CSSC	119.3132	129.2799	129.5345	119.8123	107.4374
CCSS	114.1784	124.2418	126.9562	119.4147	107.3405
CSCS	119.1939	129.2601	133.1178	136.9644	128.4247
SCSC	117.6296	129.2476	135.318	141.1146	143.7385
CSSS	102.7485	106.9623	109.5824	110.3696	103.1902
SSSC	105.2055	111.0251	113.482	110.7912	101.8345
SSCS	103.0713	107.6771	111.8604	118.8297	123.5415
SSSS	89.8159	91.6626	94.3912	89.284	83.3021
Point supported	26.4022	26.8267	27.3666	28.542	30.1836

$a/b = 1, a/h = 100, a'/b' = 1, a/h_h = 5, h_l/h_h = 0.25, E_{11}/E_{22} = 25, G_{23} = 0.2E_{22}, G_{13} = G_{12} = 0.5E_{22}$, and $\nu_{12} = \nu_{21} = 0.25$.

For cross ply shells

> Group 1: Contains CCCC, CCSC, and SCSC boundaries which exhibit relatively high frequencies.

> Group 2: Contains CSCC, CCCS, CSSC, CCSS, CSCS, SSSC, CSSS, and SSCS which exhibit intermediate values of frequencies.

Group 3: Contains SSSS and corner point supported boundaries which exhibit relatively low values of frequencies.

Similarly for angle ply shells:

> Group 1: Contains CCCC, CCSC, CSCC, CCCS, CSSC, CCSS, SCSC, and CSCS boundaries which exhibit relatively high frequencies.

> Group 2: Contains SSSC, CSSS, SSCS, and SSSS boundaries which exhibit intermediate values of frequencies.

> Group 3: Contains corner point supported shells which exhibit relatively low values of frequencies.

It is evident from the present study that the free vibration characteristics mostly depend on the arrangement of boundary constraints rather than their actual number. It can be seen from the present study that if the higher parabolic edge along $x = a$ is released from clamped to simply supported, there is hardly any change of frequency for cross ply shell. But for angle ply shells, if the edge along higher parabolic edge is released, fundamental frequency even increases more than that of a clamped shell. For cross ply shells, if the edge along $y = 0$ or $y = b$ is released, that is, along the straight edges, frequency values undergo marked decrease. The results indicate that the edge along $y = 0$ or $y = b$ should preferably be clamped in order to achieve higher frequency values, and if the edge has to be released for functional reason, the edge along $x = 0$ and $x = a$ of a conoid must be clamped to make up for the loss of frequency. But for angle ply shells, if any two edges are released, the change in fundamental frequency is not so significant.

Tables 5 and 6 show the efficiency of a particular clamping option in improving the fundamental frequency of a shell with minimum number of boundary constraints relative to that of a clamped shell. Marks are assigned to each boundary combination in a scale assigning a value of 0 to the frequency of a corner point supported shell and 100 to that of a fully clamped shell. These marks are furnished for cutouts with $a'/a = 0.2$ These tables will enable a practicing engineer to realize at a glance the efficiency of a particular boundary condition in improving the frequency of a shell, taking that of clamped shell as the upper limit.

5.1.3. Mode Shapes. The mode shapes corresponding to the fundamental modes of vibration are plotted in Figures 4 and 5 for cross-ply and angle ply shells, respectively. The normalized displacements are drawn with the shell midsurface as the reference for all the support conditions and for all the laminations used here. The fundamental mode is clearly a bending mode for all the boundary conditions for cross-ply and angle-ply shells, except for corner point supported shell. For corner point supported shells, the fundamental mode shapes are complicated. With the introduction of cutout, mode shapes remain almost similar. When the size of the cutout is increased from 0.2 to 0.4, the fundamental modes of vibration do not change to an appreciable amount.

TABLE 5: Clamping options for 0/90/0/90 conoidal shells with central cutouts having a'/a ratio 0.2.

Number of sides to be clamped	Clamped edges	Improvement of frequencies with respect to point supported shells	Marks indicating the efficiencies of number of restraints
0	Corner point supported	—	0
0	Simply supported no edges clamped (SSSS)	Good improvement	35
1	(a) Higher parabolic edge along $x = a$ (SSCS)	Marked improvement	46
	(b) Lower parabolic edge along $x = 0$ (CSSS)	Marked improvement	49
	(c) One straight edge along $y = b$ (SSSC)	Marked improvement	52
2	(a) Two alternate edges including the higher and lower parabolic edges $x = 0$ and $x = a$ (CSCS)	Marked improvement	57
	(b) 2 straight edges along $y = 0$ and $y = b$ (SCSC)	Remarkable improvement	88
	(c) Any two edges except for the above option (CSSC, CCSS)	Marked improvement	63
3	3 edges including the two parabolic edges (CSCC, CCCS)	Marked improvement	45
	3 edges excluding the higher parabolic edge along $x = a$ (CCSC)	Remarkable improvement	96
4	All sides (CCCC)	Frequency attains the highest value	100

TABLE 6: Clamping options for +45/−45/+45/−45 conoidal shells with central cutouts having a'/a ratio 0.2.

Number of sides to be clamped	Clamped edges	Improvement of frequencies with respect to point supported shells	Marks indicating the efficiencies of number of restraints
0	Corner point supported	—	0
0	Simply supported no edges clamped (SSSS)	Marked improvement	52
1	(a) Higher parabolic edge along $x = a$ (SSCS)	Marked improvement	65
	(b) Lower parabolic edge along $x = 0$ (CSSS)	Marked improvement	64
	(c) One straight edge along $y = b$ (SSSC)	Marked improvement	66
2	(a) Two alternate edges including the higher and lower parabolic edges $x = 0$ and $x = a$ (CSCS)	Remarkable improvement	81
	(b) 2 straight edges along $y = 0$ and $y = b$ (SCSC)	Remarkable improvement	83
	(c) Any two edges except for the above option (CSSC, CCSS)	Remarkable improvement	77–79
3	3 edges including the two parabolic edges (CSCC, CCCS)	Remarkable improvement	86–89
	3 edges excluding the higher parabolic edge along $x = a$ (CCSC)	Frequency attains more than a fully clamped shell.	107
4	All sides (CCCC)	Remarkable improvement	100

5.2. Effect of Eccentricity of Cutout Position

5.2.1. Fundamental Frequency. The effect of eccentricity of cutout positions on fundamental frequencies is studied from the results obtained for different locations of a cutout with $a'/a = 0.2$. The non-dimensional coordinates of the cutout centre ($\overline{x} = x/a$, $\overline{y} = y/a$) were varied from 0.2 to 0.8 along each direction, so that the distance of a cutout margin from the shell boundary was not less than one tenth of the plan dimension of the shell. The margins of cutouts were stiffened with four stiffeners. The study was carried out for all the thirteen boundary conditions for both cross ply and angle ply

shells. The fundamental frequency of a shell with an eccentric cutout is expressed as a percentage of fundamental frequency of a shell with a concentric cutout. This percentage is denoted by r. In Tables 7 and 8, such results are furnished.

It can be seen that eccentricity of the cutout along the length of the shell towards the parabolic edges makes it more flexible. It is also seen that towards the lower parabolic edge, r value is greater than that of the higher parabolic edge. This means that if a designer has to provide an eccentric cutout along the length, he should preferably place it towards the lower height boundary. The exception is there in some cases of cross ply shells. For cross ply shells with three edges simply

TABLE 7: Values of "r" for 0/90/0/90 conoidal shells.

Edge condition	$\overline{y}$	$\overline{x}$						
		0.2	0.3	0.4	0.5	0.6	0.7	0.8
CCCC	0.2	82.951	90.621	99.503	101.707	93.088	85.111	79.437
	0.3	82.557	90.021	99.158	103.034	93.743	85.175	79.190
	0.4	81.946	89.117	97.773	101.280	92.554	84.271	78.405
	0.5	81.955	88.983	97.220	100.015	91.853	83.955	78.247
	0.6	81.946	89.115	97.771	101.280	92.555	84.271	78.405
	0.7	82.557	90.020	99.159	103.036	93.743	85.175	79.189
	0.8	82.887	90.511	99.407	101.694	93.070	85.088	79.424
CSCC	0.2	93.862	100.890	106.419	103.740	95.957	89.329	85.169
	0.3	96.374	104.672	110.505	108.656	101.815	94.918	89.726
	0.4	93.858	102.085	108.275	106.920	100.353	93.653	88.508
	0.5	90.589	97.867	102.329	100	93.876	88.419	84.453
	0.6	88.150	94.257	97.223	94.591	89.294	84.754	81.462
	0.7	87.105	92.080	94.299	92.0577	87.583	83.555	80.475
	0.8	87.065	91.375	93.317	91.458	87.383	83.527	80.519
CCSC	0.2	80.918	87.869	96.510	101.953	95.020	87.039	81.087
	0.3	80.426	87.123	95.768	102.726	95.727	87.269	81.107
	0.4	79.615	85.977	94.270	101.072	94.674	86.601	80.675
	0.5	79.448	85.705	93.791	100	93.987	86.365	80.627
	0.6	79.612	85.977	94.271	101.073	94.674	86.601	80.675
	0.7	80.426	87.122	95.770	102.726	95.727	87.268	81.106
	0.8	80.837	87.743	96.390	101.907	95.007	87.017	81.075
CCCS	0.2	87.175	91.751	93.546	91.679	87.294	83.462	80.475
	0.3	87.180	92.361	94.480	92.063	87.489	83.463	80.416
	0.4	88.309	94.562	97.427	94.587	89.193	84.660	81.409
	0.5	90.816	98.202	102.571	100	93.771	88.341	84.427
	0.6	94.105	102.438	108.564	106.999	100.349	93.673	88.565
	0.7	96.611	105.025	110.815	108.872	101.974	95.062	89.875
	0.8	93.762	101.038	106.545	103.958	96.168	89.383	85.123
CSSC	0.2	91.560	99.045	106.094	105.722	98.527	91.427	86.004
	0.3	91.466	99.489	107.457	108.698	103.151	96.316	90.410
	0.4	89.314	96.955	104.290	105.667	101.025	94.902	89.407
	0.5	88.004	95.094	100.516	100	95.159	90.036	85.737
	0.6	87.238	93.625	97.473	95.882	91.217	86.821	83.288
	0.7	87.033	92.615	95.568	94.032	89.939	85.975	82.704
	0.8	87.165	92.303	94.864	93.620	89.886	86.065	82.823
CCSS	0.2	87.339	92.674	95.098	93.607	89.769	85.968	82.749
	0.3	87.146	92.875	95.730	94.015	89.811	85.845	82.608
	0.4	87.407	93.903	97.657	95.859	91.084	86.689	83.196
	0.5	88.217	95.390	100.738	100	95.034	89.925	85.675
	0.6	89.541	97.258	104.563	105.762	101.012	94.895	89.435
	0.7	91.697	99.797	107.751	108.900	103.289	96.435	90.531
	0.8	91.033	98.862	106.134	105.856	98.729	91.447	85.903
CSCS	0.2	87.176	90.165	91.941	90.278	87.528	86.028	85.511
	0.3	90.236	93.667	95.133	93.168	90.295	88.544	87.299
	0.4	93.274	97.392	98.931	97.165	94.295	92.109	89.886
	0.5	95.713	99.837	101.259	100	97.767	96.416	93.011
	0.6	93.271	97.392	98.930	97.154	94.291	92.103	89.883
	0.7	90.232	93.666	95.133	96.907	90.294	88.544	87.296
	0.8	87.045	90.057	91.800	90.237	87.504	86.015	85.510

TABLE 7: Continued.

Edge condition	$\overline{y}$	$\overline{x}$						
		0.2	0.3	0.4	0.5	0.6	0.7	0.8
	0.2	86.114	93.707	101.983	100.984	91.907	84.864	79.915
	0.3	85.570	92.996	101.670	102.014	92.324	84.978	79.910
	0.4	84.341	91.527	100.008	100.904	91.661	84.571	79.679
SCSC	0.5	83.782	90.945	99.224	100	91.321	84.472	79.663
	0.6	84.342	91.527	100.007	100.904	91.662	84.571	79.680
	0.7	85.570	92.996	101.673	102.014	92.325	84.979	79.911
	0.8	86.038	93.588	101.881	100.967	91.891	84.845	79.907
	0.2	90.957	94.019	96.237	95.319	93.007	91.535	90.574
	0.3	93.329	96.650	98.463	97.318	95.163	93.907	92.929
	0.4	94.987	98.154	99.743	99.094	97.663	96.789	95.576
CSSS	0.5	96.157	98.431	100.035	100	99.376	99.252	98.900
	0.6	94.984	98.155	99.743	99.092	97.661	96.784	95.572
	0.7	93.325	96.649	98.463	97.317	95.162	93.907	92.926
	0.8	90.826	93.898	96.032	95.227	92.974	91.510	90.575
	0.2	97.801	107.461	112.236	106.669	97.842	91.027	86.339
	0.3	100.320	109.209	113.744	109.522	101.475	94.421	89.126
	0.4	99.538	107.571	110.641	105.969	98.634	92.339	87.472
SSSC	0.5	97.804	105.127	105.975	100	93.276	88.232	84.425
	0.6	96.806	103.135	102.653	96.372	90.208	85.948	82.902
	0.7	96.657	101.965	101.055	95.197	89.518	85.604	82.855
	0.8	96.784	101.605	100.554	94.955	89.461	85.654	82.981
	0.2	93.160	96.470	95.571	91.662	88.790	87.812	87.727
	0.3	96.515	99.494	98.334	94.512	91.801	90.728	90.027
	0.4	99.292	102.150	101.314	97.900	95.242	93.801	92.363
SSCS	0.5	100.559	103.324	102.905	100	97.629	96.284	94.946
	0.6	99.290	102.155	101.314	97.896	95.240	93.802	92.362
	0.7	96.514	99.481	98.332	94.512	91.80	90.726	90.024
	0.8	93.059	96.355	95.442	91.626	88.753	87.798	87.733
	0.2	91.510	96.363	98.295	97.493	95.587	93.691	91.976
	0.3	95.252	98.790	99.925	98.895	97.114	95.560	94.407
	0.4	97.747	100.113	100.764	99.692	98.064	96.681	95.608
SSSS	0.5	98.751	100.497	101.006	100	98.510	97.233	96.154
	0.6	97.742	100.114	100.764	99.690	98.063	96.682	95.608
	0.7	95.251	98.786	99.922	98.897	97.113	95.560	94.405
	0.8	91.451	96.286	98.154	97.412	95.567	93.666	91.951
	0.2	137.892	128.050	114.295	104.647	100.791	101.361	109.077
	0.3	135.205	125.573	111.227	101.093	97.006	97.834	106.312
	0.4	132.315	124.337	110.519	100.187	95.990	97.045	105.585
CS	0.5	131.075	123.720	110.365	100	95.813	96.912	105.499
	0.6	132.292	124.356	110.550	100.211	96.017	97.088	105.603
	0.7	135.228	125.582	111.219	101.090	96.984	97.817	106.303
	0.8	135.480	125.754	112.435	102.528	98.801	99.577	107.915

$a/b = 1$, $a/h = 100$, $a'/b' = 1$, $a/h_h = 5$, $h_l/h_h = 0.25$, $E_{11}/E_{22} = 25$, $G_{23} = 0.2E_{22}$, $G_{13} = G_{12} = 0.5E_{22}$, and $\nu_{12} = \nu_{21} = 0.25$.

supported when cutout shifts towards the lower parabolic edge (including the lower parabolic edge) the shell becomes stiffer. Again for corner point supported shells, r values increase towards the parabolic edges and are maximum along lower parabolic edge. For cross ply shells, four out of thirteen boundary conditions yield the maximum value of r along $\overline{x} = 0.5$, four yield maximum values of r along $\overline{x} = 0.4$, and others show maximum values within $\overline{x} = 0.4$ to 0.5.

It is observed from Table 7 that if the eccentricity of a cutout is varied along the width, the shell becomes stiffer when the cutout shifts towards clamped edges. So, for functional purposes, if a shift of central cutout is required, eccentricity of a cutout along the width should preferably be towards the clamped straight edge. For shells having two straight edges of identical boundary condition, the maximum fundamental frequency occurs along $\overline{y} = 0.5$. For corner

TABLE 8: Values of "r" for +45/−45/+45/−45 conoidal shells.

Edge condition	$\overline{y}$	$\overline{x}$						
		0.2	0.3	0.4	0.5	0.6	0.7	0.8
CCCC	0.2	72.823	78.236	82.758	83.828	81.568	76.803	71.923
	0.3	74.884	80.770	86.548	88.893	86.815	81.249	75.551
	0.4	76.701	83.522	91.226	95.703	93.788	87.111	80.040
	0.5	77.112	84.768	94.005	100	97.403	89.411	81.517
	0.6	76.805	83.652	91.325	95.680	93.763	87.160	80.108
	0.7	75.008	80.906	86.626	88.855	86.795	81.298	75.617
	0.8	72.647	77.908	82.266	83.759	81.579	76.633	71.755
CSCC	0.2	81.077	91.056	97.803	91.446	86.032	83.235	78.655
	0.3	83.525	93.130	102.170	99.590	94.922	89.192	82.211
	0.4	84.074	93.824	101.696	103.807	101.498	94.596	86.412
	0.5	83.003	90.427	97.733	100	99.474	96.299	87.547
	0.6	81.276	87.968	94.605	96.334	95.817	92.081	84.895
	0.7	79.646	86.287	92.866	94.104	92.136	86.701	80.220
	0.8	78.630	85.619	92.366	92.585	89.172	83.321	77.139
CCSC	0.2	70.077	77.727	85.937	83.800	79.229	74.361	69.326
	0.3	71.072	78.703	87.768	88.368	83.336	77.766	72.302
	0.4	71.984	80.017	89.895	95.290	89.759	82.905	76.264
	0.5	72.468	81.214	91.396	100	93.470	85.277	77.810
	0.6	71.937	79.893	89.673	95.311	89.871	83.044	76.355
	0.7	71.086	78.669	87.710	88.508	83.481	77.891	72.371
	0.8	70.006	77.569	85.788	83.973	79.340	74.231	69.123
CCCS	0.2	78.059	85.085	91.568	92.620	90.166	84.849	78.646
	0.3	78.792	85.576	91.983	93.758	92.489	87.571	81.169
	0.4	80.502	87.433	93.909	95.951	95.851	92.337	85.557
	0.5	82.841	90.481	97.657	100	99.876	96.861	88.986
	0.6	85.401	94.488	102.330	104.306	102.177	96.765	88.819
	0.7	85.313	94.803	103.136	100.952	96.520	91.365	84.617
	0.8	81.938	91.745	98.552	93.014	87.371	84.653	80.319
CSSC	0.2	80.251	90.347	102.001	98.147	92.559	90.103	85.813
	0.3	81.688	90.911	103.714	105.159	101.546	97.766	88.928
	0.4	81.340	89.735	99.850	103.865	102.503	98.783	91.054
	0.5	81.528	89.780	96.778	100	99.635	97.691	92.230
	0.6	81.914	89.289	95.251	97.877	98.401	97.867	92.317
	0.7	81.213	88.071	94.478	97.330	98.425	94.980	87.252
	0.8	80.276	87.186	94.253	97.479	97.826	90.750	83.637
CCSS	0.2	79.271	85.734	92.215	96.177	97.750	92.440	85.311
	0.3	80.024	86.471	92.663	96.147	97.781	95.350	88.242
	0.4	81.065	87.896	94.083	97.107	98.178	97.531	92.578
	0.5	82.087	90.360	97.023	100	100.117	98.606	94.251
	0.6	82.521	90.825	101.803	104.549	103.368	100.351	93.755
	0.7	82.941	91.821	103.719	105.727	102.486	99.509	91.399
	0.8	80.346	90.089	100.955	99.195	93.707	91.345	87.507
CSCS	0.2	78.163	86.022	92.995	92.450	88.854	85.808	80.401
	0.3	79.871	86.979	93.768	95.528	94.670	90.304	83.255
	0.4	81.380	88.475	95.483	97.758	98.510	95.335	87.720
	0.5	83.417	91.209	98.288	100	101.457	99.522	90.636
	0.6	83.285	90.593	97.462	99.337	99.875	96.908	89.038
	0.7	81.807	89.179	96.019	97.206	95.825	91.360	84.354
	0.8	79.311	87.551	94.710	93.330	88.907	85.864	80.620

TABLE 8: Continued.

Edge condition	$\bar{y}$	$\bar{x}$						
		0.2	0.3	0.4	0.5	0.6	0.7	0.8
SCSC	0.2	85.127	94.329	95.136	88.093	85.184	82.426	78.466
	0.3	86.016	95.544	99.993	92.933	89.052	85.588	80.885
	0.4	86.853	97.202	105.501	97.941	92.882	88.930	83.293
	0.5	87.280	98.396	107.888	100	94.353	90.351	84.256
	0.6	86.654	96.782	104.675	97.456	92.684	88.887	83.209
	0.7	85.821	95.213	99.433	92.519	89.003	85.631	80.778
	0.8	84.798	93.899	94.867	87.932	85.323	82.385	78.165
CSSS	0.2	84.944	92.468	99.487	102.083	102.496	102.211	95.697
	0.3	87.001	94.415	100.612	101.921	103.359	105.519	99.083
	0.4	87.482	95.487	101.492	100.522	100.612	102.250	100.822
	0.5	87.931	97.254	102.629	100	99.420	100.620	99.680
	0.6	88.552	96.975	102.732	101.723	101.674	102.885	100.813
	0.7	88.608	96.436	102.875	103.969	105.120	106.825	99.475
	0.8	85.793	93.995	101.843	104.209	103.551	102.657	95.378
SSSC	0.2	90.301	101.510	103.781	96.224	91.653	89.271	87.137
	0.3	91.100	102.460	108.098	102.693	98.673	95.766	90.177
	0.4	90.295	101.404	106.028	102.757	98.698	95.253	90.038
	0.5	89.523	99.110	102.720	100	96.670	94.017	90.034
	0.6	88.793	96.857	100.664	98.472	96.351	94.739	91.563
	0.7	88.172	95.669	99.702	97.818	96.797	96.041	90.722
	0.8	87.395	94.930	98.931	96.571	95.997	95.427	88.365
SSCS	0.2	88.795	97.759	99.936	94.981	92.168	90.694	87.596
	0.3	90.150	98.974	101.928	98.665	97.216	96.201	89.404
	0.4	91.309	100.310	103.409	99.753	97.841	97.300	92.291
	0.5	92.288	101.662	104.357	100	97.487	96.678	93.643
	0.6	92.516	101.539	104.514	100.363	98.033	97.305	92.917
	0.7	91.859	100.918	103.929	99.899	98.067	97.444	90.291
	0.8	89.927	99.404	101.683	95.907	92.707	91.701	87.789
SSSS	0.2	88.805	93.985	97.537	99.443	99.34104	97.458	93.660
	0.3	90.116	95.853	98.238	99.300	99.55631	98.869	96.461
	0.4	89.799	95.332	98.006	97.770	96.847	96.012	94.848
	0.5	89.665	94.925	97.577	100	95.462	94.360	93.375
	0.6	90.247	95.512	98.509	98.722	97.792	96.672	95.013
	0.7	90.943	96.728	99.725	101.290	101.508	100.518	97.141
	0.8	88.912	94.653	99.280	101.663	101.056	99.261	95.529
CS	0.2	124.611	120.401	110.994	104.153	102.644	103.581	108.893
	0.3	120.487	117.591	108.489	101.849	100.596	101.659	107.324
	0.4	117.053	116.177	107.584	100.426	99.256	100.691	105.789
	0.5	115.794	115.786	107.429	100	98.788	100.425	105.330
	0.6	118.003	116.844	107.793	100.623	99.239	100.341	105.584
	0.7	122.490	118.779	109.260	102.384	100.470	100.978	106.534
	0.8	122.076	118.091	109.558	102.712	101.196	101.869	107.386

$a/b = 1$, $a/h = 100$, $a'/b' = 1$, $a/h_h = 5$, $h_l/h_h = 0.25$, $E_{11}/E_{22} = 25$, $G_{23} = 0.2E_{22}$, $G_{13} = G_{12} = 0.5E_{22}$, and $\nu_{12} = \nu_{21} = 0.25$.

point supported shells, the maximum fundamental frequency always occurs along the boundary of the shell. All these are true for cross ply shells only. For an angle ply shell, such unified trend is not observed, and the boundary conditions and the fundamental frequency behave in a complex manner as evident from Table 8. But for corner point supported angle-ply shells also, the maximum values of r are along the boundary.

Tables 9 and 10 provide the maximum values of r together with the position of the cutout. These tables also show the rectangular zones within which r is always greater than or equal to 95 and 90. It is to be noted that at some other points,

TABLE 9: Maximum values of r with corresponding coordinates of cutout centre and zones where $r \geq 90$ and $r \geq 95$ for 0/90/0/90 conoidal shells.

Boundary condition	Maximum values of r	Co-ordinate of cutout centre	Area in which the value of $r \geq 90$	Area in which the value of $r \geq 95$
CCCC	103.036	$\overline{x} = 0.5, \overline{y} = 0.7$	$\overline{x} = 0.6,$ $0.2 \leq \overline{y} \leq 0.8$	$0.4 \leq \overline{x} 0.5,$ $0.2 \leq \overline{y} \leq 0.8$
CSCC	110.505	$\overline{x} = 0.4, \overline{y} = 0.3$	$0.3 \leq \overline{x} 0.5,$ $0.6 \leq \overline{y} \leq 0.8$	$0.3 \leq \overline{x} 0.6,$ $0.2 \leq \overline{y} \leq 0.5$
CCSC	102.726	$\overline{x} = 0.5, \overline{y} = 0.3$ $\overline{x} = 0.5, \overline{y} = 0.7$	$\overline{x} = 0.4, 0.6,$ $0.2 \leq \overline{y} \leq 0.8$	$\overline{x} = 0.5,$ $0.2 \leq \overline{y} \leq 0.8$
CCCS	110.815	$\overline{x} = 0.4, \overline{y} = 0.7$	$0.3 \leq \overline{x} 0.5,$ $0.2 \leq \overline{y} \leq 0.4$	$0.3 \leq \overline{x} 0.6,$ $0.5 \leq \overline{y} \leq 0.8$
CSSC	108.698	$\overline{x} = 0.5, \overline{y} = 0.3$	$0.3 \leq \overline{x} 0.5,$ $0.6 \leq \overline{y} \leq 0.8$	$0.3 \leq \overline{x} 0.6,$ $0.2 \leq \overline{y} \leq 0.5$
CCSS	108.900	$\overline{x} = 0.5, \overline{y} = 0.7$	$0.3 \leq \overline{x} 0.5,$ $0.2 \leq \overline{y} \leq 0.4$	$0.3 \leq \overline{x} 0.6,$ $0.5 \leq \overline{y} \leq 0.8$
CSCS	101.259	$\overline{x} = 0.4, \overline{y} = 0.5$	$0.3 \leq \overline{x} 0.5,$ $0.2 \leq \overline{y} \leq 0.3$ $0.7 \leq \overline{y} \leq 0.8$	$0.3 \leq \overline{x} 0.5,$ $0.4 \leq \overline{y} \leq 0.6$
SCSC	102.014	$\overline{x} = 0.5, \overline{y} = 0.7$	$\overline{x} = 0.3, 0.5,$ $0.2 \leq \overline{y} \leq 0.8$	$0.4 \leq \overline{x} 0.5,$ $0.2 \leq \overline{y} \leq 0.8$
CSSS	100.035	$\overline{x} = 0.4, \overline{y} = 0.5$	$0.2 \leq \overline{x} 0.8,$ $0.2 \leq \overline{y} \leq 0.8$	$0.3 \leq \overline{x} 0.6,$ $0.3 \leq \overline{y} \leq 0.7$
SSSC	113.744	$\overline{x} = 0.4, \overline{y} = 0.3$	$0.6 \leq \overline{x} 0.7,$ $0.2 \leq \overline{y} \leq 0.4$	$0.2 \leq \overline{x} 0.5,$ $0.2 \leq \overline{y} \leq 0.8$
SSCS	103.324	$\overline{x} = 0.3, \overline{y} = 0.5$	$0.5 \leq \overline{x} 0.8,$ $0.3 \leq \overline{y} \leq 0.7$	$0.2 \leq \overline{x} 0.4,$ $0.3 \leq \overline{y} \leq 0.7$
SSSS	101.006	$\overline{x} = 0.4, \overline{y} = 0.5$	$\overline{x} = 0.8,$ $0.2 \leq \overline{y} \leq 0.8$	$0.2 \leq \overline{x} 0.7,$ $0.3 \leq \overline{y} \leq 0.7$
CS	137.892	$\overline{x} = 0.2, \overline{y} = 0.2$	nil	$0.2 \leq \overline{x} 0.8,$ $0.2 \leq \overline{y} \leq 0.8$

$a/b = 1, a/h = 100, a'/b' = 1, a/h_h = 5, h_l/h_h = 0.25, E_{11}/E_{22} = 25, G_{23} = 0.2E_{22}, G_{13} = G_{12} = 0.5E_{22},$ and $\nu_{12} = \nu_{21} = 0.25$.

r values may have similar values, but only the zone rectangular in plan has been identified. This study identifies the specific zones within which the cutout centre may be moved so that the loss of frequency is less than 5% or 10%, respectively, with respect to a shell with a central cutout. This will help a practicing engineer to make a decision regarding the eccentricity of the cutout centre that can be allowed.

5.2.2. Mode Shapes. The mode shapes corresponding to the fundamental modes of vibration are plotted in Figures 6, 7, 8, and 9 for cross-ply and angle-ply shells of CCCC and CCSC shells for different eccentric positions of the cutout. As CCCC and CCSC shells are most efficient with respect to number of restraints, the mode shapes of these shells are shown as typical results. All the mode shapes are bending modes. It is found that for different position of cutout, mode shapes are somewhat similar, only the crest and trough positions change.

The present study considers the dynamic characteristics of stiffened composite conoidal shells with square cutout in terms of the natural frequency and mode shapes. The size of the cutouts and their positions with respect to the shell centre are varied for different edge constraints of cross-ply and angle-ply laminated composite conoids. The effects of these parametric variations on the fundamental frequencies and

mode shapes are considered in details. However, the effect of the shape and orientation of the cutout on the dynamic characters of the conoid has not been considered in the present study. Future studies will evaluate these aspects.

6. Conclusions

The following conclusions are drawn from the present study.

(1) As this approach produces results in close agreement with those of the benchmark problems, the finite element code used here is suitable for analyzing free vibration problems of stiffened conoidal roof panels with cutouts. The present study reveals that cutouts with stiffened margins may always safely be provided on shell surfaces for functional requirements.

(2) The arrangement of boundary constraints along the four edges is far more important than their actual number; so far the free vibration is concerned. The relative-free vibration performances of shells for different combinations of edge conditions along the four sides are expected to be very useful in decision making for practicing engineers.

TABLE 10: Maximum values of r with corresponding coordinates of cutout centre and zones where $r \geq 90$ and $r \geq 95$ for $+45/-45/+45/-45$ conoidal shells.

Boundary condition	Maximum values of r	Co-ordinate of cutout centre	Area in which the value of $r \geq 90$	Area in which the value of $r \geq 95$
CCCC	100.000	$\overline{x} = 0.5, \overline{y} = 0.5$	$\overline{x} = 0.4, 0.6$ $0.4 \leq \overline{y} \leq 0.6$	$\overline{x} = 0.5,$ $0.4 \leq \overline{y} \leq 0.6$
CSCC	103.807	$\overline{x} = 0.5, \overline{y} = 0.4$	$\overline{x} = 0.3, 0.2 \leq \overline{y} \leq 0.5;$ $0.6 \leq \overline{x} 0.7, 0.3 \leq \overline{y} \leq 0.6$	$0.4 \leq \overline{x} \leq 0.5,$ $0.3 \leq \overline{y} \leq 0.5$
CCSC	100.000	$\overline{x} = 0.5, \overline{y} = 0.5$	$0.4 \leq \overline{x} \leq 0.6,$ $\overline{y} = 0.5$	$\overline{x} = 0.5$ $0.4 \leq \overline{y} \leq 0.6$
CCCS	104.306	$\overline{x} = 0.5, \overline{y} = 0.6$	$0.4 \leq \overline{x} \leq 0.6,$ $0.2 \leq \overline{y} \leq 0.4$	$0.4 \leq \overline{x} \leq 0.7,$ $0.5 \leq \overline{y} \leq 0.6$
CSSC	105.159	$\overline{x} = 0.5, \overline{y} = 0.3$	$0.4 \leq \overline{x} \leq 0.7,$ $0.7 \leq \overline{y} \leq 0.8;$ $\overline{x} = 0.8, 0.4 \leq \overline{y} \leq 0.6$	$0.4 \leq \overline{x} \leq 0.7,$ $0.3 \leq \overline{y} \leq 0.6$
CCSS	105.727	$\overline{x} = 0.5, \overline{y} = 0.7$	$\overline{x} = 0.5, 0.7$ $0.2 \leq \overline{y} \leq 0.4$ $\overline{x} = 0.3, 0.8$ $0.5 \leq \overline{y} \leq 0.7$	$0.5 \leq \overline{x} \leq 0.6,$ $0.2 \leq \overline{y} \leq 0.4$ $0.4 \leq \overline{x} \leq 0.7,$ $0.5 \leq \overline{y} \leq 0.7$
CSCS	101.457	$\overline{x} = 0.6, \overline{y} = 0.5$	$0.4 \leq \overline{x} \leq 0.7,$ $\overline{y} = 0.3$	$0.4 \leq \overline{x} \leq 0.7,$ $0.4 \leq \overline{y} \leq 0.7$
SCSC	107.888	$\overline{x} = 0.4, \overline{y} = 0.5$	$0.5 \leq \overline{x} \leq 0.6,$ $0.4 \leq \overline{y} \leq 0.6$	$0.3 \leq \overline{x} \leq 0.4,$ $0.3 \leq \overline{y} \leq 0.7$
CSSS	106.825	$\overline{x} = 0.7, \overline{y} = 0.7$	$\overline{x} = 0.3,$ $0.2 \leq \overline{y} \leq 0.8$	$0.4 \leq \overline{x} \leq 0.8,$ $0.2 \leq \overline{y} \leq 0.8$
SSSC	108.098	$\overline{x} = 0.4, \overline{y} = 0.3$	$0.7 \leq \overline{x} \leq 0.8,$ $0.3 \leq \overline{y} \leq 0.7$	$0.3 \leq \overline{x} \leq 0.6,$ $0.2 \leq \overline{y} \leq 0.7$
SSCS	104.514	$\overline{x} = 0.4, \overline{y} = 0.6$	$\overline{x} = 0.2, 0.8$ $0.3 \leq \overline{y} \leq 0.7$	$0.3 \leq \overline{x} \leq 0.7,$ $0.3 \leq \overline{y} \leq 0.7$
SSSS	101.663	$\overline{x} = 0.5, \overline{y} = 0.8$	$\overline{x} = 0.3, 0.8$ $0.2 \leq \overline{y} \leq 0.8$	$0.4 \leq \overline{x} \leq 0.7,$ $0.2 \leq \overline{y} \leq 0.8$
CS	124.611	$\overline{x} = 0.2, \overline{y} = 0.2$	nil	$0.2 \leq \overline{x} 0.8,$ $0.2 \leq \overline{y} \leq 0.8$

$a/b = 1, a/h = 100, a'/b' = 1, a/h_h = 5, h_l/h_h = 0.25, E_{11}/E_{22} = 25, G_{23} = 0.2E_{22}, G_{13} = G_{12} = 0.5E_{22},$ and $\nu_{12} = \nu_{21} = 0.25$.

(3) The information regarding the behaviour of stiffened conoids with eccentric cutouts for a wide spectrum of eccentricity and boundary conditions for cross ply and angle ply shells may also be used as design aids for structural engineers.

Notations

a, b:	Length and width of shell in plan
a', b':	Length and width of cutout in plan
b_{st}:	Width of stiffener in general
b_{sx}, b_{sy}:	Width of X- and Y-stiffeners, respectively
B_{sx}, B_{sy}:	Strain-displacement matrix of stiffener elements
d_{st}:	Depth of stiffener in general
d_{sx}, d_{sy}:	Depth of X- and Y-stiffeners, respectively
$\{d_e\}$:	Element displacement
e:	Eccentricities of both x- and y-direction stiffeners with respect to shell midsurface
E_{11}, E_{22}:	Elastic moduli

G_{12}, G_{13}, G_{23}:	Shear moduli of a lamina with respect to 1, 2, and 3 axes of fibre
h:	Shell thickness
h_h:	Higher height of conoid
h_l:	Lower height of conoid
M_x, M_y:	Moment resultants
M_{xy}:	Torsion resultant
np:	Number of plies in a laminate
N_1–N_8:	Shape functions
N_x, N_y:	Inplane force resultants
N_{xy}:	Inplane shear resultant
Q_x, Q_y:	Transverse shear resultant
R_y, R_{xy}:	Radii of curvature and cross curvature of shell, respectively
u, v, w:	Translational degrees of freedom
x, y, z:	Local coordinate axes
X, Y, Z:	Global coordinate axes
z_k:	Distance of bottom of the kth ply from midsurface of a laminate
α, β:	Rotational degrees of freedom

ε_x, ε_y: Inplane strain component
ϕ: Angle of twist
γ_{xy}, γ_{xz}, γ_{yz}: Shearing strain components
ν_{12}, ν_{21}:   Poisson's ratios
ξ, η, τ: Isoparametric coordinates
ρ: Density of material
σ_x, σ_y: Inplane stress components
τ_{xy}, τ_{xz}, τ_{yz}: Shearing stress components
ω: Natural frequency
$\overline{\omega}$:   Nondimensional natural
             frequency $= \omega a^2 (\rho/E_{22}h^2)^{1/2}$.

References

[1] H. A. Hadid, *An analytical and experimental investigation into the bending theory of elastic conoidal shells [Ph.D. thesis]*, University of Southampton, 1964.

[2] C. Brebbia and H. Hadid, Analysis of plates and shells using rectangular curved elements, CE/5/71 Civil engineering department, University of Southampton, 1971.

[3] C. K. Choi, "A conoidal shell analysis by modified isoparametric element," *Computers and Structures*, vol. 18, no. 5, pp. 921–924, 1984.

[4] B. Ghosh and J. N. Bandyopadhyay, "Bending analysis of conoidal shells using curved quadratic isoparametric element," *Computers and Structures*, vol. 33, no. 3, pp. 717–728, 1989.

[5] B. Ghosh and J. N. Bandyopadhyay, "Approximate bending analysis of conoidal shells using the galerkin method," *Computers and Structures*, vol. 36, no. 5, pp. 801–805, 1990.

[6] A. Dey, J. N. Bandyopadhyay, and P. K. Sinha, "Finite element analysis of laminated composite conoidal shell structures," *Computers and Structures*, vol. 43, no. 3, pp. 469–476, 1992.

[7] A. K. Das and J. N. Bandyopadhyay, "Theoretical and experimental studies on conoidal shells," *Computers and Structures*, vol. 49, no. 3, pp. 531–536, 1993.

[8] D. Chakravorty, J. N. Bandyopadhyay, and P. K. Sinha, "Free vibration analysis of point-supported laminated composite doubly curved shells-A finite element approach," *Computers and Structures*, vol. 54, no. 2, pp. 191–198, 1995.

[9] D. Chakravorty, J. N. Bandyopadhyay, and P. K. Sinha, "Finite element free vibration analysis of point supported laminated composite cylindrical shells," *Journal of Sound and Vibration*, vol. 181, no. 1, pp. 43–52, 1995.

[10] D. Chakravorty, J. N. Bandyopadhyay, and P. K. Sinha, "Finite element free vibration analysis of doubly curved laminated composite shells," *Journal of Sound and Vibration*, vol. 191, no. 4, pp. 491–504, 1996.

[11] D. Chakravorty, P. K. Sinha, and J. N. Bandyopadhyay, "Applications of FEM on free and forced vibration of laminated shells," *Journal of Engineering Mechanics*, vol. 124, no. 1, pp. 1–8, 1998.

[12] A. N. Nayak and J. N. Bandyopadhyay, "On the free vibration of stiffened shallow shells," *Journal of Sound and Vibration*, vol. 255, no. 2, pp. 357–382, 2003.

[13] A. N. Nayak and J. N. Bandyopadhyay, "Free vibration analysis and design aids of stiffened conoidal shells," *Journal of Engineering Mechanics*, vol. 128, no. 4, pp. 419–427, 2002.

[14] A. N. Nayak and J. N. Bandyopadhyay, "Free vibration analysis of laminated stiffened shells," *Journal of Engineering Mechanics*, vol. 131, no. 1, pp. 100–105, 2005.

[15] A. N. Nayak and J. N. Bandyopadhyay, "Dynamic response analysis of stiffened conoidal shells," *Journal of Sound and Vibration*, vol. 291, no. 3–5, pp. 1288–1297, 2006.

[16] H. S. Das and D. Chakravorty, "Design aids and selection guidelines for composite conoidal shell roofs—a finite element application," *Journal of Reinforced Plastics and Composites*, vol. 26, no. 17, pp. 1793–1819, 2007.

[17] H. S. Das and D. Chakravorty, "Natural frequencies and mode shapes of composite conoids with complicated boundary conditions," *Journal of Reinforced Plastics and Composites*, vol. 27, no. 13, pp. 1397–1415, 2008.

[18] S. S. Hota and D. Chakravorty, "Free vibration of stiffened conoidal shell roofs with cutouts," *JVC/Journal of Vibration and Control*, vol. 13, no. 3, pp. 221–240, 2007.

[19] M. S. Qatu, E. Asadi, and W. Wang, "Review of recent literature on static analyses of composite shells: 2000–2010," *Open Journal of Composite Materials*, vol. 2, pp. 61–86, 2012.

[20] M. S. Qatu, R. W. Sullivan, and W. Wang, "Recent research advances on the dynamic analysis of composite shells: 2000–2009," *Composite Structures*, vol. 93, no. 1, pp. 14–31, 2010.

[21] V. V. Vasiliev, R. M. Jones, and L. L. Man, *Mechanics of Composite Structures*, Taylor and Francis, New York, NY, USA, 1993.

[22] M. S. Qatu, *Vibration of Laminated Shells and Plates*, Elsevier, London, UK, 2004.

[23] S. Sahoo and D. Chakravorty, "Finite element bending behaviour of composite hyperbolic paraboloidal shells with various edge conditions," *Journal of Strain Analysis for Engineering Design*, vol. 39, no. 5, pp. 499–513, 2004.

14

Influence of the Constitutive Flow Law in FEM Simulation of the Radial Forging Process

Olivier Pantalé and Babacar Gueye

Université de Toulouse, INP/ENIT, Laboratoire Génie de Production, 47 Avenue d'Azereix, 65016 Tarbes, France

Correspondence should be addressed to Olivier Pantalé; pantale@enit.fr

Academic Editor: Fabio Galbusera

Radial forging is a widely used forming process for manufacturing hollow products in transport industry. As the deformation of the workpiece, during the process, is a consequence of a large number of high-speed strokes, the Johnson-Cook constitutive law (taking into account the strain rate) seems to be well adapted for representing the material behavior even if the process is performed under cold conditions. But numerous contributions concerning radial forging analysis, in the literature, are based on a simple elastic-plastic formulation. As far as we know, this assumption has yet not been validated for the radial forging process. Because of the importance of the flow law in the effectiveness of the model, our purpose in this paper is to analyze the influence of the use of an elastic-viscoplastic formulation instead of an elastic-plastic one for modeling the cold radial forging process. In this paper we have selected two different laws for the simulations: the Johnson-Cook and the Ludwik ones, and we have compared the results in terms of forging force, product's thickness, strains, stresses, and CPU time. For the presented study we use an AISI 4140 steel, and we denote a fairly good agreement between the results obtained using both laws.

1. Introduction

Metal forming is a widely used tool in the industry to manufacture a large range of parts in different sectors. Researchers and engineers have always resorted to some techniques in order to improve their understanding of the process on the one hand, and on the other hand to properly control the process parameters and their influence on the produced workpieces. The radial forging process is used to reduce the diameter of tubes and bars. The final shape of the workpiece is obtained thanks to a large number of strokes achieved by four anvils radially arranged around the workpiece as shown in Figure 1. The preform is maintained in the proper position and is pushed into the dies using a mandrel which grips it securely. The preform is subjected to a combination of an axial translation and a circumferential rotation.

For those studies, analytical methods have been originally developed. Hosford and Caddell [1] in their contribution have detailed many of such methods. One of these methods is the work balance method is used to estimate the amount of force required to achieve a metal forming operation. In the proposed approach, the global work is divided into three parts: (a) an ideal work that would be required for only the shape change, (b) a friction work, and (c) an unwanted redundant work. The force required is estimated from the computation of the global energy involved in the forming operation. Other approaches have also been used such as the one proposed by Ghaei et al. [2], based on the slab analysis method, to compute the deformation versus the geometry of the die in a radial forging process. Choi et al. [3] employed the upper-bound method to forecast the forging force. Their results were compared to experimental data, and a good agreement was denoted for the proposed analytical model. Pitt-Francis et al. [4] have proposed a three-dimensional formulation, also based on the upper bound method, to analyze the forging process. In their contribution, Donald and Chen [5] have developed an upper bound approach to analyze the stability in soils and rocks. Even if those methods are very useful and simple to use, they are limited as soon as we need to get more detailed results concerning the material state. For

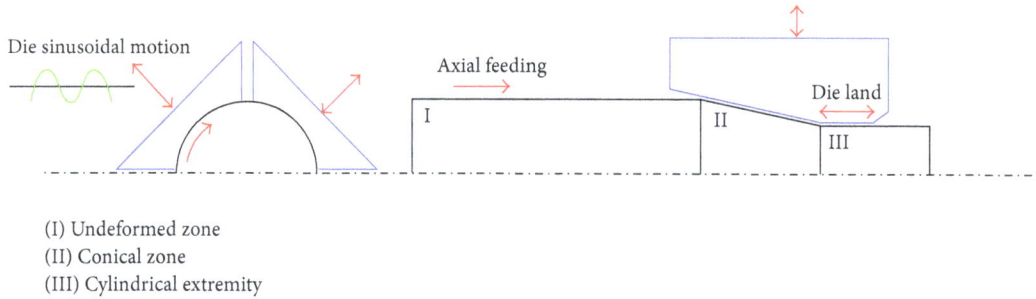

(I) Undeformed zone
(II) Conical zone
(III) Cylindrical extremity

FIGURE 1: Radial forging process used to reduce the diameter of a bar.

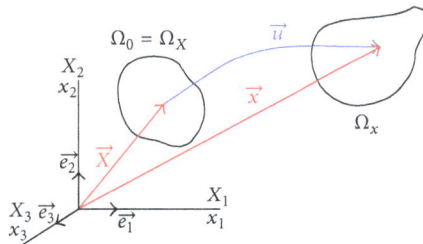

FIGURE 2: Initial and final configuration during a transformation.

those reasons the finite element method (FEM) is now widely used to simulate and analyze the forming processes.

Finite element modeling was originally used in structural mechanics for managing the parts design cycle or for evaluating the performance of an existing system. The development of computational capabilities during the last three decades allows the extension of its use to many engineering fields including the metal forming processes and in particular the radial forging. Numerous authors have used FEM to study the radial forging process. In 1992, Piela [7] has studied the applicability of the finite element method for simulating radial forging process and has proposed his own model in a latter paper in 1997 [8]. In the same period, Domblesky et al. [9, 10] have proposed their operating strategy to model multiple-pass radial forging using the finite element method. Their approach is based on a stroke by stroke axisymmetric simulation with an automatic update of the die and the workpiece positions between each stroke. Due to some convergence difficulties and an excessive CPU time, they modeled the chuckhead as a surrounded ring. Ameli and Movahhedy [11] have also proposed a parametric study in cold radial forging process using a finite element model where they analyzed the influence of three parameters (the axial feed, the preform thickness, and the friction coefficient) on the residual stresses. The effect of the die inlet angle and the die land length on the forging force has also been evaluated. In their contribution, Ghaei and Movahhedy [12] have used the finite element method for the design of the dies. In a later work [13] they have proposed their modeling results in terms of axial stress distribution within the tube.

As presented before, the large deformation of the preform during radial forging process results from numerous strokes of the dies. The dies are hence driven in such a way that they are subjected to a high-frequency sinusoidal motion. In such conditions (large strain and large strain rate), the Johnson-Cook dynamic constitutive law [14] seems to be more adapted than a classical power law like the Hollomon or Ludwik laws usually used in most of the proposed analytical and numerical models described above. Usually, the authors make the assumption that the material presents an elastic-plastic behavior in cold conditions; therefore, they have used a classical power law to describe the hardening. As far as we know, in the literature, there are no papers which deal with the validation of this assumption when the simulation of the radial forging process is concerned. Knowing the importance of the flow law choice on the effectiveness of the model, we propose in this paper to compare the results of the same simulated process with regard to the use of the Johnson-Cook and the Ludwik laws, respectively.

In this paper, theoretical bases will be first briefly presented. The mechanical problem in its local form and the bases of the finite element method used to transform the local differential equations to a set of nonlinear equations will be analyzed in the second section of the paper. In the third section the model used in this study will be detailed, and its accuracy will be presented by comparing the predicted forging force with experimental results available in the literature. Finally we will compare the Johnson-Cook and the Ludwik constitutive laws in terms of forging force, product thickness, strains, stresses, and CPU time in the last section.

2. Theoretical Bases

In this section, we present the mechanical problem in its local form, the discretization of the problem and the construction of the weak form associated. The constitutive law integration based on the radial return algorithm is presented in a second part of this section with the details concerning the elastic prediction and the plastic correction. As the problem is highly nonlinear, an explicit integration scheme has been retained for the time integration of the proposed equations.

2.1. Governing Equations. From the knowledge of the initial configuration and the imposed boundary conditions, the mechanical problem consists in predicting the final configuration depending on the initial configuration as depicted in Figure 2. As presented in Figure 2 and according to

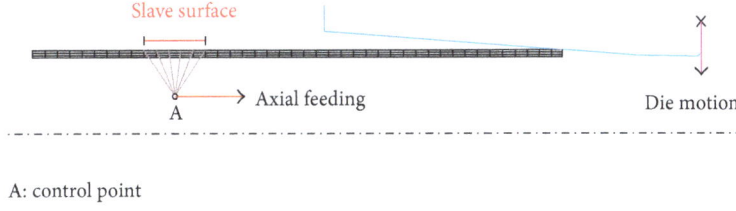

A: control point

FIGURE 3: Axisymmetric finite element model.

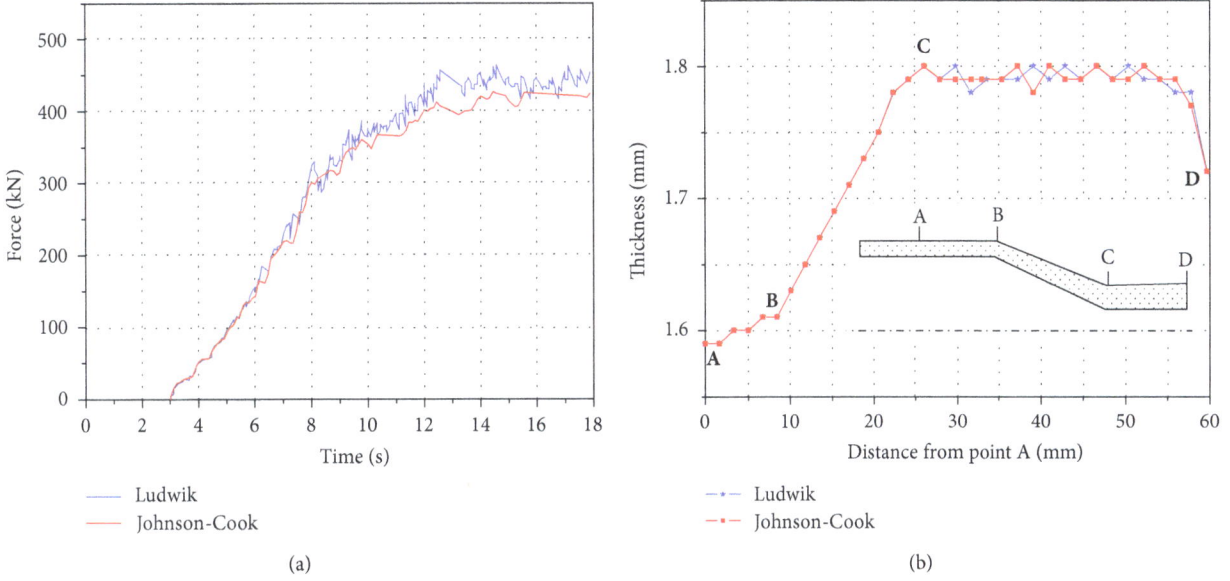

(a)

(b)

FIGURE 4: Numerically predicted forging force and geometry.

the continuum mechanics approach [15], we consider an arbitrary domain Ω with boundary Γ leading to the mechanical problem presented in (1) that must be solved at each increment. Consider,

$$\vec{\nabla} \cdot \sigma + \rho \vec{b} = \rho \vec{\gamma} \quad \text{in } \Omega$$

$$\vec{u} = \vec{u}^d \quad \text{on } \Gamma^u \quad\quad (1)$$

$$\sigma \cdot \vec{n} = \vec{t}^d \quad \text{on } \Gamma^d,$$

where $\vec{\nabla}$ is the divergence operator, σ is Cauchy stress tensor, $\rho \vec{b}$ is the body forces vector, ρ is the mass density of the material, $\vec{\gamma}$ is the acceleration vector, $\vec{u}$ is the displacement vector, $\vec{n}$ is the surface normal vector, $\cdot$ is the contraction of inner indices operator,: is the double contractor of inner indices operator, and Γ^u and Γ^d are a partition of the domain's boundary Γ, where displacements $\vec{u}^d$ and external loads $\vec{t}^d$ are imposed, respectively. In this state, the mechanical problem cannot be solved because of a lack of equations compared to the unknown variables. Therefore the additional equations given below are added to the previous system:

(i) the geometrical compatibility equation which is written in a general way as folows:

$$\varepsilon = \frac{1}{2} \left[\vec{\nabla}\vec{u} + \left(\vec{\nabla}\vec{u}\right)^T + \vec{\nabla}\vec{u} \cdot \left(\vec{\nabla}\vec{u}\right)^T \right], \quad\quad (2)$$

(ii) the constitutive equation used to represent the material behavior as a relation between different variables $f(\sigma, \varepsilon, \dot{\varepsilon}, T) = 0$, where ε is the strain tensor $\dot{\varepsilon}$ is the strain rate tensor, and T is the temperature.

2.2. Finite Element Discretization. The finite element method is neither more or less than a mathematical way to resolve differential equations. It is an approximate method based on the discretization of the problem's equations and the domain in which a solution is looked for. In radial forming, the mechanical problem is given by (1). Before resolution, this equation is turned into a weak form by multiplying (1) with an admissible virtual displacement $\delta\vec{u}$ and integrating in the hole domain Ω. So we obtain the following form on the whole domain:

$$\int_{\Omega} \delta\vec{u}\rho\vec{b}d\Omega + \int_{\Gamma^d} \delta\vec{u}\vec{t}^d d\Gamma - \int_{\Omega} \delta\varepsilon : \sigma d\Omega$$

$$= \int_{\Omega} \delta\vec{u}\rho\vec{\gamma}d\Omega. \quad\quad (3)$$

(a)

(b)

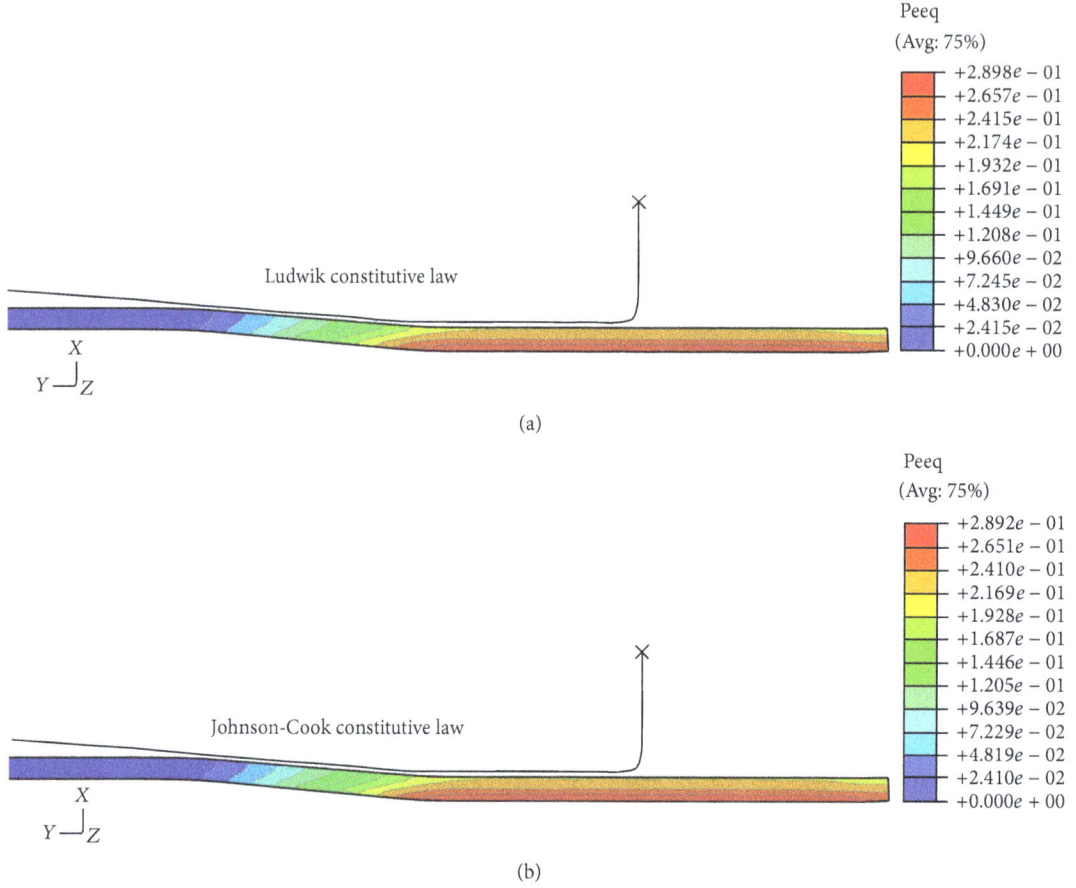

FIGURE 5: Equivalent plastic strain after forming.

TABLE 1: Comparison between experimental [6] and computed loads.

Preform diameter (mm)	Product diameter (mm)	Sizing zone (mm)	Experimental load (KN)	FEM load (KN)
15.97	13.18	18.00	172.00	181.63

TABLE 2: AISI 4140 Johnson-Cook parameters.

A (MPa)	B (MPa)	n	C	m
806	614	0.168	0.0089	1

TABLE 3: AISI 4140 Ludwik parameters.

A (MPa)	K (MPa)	n
817	699	0.156

The finite element method allows writing the approximation of the displacement vector $\vec{u}_e$ for each element of the decomposed domain Ω_e and the ponderation vector $\delta\vec{u}_e$ using the following form:

(i) approximation of the displacement: $\vec{u}_e = \mathbf{N}_e\vec{u}$, where $\mathbf{N}_e$ is the matrix of the shape functions and $\vec{u}$ is the displacement vector of the nodes involved in the current element e,

(ii) approximation of the admissible displacement: $\delta\vec{u}_e = \mathbf{N}_e\delta\vec{u}$.

Using the above approximation, (1) leads to the following system:

$$\mathbf{M}_e\ddot{\vec{u}}_e + \vec{F}_e^{\text{ext}} = \vec{F}_e^{\text{int}}, \qquad (4)$$

where,

(i) $\mathbf{M}_e = \int_{\Omega_e} \rho\mathbf{N}_e^T\mathbf{N}_e d\Omega_e$ is the elementary mass matrix.

(ii) $\vec{F}_e^{\text{int}} = \int_{\Omega_e} \mathbf{B}^{e^T}\sigma d\Omega_e$ is the elementary internal force vector.

(iii) $\vec{F}_e^{\text{ext}} = \int_{\Omega_e} \rho\mathbf{N}_e^T\vec{b}d\Omega_e + \int_{\Gamma_e^d} \mathbf{N}_e\vec{t}^d d\Gamma_e$ is the elementary external force vector.

The global problem is obtained by assembling the matrices above, and subsequently the system is resolved using an iterative method such as the Newton-Raphson algorithm.

(a)

(b)

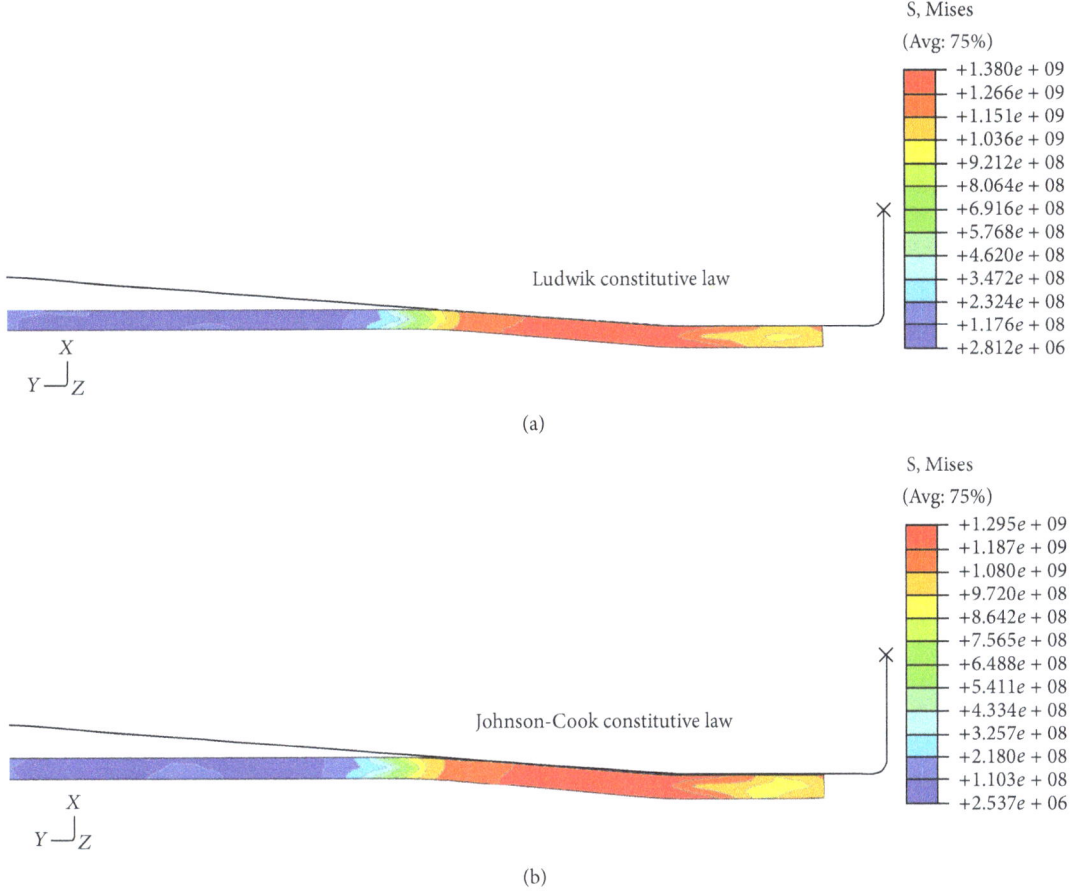

FIGURE 6: Von Mises stress when the die strokes.

(1) Initial conditions and initialization: $n = 0$; $\sigma_0 = \sigma(t_0)$; $\vec{x}_0 = \vec{x}(t_0)$; $\vec{v}_0 = \vec{v}(t_0)$
(2) Update quantities: $n \leftarrow n + 1$; $\sigma_n = \sigma_{n-1}$; $\vec{x}_n = \vec{x}_{n-1}$; $\vec{v}_{n+1/2} = \vec{v}_{n-1/2}$
(3) Compute the time-step and update current time: $t_n = t_{n-1} + \Delta t$
(4) Update nodal displacements: $\vec{x}_n = \vec{x}_{n-1} + \Delta t \vec{v}_{n-1/2}$
(5) Compute internal and external force vector $\vec{F}_n^{\text{int}}, \vec{F}_n^{\text{ext}}$
(6) Integrate the conservative equations and compute accelerations: $\dot{\vec{v}}_n = \mathbf{M}^{-1}(\vec{F}_n^{\text{ext}} - \vec{F}_n^{\text{int}})$
(7) Update nodal velocities: $\vec{v}_{n+1/2} = \vec{v}_{n-1/2} + \Delta t \dot{\vec{v}}_n$
(8) Enforce essential boundary conditions: if node I on Γ_v
(9) Output; if simulation not complete go to (2).

ALGORITHM 1: Flowchart for explicit time integration.

2.3. *Constitutive Law.* The problem to be solved here is a large strain simulation; therefore we must ensure the objectivity of all the terms appearing in the constitutive law in order to have correct responses of the model. It is therefore necessary to maintain correct rotational transformation properties all along a finite time step. The symmetric part of the spatial velocity gradient $\mathbf{L}$, denoted by $\mathbf{D}$, is objective while its skew-symmetric part $\mathbf{W}$, called the spin tensor, is not objective as reported, for example, in [16]. Assuming that the Cauchy stress tensor σ is objective, its classical material time derivative $\dot{\sigma}$ is nonobjective, so one must introduce an objective rate notion $\overset{\triangledown}{\sigma}$ which is a modified time derivative form of the

Cauchy stress tensor as the Jaumann or the Green-Naghdi rates. The incremental formulation of the constitutive law is therefore given by $\overset{\triangledown}{\sigma} = f(\mathbf{D}, \ldots)$.

One of the solutions to this problem consists in transporting the Cauchy stress σ in a corotational frame defined using a rotation tensor $\mathbf{w}$ based on the following set of equations:

$$\dot{\mathbf{w}} = \omega \mathbf{w},$$
$$\mathbf{w}(t = t_0) = \mathbf{I}.$$

(5)

Defining any quantity () in this rotating frame as a corotational one denoted by $()^c$, one may obtain in these axes, when transformed by $\mathbf{w}$, the following expressions for the Cauchy stresses:

$$\sigma^c = \mathbf{w}^T \sigma \mathbf{w}, \qquad \dot{\sigma}^c = \mathbf{w}^T \dot{\sigma} \mathbf{w}. \qquad (6)$$

In fact, the choice of $\omega = \mathbf{W}$ corresponds to the Jaumann rate. The major consequence of corotational rates is that if we choose the local axis system as the corotational one, constitutive laws integration can be performed as in small deformation, leading to a simplified formulation. According to the decomposition of the Cauchy stress tensor σ into a deviatoric term $\mathbf{s}$ and a hydrostatic term p, one may obtain:

$$\dot{\mathbf{s}}^c = \mathbb{C}^c : \mathbf{D}^c,$$
$$\dot{p} = K \operatorname{tr} \left[\mathbf{D}^c \right], \qquad (7)$$

where K is the bulk modulus of the material, $\mathbf{D}^c = \mathbf{w}^T \mathbf{D} \mathbf{w}$, $\mathbb{C}^c = \mathbf{w}^T [\mathbf{w}^T \mathbb{C} \mathbf{w}] \mathbf{w}$, and $\mathbb{C}$ is the fourth order constitutive tensor. In this application, we use a J_2 plasticity model with a nonlinear isotropic hardening law. It is generally assumed that the rate of deformation can be additively decomposed into a elastic and a inelastic parts; therefore the approach consisting in decomposing the stress computation into an elastic predictor and a plastic corrector can be used. The associated von Mises yield criterion allows the use of the radial-return mapping strategy, briefly summarized hereafter, to integrate the constitutive behavior of the material along the time increment $\Delta t = t_{n+1} - t_n$.

2.3.1. Elastic Prediction. Due to the objectivity and the use of a corotational system, all the terms of the constitutive equation are co-rotational ones, so we can drop the upperscript c in the following equations for simplicity. The predicted elastic stresses at increment $n + 1$ are calculated from the current known values at increment n using the Hooke's law, according to (7), by the following equations:

$$p_{n+1}^{\text{trial}} = p_n + K \operatorname{tr} \left[\Delta \mathbf{e} \right],$$
$$\mathbf{s}_{n+1}^{\text{trial}} = \mathbf{s}_n + 2G \operatorname{dev} \left[\Delta \mathbf{e} \right], \qquad (8)$$

where $\Delta \mathbf{e} = \ln[\mathbf{U}]$ is the co-rotational strain increment tensor between increment n and increment $n + 1$ and G is the Lamé coefficient. Hence, the von Mises criterion $f(\sigma, \sigma_v)$ is then defined by

$$f = \sqrt{\frac{3}{2} \mathbf{s}_{n+1}^{\text{trial}} : \mathbf{s}_{n+1}^{\text{trial}}} - \sigma_v, \qquad (9)$$

where σ_v is the current yield stress depending on the history of the deformation in case of a plastic behavior. Hence, if $f \leq 0$, the predicted solution is physically admissible and the whole increment is assumed to be elastic. If $f > 0$, then a plastic correction must be taken into account in order to restore computed stresses to a physically admissible value.

2.3.2. Plastic Correction. If the predicted elastic stresses do not correspond to a physically admissible state, a plastic correction has to be performed. The previously trial stresses

serve as the initial condition for the so-called return mapping algorithm. This one is summarized by the following equation:

$$\mathbf{s}_{n+1} = \mathbf{s}_{n+1}^{\text{trial}} - 2G\Gamma\mathbf{n}, \qquad (10)$$

where $\mathbf{n} = \mathbf{s}_{n+1}^{\text{trial}} / \sqrt{\mathbf{s}_{n+1}^{\text{trial}} : \mathbf{s}_{n+1}^{\text{trial}}}$ is the unit outward normal to the yield surface and Γ is the consistency parameter defined as the solution of the one scalar parameter nonlinear equation below:

$$f(\Gamma) = \left(\mathbf{s}_{n+1}^{\text{trial}} - 2G\Gamma\mathbf{n} \right) : \left(\mathbf{s}_{n+1}^{\text{trial}} - 2G\Gamma\mathbf{n} \right) - \frac{2}{3} (\sigma_v(\Gamma))^2 = 0. \qquad (11)$$

Equation (11) is effectively solved by a local Newton iterative procedure [17]. Since $f(\Gamma)$ is a convex function, convergence is guaranteed.

2.4. Time Integration. All above equations are integrated by an explicit scheme associated with lumped mass matrices. The flowchart for the explicit time integration of the Lagrangian mesh is given in Algorithm 1 as proposed, for example, by Belytschko et al. [15].

3. Finite Element Modeling and Validation

In a general way, the radial forging process does not exhibit an axial symmetry property. So the modeling should be performed under three-dimensional conditions. But this is excessively time consuming as we need a couple of weeks to complete a single simulation. For this reason we have made some assumptions that allow the use of a 2D model (see Figure 3). Therefore the following points are not considered in the proposed model:

  (i) the tube's rotation,

  (ii) the clearances between the hammers,

  (iii) variation of the dies' shape along their cross section.

These assumptions lead to a significant computational cost saving while providing relatively good results as demonstrated in the literature. In this work, we used the Abaqus Explicit [18] finite element commercial code for the analysis, and the preform is modeled using CAX4R elements, which are 4 node quadrilateral elements with reduced integration and hourglass control. On the other hand, the dies are assumed to be rigid, and only the contacting surface with the tube is taken into account in the model. So going under axisymmetric conditions the die is discretized by 2 nodes axisymmetric rigid elements.

In the literature the chuckhead is systematically represented, including thereby a supplementary contact problem. So the solving efficiency, in terms of time consumption, can be collapsed. In our work, the chuckhead is suppressed, and the tube feeding is supported by a coupling constraint between a part of the preform's outer surface and a control point as illustrated in Figure 3. For the die-preform interaction, we use a penalty contact formulation with a Coulomb friction coefficient $\mu = 0.2$. Furthermore, contrary to previous studies, our strategy does not consist in a stroke by stroke simulation. In fact we perform one single simulation for the entire process, taking into account the dwell time between each consecutive stroke.

The validation of the proposed model is made by comparing our forecasted computed forging load with the experimental values available in the literature and proposed by Uhlig [6]. This way of evaluating the effectiveness of the model was inspired by Ghaei et al. [13] and Ameli and Movahhedy [11]. The initial and the final diameter, are reported in Table 1 as well as the experimental and numerically computed loads. A reasonable good agreement (around 5%) is observed; therefore, our model can then be used to study the influence of the flow law on the results as proposed in the next section of this paper.

4. Results and Discussion

We propose here to compare the influence of the use of the Ludwik or the Johnson-Cook laws on the numerical results in a radial forming simulation. For the Ludwik law (12), the equivalent plastic stress σ_{eq} depends only on the plastic strain $\bar{\varepsilon}$ whereas the Johnson-Cook law (13) considers in addition the influence of the strain rate $\dot{\bar{\varepsilon}}$ and the effect of the temperature T during the forming process. According to the solicitations, the Johnson-Cook law seems to suit better to represent the material behavior. But, in the case of cold forming, the Ludwik law may be relevant and more practical as one must only identify three parameters $(A, K, \text{and } n)$ versus five $(A, B, C, n, \text{and } m)$ for the Johnson-Cook law. In this section we will compare the numerical results of the simulations, such as the forging force, the predicting tube thickness, strains, and stresses, obtained by using the two distinct laws:

$$\sigma_{eq} = A + K\bar{\varepsilon}^n, \qquad (12)$$

$$\sigma_{eq} = \left[A + B\bar{\varepsilon}^n\right]\left[1 + C\ln\frac{\dot{\bar{\varepsilon}}}{\dot{\bar{\varepsilon}}_0}\right]\left[1 + \left(\frac{T - T_0}{T_m - T_0}\right)^m\right]. \qquad (13)$$

We have employed an AISI 4140 material for this study. The Johnson-Cook and Ludwik parameters [16] are, respectively reported in Tables 2 and 3. The Ludwik parameters have been obtained using a classical traction test whereas the Johnson-Cook ones have been identified by using dynamic tests [16] on the same material. For the numerical simulation we used an initial diameter $D_0 = 16$ mm, a percentage of reduction in diameter $\Delta D = 17\%$, and a feeding rate $V_r = 0.37$ mm/stroke. Concerning the numerical simulation, similar CPU times are denoted for both laws. In fact, a CPU time of 13 min 08 s is required when using the Johnson law, whereas a CPU time of 12 min 52 s is necessary for the Ludwik law, that is, a deviation of 2%.

4.1. Forging Force Evolution. In Figure 4 the evolution of the forging force during forming is plotted for both constitutive laws. We can denote the existence of a transient and a stationary phase. From the beginning to the exit of the die land, the forging force increases because the amount of material being deformed and the frictional work increase continuously. Once the die land exit is reached, deformation and friction energies remain stationary and the load is stabilized around $F = 425$ kN. Predicting the forging force can be very interesting as it allows knowing whether the capabilities required to manufacture a given product are available in a workshop or not. On the other hand, a parametric or optimization study can help to minimize the force and consequently to reduction cost by saving energy. For the proposed simulation, both laws exhibit the same evolution and a good concordance of results is denoted.

4.2. Geometry and Product Thickness. Knowing that our study focuses on the radial forging without a mandrel, the inner surface of the tube can be radially deformed when the dies stroke. Consequently, the tube thickness is greater after than before deformation. In Figure 4, we have plotted the variation of the product thickness versus the distance from the exit. We denote a maximum thickening of 10% with regard to the initial thickness. Furthermore we can see that this greatest thickness concerns the cylindrical extremity of the forged product. At the product end a little bit pinch is depicted. That leads to a thickness decrease of around 3.5% compared to the neighboring points. The conical region represents the transition between the undeformed cylindrical domain and the extremity. Hence, the thickness in this conical zone increases from the initial value up to the maximum one. Johnson-Cook and Ludwik diagrams follow the same trend, and the maximum deviation is about 1.12%. So, once again, we obtain a fairly good agreement between both laws.

4.3. Plastic Strain Distribution. In Figure 5, the equivalent plastic strain after forming is shown. As we expect, plastic strain is more important at the extremity of the tube as this region has the biggest amount of diameter reduction. So the maximum strain is about 29% for both Johnson-Cook and Ludwik laws. In addition, the non deformed region is highlighted by a zero value of the plastic strain. Within the transition between the two regions, represented by the conical shape, the plastic strain increases from zero up to the maximum value. The two constitutive equations used in this document have the same results on both the contourplot strain and the numerical values. The deviations in absolute values denoted are almost equal to zero.

4.4. Stress Distribution. In Figure 6 the von Mises stress distribution when the die strokes is depicted. The maximum value is reached under the die land and is about 1300 MPa. We can also see in Figure 6 that, in the die inlet, the change in the material flow direction causes a shift on the stress compared with the back and front neighboring regions.

Even if the Ludwik law does not account for the strain rate, the forecasted stresses are close to those given by the Johnson-Cook law. We can see in Figure 6 similar stress distributions. The comparison made up till now is thereby confirmed.

5. Conclusion

In this work, a 2D axisymmetric finite element model was presented to study the radial forging process. Unlike some previous works, our strategy was not based on a stroke by stroke simulation, but we perform a single simulation accounting for the dwell time between two successive strokes. The comparison of the predicting forging force with experimental results available in the literature shows the accuracy of our model. The numerical model proposed in this paper has provided some insights that can be summarized as follows:

(i) the forging force evolution during the process exhibits a transient and a stationary phase. The limit between the two phases is reached when the part gets out of the die land;

(ii) when the process is performed without a mandrel, the maximum 10% thickening of the tube is denoted in a case of 17% diameter reduction.

Our purpose in this paper is to compare the Johnson-Cook and the Ludwik constitutive laws in modeling the radial forging process. In the proposed approach, we have shown a very good accordance of the results between the two laws in terms geometry, forging forces, strains, and stresses. We can therefore conclude that even if the preform undergoes high-speed stroke, the strain rate is not great enough in the case of cold forming to request the use of an elasto-viscoplastic formulation. Therefore a simple Ludwik or Hollomon law can be used to represent the material behavior and lead to quite good results. Concerning the CPU time, we have denoted a deviation of 2% between the two laws with the advantage to the simplest one.

References

[1] W. F. Hosford and R. M. Caddell, *Metal Forming: Mechanics and Metallurgy*, Cambridge University Press, 3rd edition, 2007.

[2] A. Ghaei, M. R. Movahhedy, and A. K. Taheri, "Study of the effects of die geometry on deformation in the radial forging process," *Journal of Materials Processing Technology*, vol. 170, no. 1-2, pp. 156–163, 2005.

[3] S. Choi, K. H. Na, and J. H. Kim, "Upper-bound analysis of the rotary forging of a cylindrical billet," *Journal of Materials Processing Technology*, vol. 67, no. 1–3, pp. 78–82, 1997.

[4] J. M. Pitt-Francis, A. Bowyer, and A. N. Bramley, "A simple 3D formulation for modeling forging using the upper bound method," *Analysis of the CIRP*, vol. 45, no. 1, pp. 245–248, 1996.

[5] I. B. Donald and Z. Chen, "Slope stability analysis by the upper bound approach: fundamentals and methods," *Canadian Geotechnical Journal*, vol. 34, no. 6, pp. 853–862, 1997.

[6] A. Uhlig, *Investigation on the motions and the forces in radial swaging [Ph.D. thesis]*, Technical University of Hannover, 1964.

[7] A. Piela, "Studies on the applicability of the finite element method to the analysis of swaging process," *Archives of Metallurgy*, vol. 37, pp. 425–443, 1992.

[8] A. Piela, "Analysis of the metal flow in swaging-numerical modelling and experimental verification," *International Journal of Mechanical Sciences*, vol. 39, no. 2, pp. 221–231, 1997.

[9] J. P. Domblesky and R. Shivpuri, "Development and validation of a finite-element model for multiple-pass radial forging," *Journal of Materials Processing Technology*, vol. 55, no. 3-4, pp. 432–441, 1995.

[10] J. P. Domblesky, R. Shivpuri, and B. Painter, "Application of the finite-element method to the radial forging of large diameter tubes," *Journal of Materials Processing Technology*, vol. 49, no. 1-2, pp. 57–74, 1995.

[11] A. Ameli and M. R. Movahhedy, "A parametric study on residual stresses and forging load in cold radial forging process," *International Journal of Advanced Manufacturing Technology*, vol. 33, no. 1-2, pp. 7–17, 2007.

[12] A. Ghaei and M. R. Movahhedy, "Die design for the radial forging process using 3D FEM," *Journal of Materials Processing Technology*, vol. 182, no. 1–3, pp. 534–539, 2007.

[13] A. Ghaei, M. R. Movahhedy, and A. Karimi Taheri, "Finite element modelling simulation of radial forging of tubes without mandrel," *Materials & Design*, vol. 29, no. 4, pp. 867–872, 2008.

[14] G. R. Johnson and W. H. Cook, "A constitutive model and data for metals subjected to large strains, high strain rates and high temperatures," in *Proceedings of the 7th International Symposium on Balistics*, pp. 541–547, 1983.

[15] T. Belytschko, W. K. Liu, and B. Moran, *Nonlinear Finite Element For Continua and Structures*, John Wiley & Sons, 2000.

[16] O. Pantalé, *Virtual prototyping platform for numerical simulation in large thermomechanical transformations [Ph.D. thesis]*, Institut National Polytechnique de Toulouse, 2005.

[17] J. C. Simo and T. J. R. Hugues, *Computational Inelasticity*, Springer, 1998.

[18] Simulia, *Abaqus Analysis User Manual*, Rising Sun Mills, Providence, RI, USA.

15

A Fall Protection System for High-Rise Construction

Haluk Çeçen[1] and Begüm Sertyeşilişık[2,3]

[1] *Division of Construction Management, Department of Civil Engineering, Yildiz Technical University, Esenler, 34220 İstanbul, Turkey*
[2] *Liverpool John Moores University, Liverpool L3 2AJ, UK*
[3] *Department of Architecture, Istanbul Technical University, Taşkışla Taksim, 34437 Istanbul, Turkey*

Correspondence should be addressed to Begüm Sertyeşilişık; b.sertyesilisik@ljmu.ac.uk

Academic Editor: Brian Uy

In construction industry, the number of fatal and nonfatal occupational injuries is higher than other industries. Among causes of these accidents, "falls" play a key role. This situation reveals the importance for carrying out research in fall protection systems. In this paper, a practical, economical, and functional fall protection system is introduced. Following determination and evaluation of existing solutions, weekly brainstorming meetings were held among the responsible project staff (general coordinator, project coordinator, project manager, site manager, and health and safety manager). As a result of these meetings, design criteria were developed. Based on these criteria, the fall protection system for high-rise construction (FPSFHC) was developed which satisfied all the specified design criteria. Required materials were procured from local dealers. In this paper, criteria used in design and details of the final design are presented. Field performance of the system is evaluated, and recommendations for further development and standardization of the system are added.

1. Introduction

The fall protection system for high-rise construction (FPSFHC) was developed in June 2006 during the construction of the Federation Tower, Moscow, Russia (a 63-storey-243 m high reinforced concrete skyscraper built by the Turkish contractor ANT YAPI, the tallest building in Russia in 2006). Due to the high-rise nature of the building construction, construction safety was an important concern. Thus, a special safety team was formed, and conventional safety precautions were taken. Nevertheless, a fatal accident which involved the fall of a worker from upper stories could not be prevented. This was due to a sudden gust of unprecedented strength. That accident has triggered an urgent on-site research and development process. The target was to develop a fall protection system for the upper working levels of the high-rise building construction.

2. A Literature Survey

In construction industry, rates of fatal and nonfatal occupational accidents are higher compared to other industries. This fact was emphasized in researches of different authors: Rivara and Thompson [1], Saloniemi and Oksanen [2], Abudayyeh et al. [3], Waehrer et al. [4], Sorock et al. [5], Jeong [6], Chi et al. [7], and Mohamed [8]. "The average cost per case of fatal or nonfatal injury is $27,000 in construction, almost double the per-case cost of $15,000 for all industry in 2002... Injury rates and cost rates are higher for construction than for the average of all industries" (Waehrer et al. [4], 1258-1259).

Various researches pointed out the main cause of accidents as "falls": Hinze and Russell [9], Sorock et al. [5], Jeong [6], Rivara and Thompson [1], Huang and Hinze [10], and Hinze [11]. Among them, Jeong [6] evaluated the percentage distribution of deaths and injuries by accident type. According to Jeong's [6] study slip and falls cause 14.50% of nonfatal injury resulted accidents and 6.80% of the fatal accidents. Kines [12] indicated that according to studies from many countries preventing falls from heights in construction is necessary to prevent fatal injuries.

Ale et al. [13], who carried out a research on accidents in the construction industry in the Netherlands, pointed out that the percentage of death drastically increases as the height of fall increases and that roof edge protection failure (48%)

and failure of fall arrests (28%) were two of the major causes of fatal fall accidents. Similarly, based on investigation of 508 fatalities due to falls, Hinze and Russell [9] identified the most common causes of falls as follows: off roof, collapse of scaffolding, off scaffolding, collapse of structure, through floor opening, off ladder, off structure, through roof opening, off edge of open floor, and off beam support.

Johnson et al. [14] presented criteria for a fall protection system to be implemented on a wide scale in roof construction. They suggested that such a fall protection system should be economical, flexible, passive, feasible, simple, and protective (Johnson et al. [14]).

In order to increase the efficiency of fall protection system, it must be supported by management. The effect of management commitment to safety on rates of injury and illnesses was emphasized by Heinrich [15], Komaki [16], and Abudayyeh et al. [3].

3. Methodology Used in Development of Fall Protection System

The fall protection system developed and presented in this paper satisfies the requirements underlined in the above-summarized research.

The research to develop a satisfactory fall protection system was carried out in two avenues.

(1) Determination and evaluation of existing solutions upon a swift search of available sources including internet, and one well-known international company (producing formwork systems for reinforced concrete) was contacted and invited to make a proposal. They proposed a fall protection system which used their standard elements coupled with special elements and mechanisms. However, their proposal was not acceptable due to the following drawbacks:

(a) it would involve a heavy and bulky system of steel framework,

(b) it would be on the way of other construction activities,

(c) it would be hard to move up to the next floor,

(d) it would require crane support,

(e) it would take long time to produce and bring on site,

(f) it would require major modifications to be used in construction of other buildings,

(g) it would cost around US$150.000–200.000.

Several other solutions used in international construction projects were evaluated too. They were not satisfactory either, due to reasons similar to the ones listed above.

(2) Development of a solution on site: the target was the development of a practical, economical, and functional system. For this purpose, weekly brainstorming meetings were held among the responsible project staff (general coordinator, project coordinator,

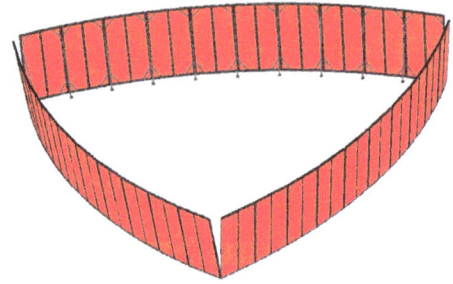

FIGURE 1: General configuration of FPSFHC.

project manager, site manager, and health and safety manager). To start with, a set of design criteria were developed. Eventually (within a month), the FPSFHC was developed which satisfied all the specified design criteria. Procurement of required materials from local dealers and production of the FPSFHC on site took only two weeks. The production cost of FPSFHC remained within US$10 000. The proposed system is new due to the fact that it was the first economical, light, and practical modular FPSFHC system with 15-degree (with vertical) inclined modules developed in a way to eliminate the above-mentioned drawbacks of the other existing systems.

4. Design Criteria

A practical, economical, and functional solution was sought. Reusability was another major criterion. Design criteria used for FPSFHC, presented in Table 1, were determined based on the evaluation of the following:

(i) construction site conditions (availability of materials and site production capabilities),

(ii) the related literature,

(iii) the existing solutions,

(iv) the discussions in the brainstorming meetings among the responsible project staff.

5. Details of the Design

FPSFHC is a modular system consisted of a series of panels. Each panel is 3×5 m in dimension. The enough number of panels is mounted side by side in order to cover the circumference of working floors completely (Figure 1). Details of a panel are shown in Figures 2, 3, and 4. The main frame of a panel is made up of standard pipes ($\emptyset = 40$ mm). A backbone running along in the middle and four braces ($45°$) of each corner strengthens the rectangular frame. Two segments (1.5×5.0 m) of steel mesh are spread over each frame and stitched onto the pipes all along the perimeter and the back bone (steel cables $\emptyset = 1$ mm were used for stitching). The steel mesh is made up of steel wires ($\emptyset = 1.0$ mm) with openings 15 mm $\times$ 15 mm in dimension.

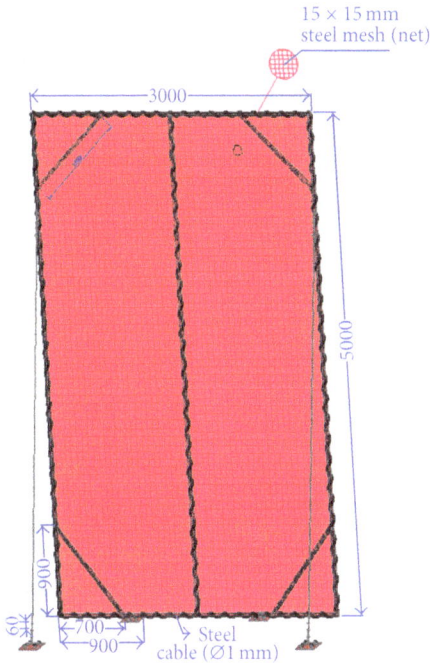

FIGURE 2: Details of panels.

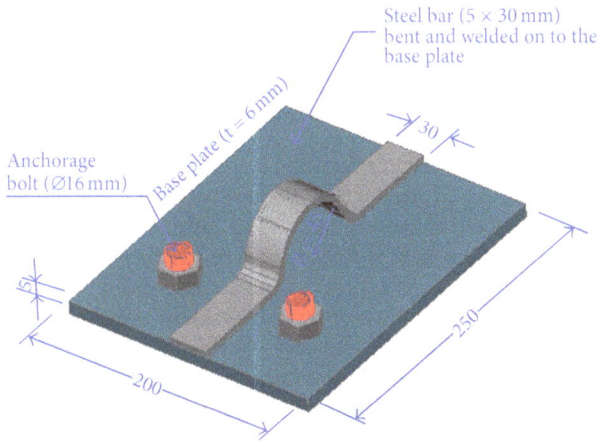

FIGURE 3: Base plate of the panels.

The modular feature of the FPSFHC system provides the following solutions as required due to unusual/irregular plan dimensions of buildings.

(a) The modules may be placed in an overlapping way.

(b) Special modules may easily be produced locally at the construction site.

Base plate details of the panel and that of the connection cables are shown in Figures 3 and 4, respectively.

As seen from Table 1, FPSFHC system was developed based on the following criteria.

Design criterion: strong enough to hold falling man and material.

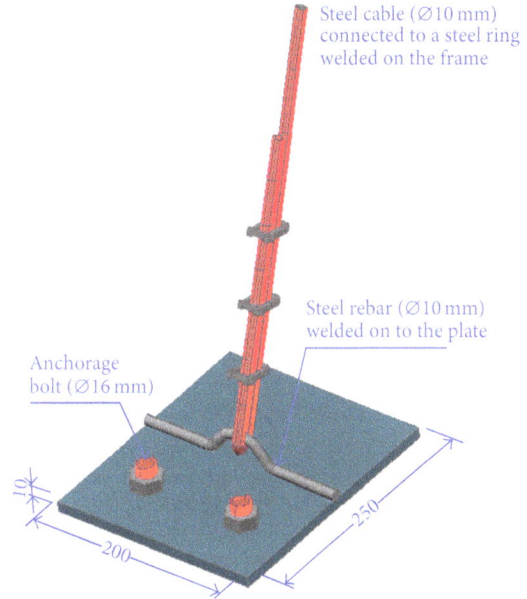

FIGURE 4: Base plate of the steel cables.

Performance criterion: should hold falling items weighting up to 1 tonne.

Thus, the FPSFHC system was produced to satisfy the above criteria.

Figure 5 shows that panels are mounted in a position that would make an angel about 15° from the vertical. This way the formwork extensions (approximately 400 mm) of the upper floor slab are engulfed in FPSFHC.

Since the falling items would hit the system modules along a 15-degree angle (Figure 5), the angled mounting position of the modules helps to reduce the impact of loads on the modules considerably.

6. Assembling and Disassembling

A specially trained team of four workers guided by a health and safety (H & S) foreman handled assembling and disassembling operations. Wearing of proper work outfit, especially the use of safety belts, was strictly enforced.

Each time, two of the upper most working floors of the building construction were enclosed in FPSFHC. Once the first segment of the concrete floor slab of the upcoming floor was casted (which coincided with the striking off the formwork of the lower floor), the panel of the lower floor was started to be disassembled and carried up to the new floor level.

(a) Disassembling steps of the FPSFHC

(1) The anchorage bolts holding a cable are unfastened, and the end of the cable is securely tied onto a column with a rope.

(2) Second cable is secured by the same way.

(3) The anchorage bolts of the anchorage plates at the bottom of the panel are unfastened.

TABLE 1: Design and performance criteria of FPSFHC.

Design criteria	Design	Performance criteria	Performance
Economical solution	Inexpensive materials	Production cost $\leq$ \$30 000	Satisfied (cost = \$10 000)
Practical solution	Modular system avoiding bulky solutions	Easy to handle	Satisfied
Functional solution	Fall prevention barriers	Protection of working levels	Satisfied
No procurement problem	Standard materials	Local availability	Satisfied
Easy to produce	Simple rectangular panels	On-site production	Satisfied
No need for crane	Light weight panels	Few men to handle	Satisfied (4-man team)
Speedy assembling and disassembling	Light weight modular system	One day per floor	Satisfied (8 hours/floor)
Not to hinder construction activities	Mounted in a way that would not interfere with other construction activities	No obstacle and delay	Satisfied
Fall protection for the two upper most floors	3×5 m panels extending and protecting the top floor too	Gapless protection all along the perimeters of the top two floors	Satisfied (special corner modules)
Strong enough to hold falling man and material	Steel frames (diameter = 40 mm pipes) covered with steel mesh (15×15 mm)	Should hold falling items weighting up to 1 ton	Satisfied Held a load > 1 ton (2-man + steel formwork)
No additional wind load	Panels covered with steel mesh	Should stay in place under wind	Satisfied
Physiological confidence	Gapless protection all along the perimeter of the working floors	Confidence of the construction workers	Satisfied
Reusable	Painted steel elements	No damage after repeated usage	Satisfied
Encompass the extensions of floor formworks	Panels placed 15° from vertical	Sufficient space for floor formwork extensions	Satisfied

(4) The untied panel is pulled inside by three men while the other two helped them by pulling on each cable.

(5) Each recovered panel is carried onto and stored on the material handling platform (a temporary balcony made up of a steel framework) of the same floor.

(b) Assembling steps of FPSFHC

(1) Groups of panel are carried up to the new floor level by crane (the only time crane is used).

(2) Each panel is carried in its place along the rim and is laid on the slab.

(3) Anchorage bolts of the anchorage plates at the bottom of the panel are fastened.

(4) Anchorage bolts of the plates holding the cables are fastened.

(5) The panel is lifted up (by pivoting around the hinges at the bottom of the panel) and is placed in its final position having an angle of 15° with the vertical.

7. Performance of the FPSFHC

FPSFHC had proved its functionality by catching countless falling materials on many occasions. Once it caught a large and heavy piece of steel formwork with two workers on top of it. On-site performance of the system satisfied the design criterion and performance criterion stated in Table 1.

As Moscow has severe weather conditions, the proposed FPSFHC system has been observed to perform well under snowy weather as well. The workers could easily clean off the excessive snow when observed. Besides, additional snow load on the roof at shell-and-core construction stage has not been critical on a high-rise building designed to carry all the dead loads partitions, furniture, and live loads of all floors. It is worth noting that the FPSFHC steel formwork was used for the major columns of the building. The weight of the formwork plus the two workers on top of it was more than 1 ton.

Upon observing this successful site performance, other neighboring constructors started to use similar fall protection systems.

8. Recommendations for Further Development

As explained in this paper, the FPSFHC was developed on a construction site within a couple of weeks following a fatal accident. Only locally (Moscow, Russia) available materials were used to manufacture the panel forming the FPSFHC.

Thus, every part of the FPSFHC is open to improvement through research and development. Especially, the connection pieces and details can be improved in order to facilitate quick assembling and disassembling operations. It is believed that well-planned R & D projects including laboratory tests will produce functional improvements and standardization.

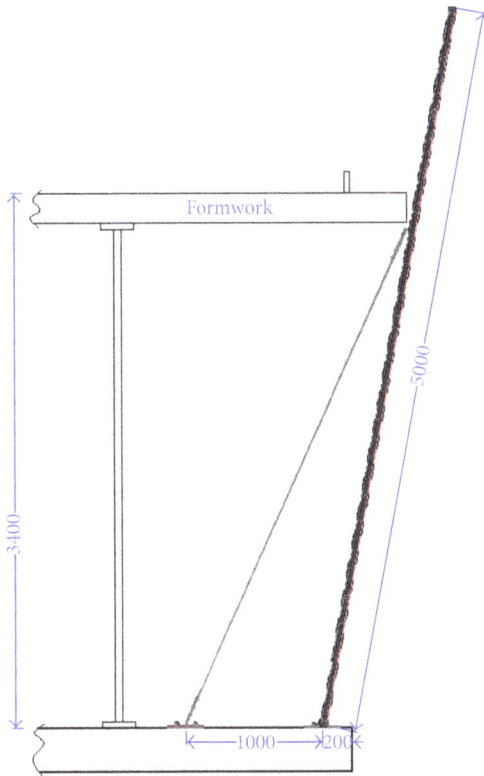

FIGURE 5: Mounting position of the FPSFHC.

9. Summary and Conclusion

FPSFHC was developed to be a practical, functional, and economical fall protection system for upper most working levels of high-rise building constructions. The FPSFHC proved to be as such in the construction of a reinforced concrete skyscraper in Moscow, Russia.

Now that the details of the FPSFHC are shared with the construction industry, the authors sincerely hope that the system will be used throughout the world in order to prevent thousands of fatal and nonfatal accidents that take place each year in constructions of high-rise buildings.

Acknowledgments

The main idea and basic features of the FPSFHC were developed by Professor Dr. Haluk Cecen then the Project Coordinator of construction of the Federation Tower-B in Moscow, Russia. The authors wish to thank the managers of ANT YAPI for their support for the research and development of FPSFHC. Valuable contributions of the colleagues are greatly appreciated. The following professionals contributed during the final evaluation, approval, production, and execution stages of the FPSFHC: Messers Kadir Tokman, Ertan Yilmaz, Argun Koculu, Ercan Gurbuz, Gurkan Bostanci, Bahtiyar Azeri, Hayati Kalfa, Timur Usta, Nazim Ahiskali (all were Members of ANT YAPI construction company), and Mr. Gleen A. Collins (TURNER Construction Company).

References

[1] F. P. Rivara and D. C. Thompson, "Prevention of falls in the construction industry: evidence for program effectiveness," *American Journal of Preventive Medicine*, vol. 18, supplement 4, pp. 23–26, 2000.

[2] A. Saloniemi and H. Oksanen, "Accidents and fatal accidents—some paradoxes," *Safety Science*, vol. 29, no. 1, pp. 59–66, 1998.

[3] O. Abudayyeh, T. K. Fredericks, S. E. Butt, and A. Shaar, "An investigation of management's commitment to construction safety," *International Journal of Project Management*, vol. 24, no. 2, pp. 167–174, 2006.

[4] G. M. Waehrer, X. S. Dong, T. Miller, E. Haile, and Y. Men, "Costs of occupational injuries in construction in the United States," *Accident Analysis and Prevention*, vol. 39, no. 6, pp. 1258–1266, 2007.

[5] G. S. Sorock, E. O. Smith, and M. Goldoft, "Fatal occupational injuries in the New Jersey construction industry, 1983 to 1989," *Journal of Occupational Medicine*, vol. 35, no. 9, pp. 916–921, 1993.

[6] B. Y. Jeong, "Occupational deaths and injuries in the construction industry," *Applied Ergonomics*, vol. 29, no. 5, pp. 355–360, 1998.

[7] C. F. Chi, T. C. Chang, and H. I. Ting, "Accident patterns and prevention measures for fatal occupational falls in the construction industry," *Applied Ergonomics*, vol. 36, no. 4, pp. 391–400, 2005.

[8] S. Mohamed, "Safety climate in construction site environments," *Journal of Construction Engineering and Management*, vol. 128, no. 5, pp. 375–384, 2002.

[9] J. Hinze and D. B. Russell, "Analysis of fatalities recorded by OSHA," *Journal of Construction Engineering and Management*, vol. 121, no. 2, pp. 209–214, 1995.

[10] X. Huang and J. Hinze, "Analysis of construction worker fall accidents," *Journal of Construction Engineering and Management*, vol. 129, no. 3, pp. 262–271, 2003.

[11] J. Hinze, *Construction Safety*, Prentice-Hall, Upper Saddle River, NJ, USA, 1997.

[12] P. Kines, "Construction workers' falls through roofs: fatal versus serious injuries," *Journal of Safety Research*, vol. 33, no. 2, pp. 195–208, 2002.

[13] B. J. M. Ale, L. J. Bellamy, H. Baksteen et al., "Accidents in the construction industry in the Netherlands: an analysis of accident reports using Storybuilder," *Reliability Engineering and System Safety*, vol. 93, no. 10, pp. 1523–1533, 2008.

[14] H. M. Johnson, A. Singh, and R. H. F. Young, "Fall protection analysis for workers on residential roofs," *Journal of Construction Engineering and Management*, vol. 124, no. 5, pp. 418–428, 1998.

[15] H. Heinrich, *Industrial Accident Prevention*, McGraw-Hill, New York, NY, USA, 1931.

[16] J. L. Komaki, "Toward effective supervision: an operant analysis and comparison of managers at work," *Journal of Applied Psychology*, vol. 71, no. 2, pp. 270–279, 1986.

16

A Morphological Model for Extracting Road Networks from High-Resolution Satellite Images

Mohamad M. Awad

National Council for Scientific Research, National Center for Remote Sensing, P.O. Box 11-8281, Beirut 11072260, Lebanon

Correspondence should be addressed to Mohamad M. Awad; mawad@cnrs.edu.lb

Academic Editor: João M. Tavares

Urban planning depends strongly on information extracted from high-resolution satellite images such as buildings and roads features. Nowadays, most of the available extraction techniques and methods are supervised, and they require intensive labor work to clean irrelevant features and to correct shapes and boundaries. In this paper, a new model is implemented to overcome the limitations and to correct the problems of the known and conventional techniques of urban feature extraction specifically road network. The major steps in the model are the enhancement of the image, the segmentation of the enhanced image, the application of the morphological operators, and finally the extraction of the road network. The new model is more accurate position wise and requires less effort and time compared to the traditional supervised and semi-supervised urban extraction methods such as simple edge detection techniques or manual digitization. Experiments conducted on high-resolution satellite images prove the high accuracy and the efficiency of the new model. The positional accuracy of the extracted road features compared to the manual digitized ones, the counted number of detected road segments, and the percentage of completely closed and partially closed curves prove the efficiency and accuracy of the new model.

1. Introduction

Urban areas are the inhabited areas on earth where most dynamic changes can be observed. They are characterized by a trend of permanent expansion and growth. For the first time ever more than 60% of earth's entire human population can be found in urban areas.

These urban areas are characterized by the existence of diversity of complex features such as buildings, roads, and bridges. One important urban feature is the road network which is extracted from up-to-date high-resolution images such as Ikonos, Quickbird, WorldView, and GeoEye. Many modern applications in many disciplines use these networks. These applications are widely used by a large number of individuals, organizations, and institutes such as car navigation system, emergency rescue system, and urban and environmental planning. There are two types of approaches for extraction and identification of road network features in a remotely sensed image: manual and task-specific automated approaches [1]. In the past, most feature extraction methods are usually done by visual interpretation and manual digitizing from aerial photographs or satellite imagery. Although this is still the predominant approach to geospatial data production, but it consumes labor and time for manual feature extraction or identification. Successful development of feature extraction technologies from high-resolution satellite can greatly increase its usability in geographic databases updating and remote sensing application. At the present, there are many algorithms, semiautomatic [2, 3] and manual for the extraction of road network features. Automated Feature Extraction (AFE) methods have been the long-term goal of geospatial data production workflows for the past 30 years; extracted features over small training sets can then be applied to larger areas, reducing the extraction time required by several orders of magnitude [4]. Automated Feature Extraction applications use spectral, ancillary information, and feature characteristics such as spatial association, size, shape, texture, pattern, and shadow. Machine-learning algorithms and techniques serve to automate the feature recognition process [1].

Many researchers concentrated on developing automated systems to detect the urban area. Karathanassi et al. [5] used building density information to classify residential regions. They benefit from texture information and segmentation to extract the residential areas. Unfortunately, they had several parameters to be adjusted manually. Mathematical morphological operators are used widely in the extraction process of urban features within the AFE methods. Mura et al. [6], Bellens et al. [7], and Akcay and Aksoy [8] used mathematical morphological operators for automated extraction of multiscale urban features, such as buildings, shadows, roads, and other man-made objects. Benediktsson et al. [9] used mathematical morphological operations to extract structural information to detect the urban area boundaries in satellite images. This method is based on neural network architecture. Valero et al. [10] deployed the mathematical morphological operators of opening and closing to extract road features. The method is based on advanced directional morphological operators, namely, path opening and path closing [11]. The above research showed success in extracting different urban features, but with limitations due to the complex content and structure of the high-resolution satellite images such as the road-width can vary considerably, presence of lane markings, vehicles, shadows cast by buildings and trees, and changes in surface material.

Although AFE-existing techniques offer several advantages, such as saving hours of labor time and reducing budgets for heads-up digitizing to create or maintain GIS data, AFE requires posttreatment or guidance to overcome some deficiencies which exist in some AFE methods. Other AFE methods require training and supervision before they automate the process of urban feature extraction such as Feature Analyst (Feature Analyst for ArcMap http://www.esri.com/). Finally others require the image to be converted to binary format and to be clear of noises and other existing features. In addition, they require the scanned map to be large scale before digitizing such as ArcScan (ArcScan for ArcMap http://www.esri.com/) which means that it works for utility and cadastral maps only.

To overcome the above problems and to create a robust toward an unsupervised road features extraction model, several methods are combined. The new model consists of several processing steps such that each step is a prerequisite for the success of the next one. The initial step is to improve the extraction of information by applying many different image enhancement techniques. The next step in the model requires the use of an unsupervised classification technique to separate the road features from other features. Then morphological operators are used to improve the feature extraction. Finally, a reliable edge detection technique is used to extract the features which are provided to a raster to vector conversion algorithm. All the above details will be explained later in the following sections.

The remaining of this paper is divided into two more sections and conclusion. After the introduction above, Section 2 describes in detail the new model including all its components. Section 3 is a complete explanation and details about the experimental results to extract road network features from high-resolution satellite images using the new model. Finally, conclusion and future perspectives provide a summary of the completed steps in the model and the

planned ones. The reader is urged to further investigate the important use of the extracted urban features in the planning and security reasons in many literatures.

2. The New Urban Feature Extraction Model

The complexity of the high-resolution satellite images requires sophisticated techniques to handle different tasks which are necessary to extract information such as urban objects. Due to the problems which are inherited by every satellite image such as radiometric, atmospheric, and sensor malfunction, there is need to eliminate these problems. The noises, missing information, and atmospheric effects such as haziness in the images require preprocessing using restoration and filtering techniques to improve information extraction from the image. So, enhancement of the image is the first step in the model. The next step is to classify the image in a semisupervised or unsupervised way using existing or newly implemented segmentation/classification methods.

2.1. Classification/Segmentation of the Enhanced Image. There are several commercial-known clustering algorithms such as Fuzzy C-means (FCM) [12] and ISODATA [13]. Moreover, there are clustering methods which consist of a combination of Artificial Neural Network and evolutionary methods such as Self-Organizing Maps (SOMs) [14] and Hybrid Dynamic Genetic Algorithm (HDGA) [15]. The selection of the segmentation method is very important in eliminating any extra information which can represent nonroad network features. The literature has been investigated, and several approaches which are combinations of two or more methods and which show high accuracy in satellite image segmentation are examined. These approaches are Self-Organizing Maps and Hybrid Genetic Algorithm (SOMs-HGA) [16], Fuzzy C-Means and Hybrid Dynamic Genetic Algorithm (FCM-HDGA) [17], and the combination of both SOMs and FCM [18]. In this paper the unsupervised SOMs-FCM approach is used because of the efficiency in the segmentation of high-resolution images which includes speed and accuracy. In brief the standard Fuzzy C-Means clustering algorithm works as follows. Given a set of n data patterns, $x = xi, \ldots, xn$, the FCM algorithm minimizes the weights within the group sum of the squared error objective function $J(U, V)$:

$$J(U, V) = \sum_{k=1}^{n} \sum_{i=1}^{c} u_{ik}^m d^m (\mathbf{x}_k, v_i), \qquad (1)$$

where $\mathbf{x}_k$ is the kth p-dimensional data vector, v_i is the prototype of the cluster center i, u_{ik} is the degree of membership of $\mathbf{x}_k$ in the ith *cluster*, and m is a weighting exponent on each fuzzy membership. The function $d_{ik}(\mathbf{x}_k, v_i)$ is a distance measure between object $\mathbf{x}_k$ and cluster centre v_i, n is the number of objects (pixels of an image), and c is the number of clusters.

The Self-Organizing Maps (SOMs) algorithm is used to modify FCM by providing it with the initial cluster centers which reduce the required time for FCM to converge to a solution and optimize and stabilize the final solution.

The new FCM objective function is explained in the following equation:

$$J(U, W) = \sum_{k=1}^{n} \sum_{i=1}^{c} u_{ik}^{m} \frac{\sum_{j=1}^{3} d^{m}\left(\mathbf{x}_{kj}, \mathbf{W}_{ij}\right)}{3}, \qquad (2)$$

where W_{ij} is the vector of weights representing the cluster centers obtained by SOMs final iteration and multiplied by 255 (grey level values). One should notice that i represents the number of neurons defined as the size of SOMs network (i.e., 16×16) and the final cluster numbers (this varies if threshold is used). In addition, every neuron has three weights for every band which means that a distance must be computed between every pixel and center in a specific band. SOMs-FCM is very efficient compared to some commercial segmentation/classification methods such as ISODATA which stands for Iterative Self-Organizing Data algorithm. The ISODATA algorithm has some further refinements by splitting and merging of clusters. Clusters are merged if either the number of pixels in a cluster is less than a certain threshold or if the centers of two clusters are closer than a certain threshold. Clusters are split into two different clusters if the cluster standard deviation exceeds a predefined value and the number of members (pixels) is twice the threshold for the minimum number of members. The algorithm is similar to the k-means algorithm with the distinct difference that the ISODATA algorithm allows for different number of clusters while the k-means algorithm assumes that the number of clusters is known a *priori*. The efficiency of the selected segmentation algorithm is due to the instability of the existing clustering algorithms such as ISODATA algorithm which provides different results each time the threshold value and the number of iterations are changed even when the number of clusters is fixed.

The comparison between SOMs-FCM and ISODATA is completely explained in [18], and the superiority of SOMs-FCM over ISODATA is empirically proved.

2.2. Enhancement of the Road Extraction Process Using Morphological Operators. After the segmentation of the image, several classes are obtained which represent urban and nonurban objects. The road network objects are selected, and they are subjected to a repetitive and equal number of morphological operators [11] of *opening* and *closing*. These operators depend on other operators which are the *erosion* and *dilation* operators. The last two operations are fundamental to the morphological image processing. *Dilation* is an operation that "grows" or "thickens" objects in a binary image. The specific manner and extent of this thickening is controlled by a shape referred to as a *structuring element*. Mathematically, *dilation* is defined in terms of set operations. The *dilation* of A by B, denoted $A \oplus B$, is defined as

$$A \oplus B = \{z \mid (B)_z \cap A \neq \phi\}. \qquad (3)$$

Here ϕ is the empty set, and B is the *structuring element*. In other words, the *dilation* of A by B is the set consisting of all the *structuring element* origin locations, where the *reflected* and *translated* B overlaps at least some portion of A.

Erosion "shrinks" or "thins" objects in a binary image. As in dilation, the manner and extent of shrinking is controlled by the *structuring element*.

The mathematical definition of *erosion* is similar to that of *dilation* except that the intersection is equal to an empty set. The *erosion* of A by B, denoted $A \ominus B$, is defined as

$$A \ominus B = \{z \mid (B)_z \cap A^c = \phi\}. \qquad (4)$$

In other words, *erosion* of A by B is the set of all *structuring element* origin locations, where the translated B has no overlap with the background of A.

The *opening* operator $\circ$ is the combination of *erosion* followed by *dilation* such that

$$A \circ B = \{A \ominus B\} \oplus B. \qquad (5)$$

On the other hand, *closing* $\bullet$ is the combination of *dilation* followed by *erosion* such that

$$A \bullet B = \{A \oplus B\} \ominus B. \qquad (6)$$

The morphological process requires the *structuring element* to be set before any operation takes place. There are many shapes and sizes for the *structuring element* such as disk, ball, square, and line. The selection of the shape and size depends on the characteristics of the information in the image and the speed requirements of the process. In this paper several *structuring element* shapes have been tested in order to obtain the best combination in order to reduce irrelevant information and to close gaps (missing information) in every road segment shape.

2.3. Edge Detection of the Road Segments. After the morphological operator, the edges of the urban objects are delineated using an edge detection method. Edge detection is one of many efficient techniques in image segmentation. Edge detection plays an important role in reducing significantly the amount of data and filters out information that may be regarded as less relevant, yet preserving the important structural properties of an image. There are many methods for edge detection, such as search-based, zero-crossing-based, and active-based contours and deformable models [19–22]. However, these methods suffer from sensitivity to noise which does not represent edges. In addition, they are inefficient in providing complete edge boundaries.

To overcome these problems, Canny-Deriche [23] edge detection technique is considered to extract the boundaries of different road networks from the processed high-resolution images.

Finally, the conversion of the raster edges of the road networks to vector format is done using any available software such as ArcMap (a trademark of ESRI http://www.esri.com/) toolbox. The following graph (Figure 1) shows the hierarchical steps of the model. The number of morphological operators of open and close depends on how successful the previous steps. The existence of noise and missing information due to the existence of clouds or malfunction of the sensor may require more work before the morphological operators.

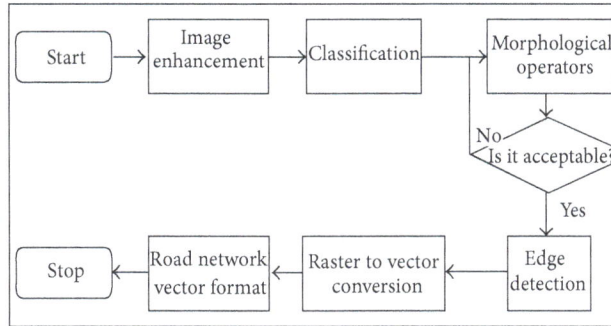

FIGURE 1: Graph showing the steps necessary to extract road network from high-resolution image.

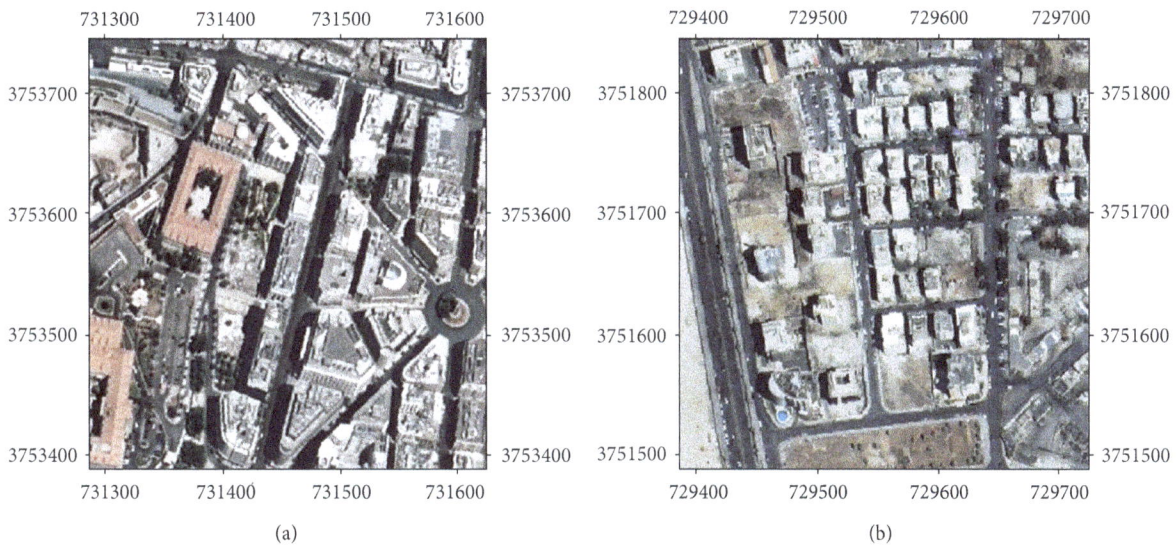

(a)

(b)

FIGURE 2: QuickBird-enhanced original images: (a) first experiment, (b) second experiment.

3. Experimental Results

In order to prove the efficiency of the model, QuickBird satellite images are used in the experiments. QuickBird has 5 bands one panchromatic with 0.6 m, three true color bands, and one near infrared with spatial resolution of 2.4 m and spectral wavelength between 0.45 μm and 0.85 μm. The images are projected using the Universal Transverse Mercator (UTM) with zone 36 north.

In the first experiment Figure 2(a) is used to obtain road networks. The figure with a size of 573×540 represents a touristic, government, and financial area in the Center of Beirut, capital of Lebanon. The figure was enhanced to remove noise and to improve the scene illumination. The enhancements include adjusting the contrast and brightness and applying smoothing filter to remove noise. The same enhancements are applied to the second experiment which is a QuickBird image of size 450×450 pixels. The image represents a coastal area in Beirut "Ramleh El-Baida—White Sand area" (Figure 2(b)). The diversity of the themes in the images makes it more complex for processing them in order to extract roads, and this is intentionally done.

After enhancement, the images are classified using SOMs-FCM algorithm into several classes including roads in red color and shadow in blue (Figures 3(a) and 3(b)). Sometime, different road network classes may exist; this richness in the number of road classes depends on the material (type of asphalt) used to pave the roads and the surrounding environment (dusty or sandy area). Moreover, classification indicates the status of road: if the classification is homogeneous (one simple class), then the road is made of the same material or the road does not have any structural problem. If the homogeneity is intermittent, this indicates that the quality of the pavement material is not the same.

The road network class is extracted from the classified images, and it is converted into binary image where the black indicates the background and white color indicates roads (Figures 4(a) and 4(b)). In the first image the shadow class is included (grey color) because the majority of the roads are covered by the shadow of the buildings.

One can see clearly that homogeneity is not complete with some roads. Some black gaps exist which represent different classes and which indicate the condition or the material property of the asphalt and pavements.

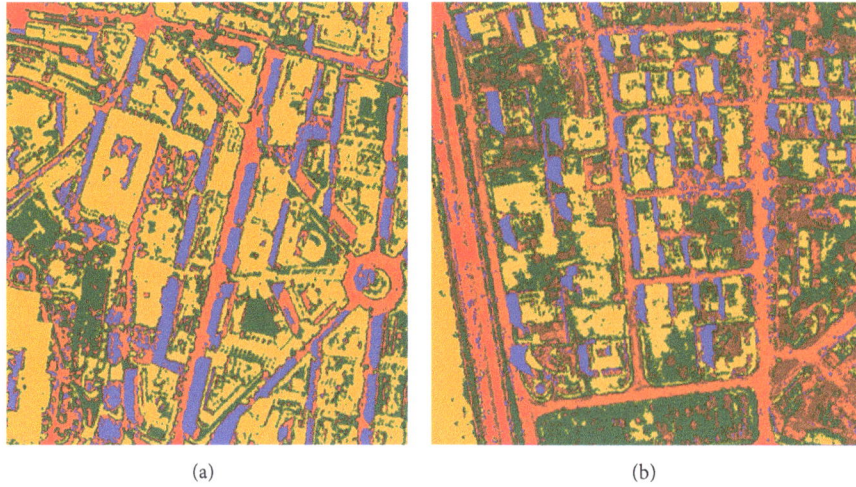

FIGURE 3: Classified images using SOMs-FCM: (a) first experiment, (b) second experiment.

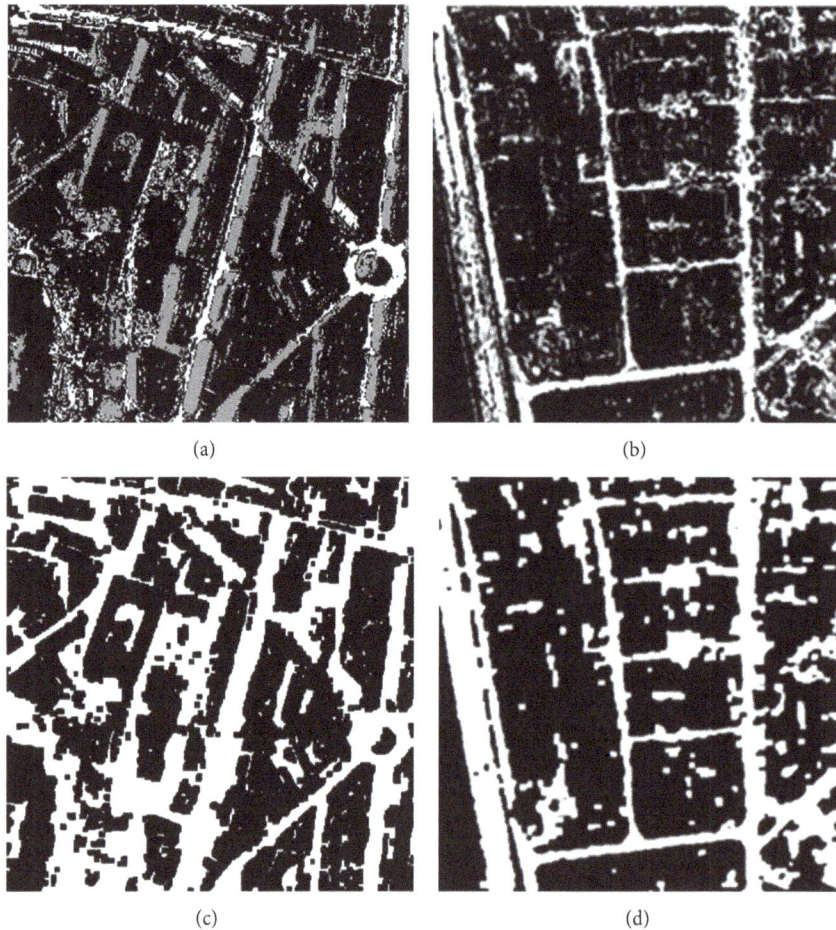

FIGURE 4: Road networks ((a) and (b)) extracted from classified images and ((c) and (d)) fixed by the morphological operators.

In order to extract linear and continuous roads features the regions must be homogeneous and continuous; in other words, the following should be applied with respect to the background such that the partition of image I into n road regions $R1, R2, \ldots, Rn$ must satisfy the following

conditions: R_i is a connected region, $i = 1, 2, \ldots, n$ (homogeneous).

However, if this condition is not satisfied, then the morphological operators must be used to reduce the heterogeneity. The open and close operator which depends on the

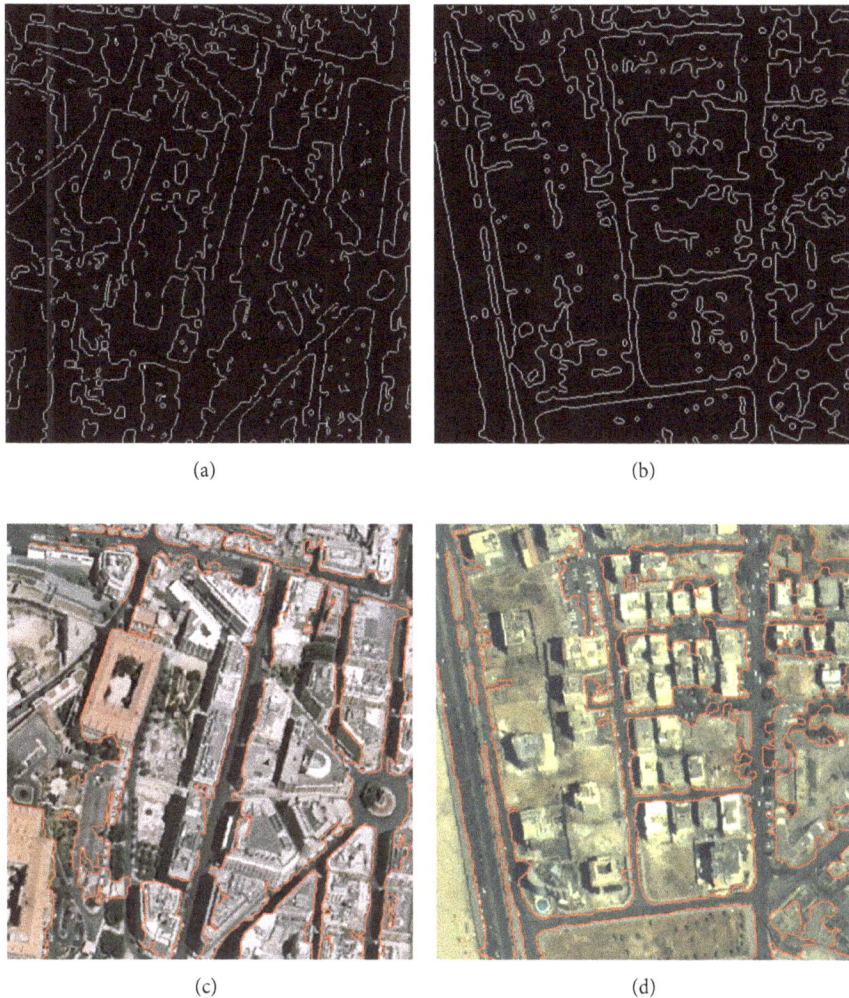

(a)

(b)

(c)

(d)

FIGURE 5: Final results of ((a) and (b)) applying Canny-Deriche and ((c) and (d)) conversion and elimination.

famous morphological operators of dilation and erosion as explained before are used several times based on trials and errors outcomes. The results of the morphological operators for both images in the two experiments are shown in Figures 4(c) and 4(d). The reader can see that several road segments are fixed by filling gaps with more white pixels and removing isolated ones and leaving others which may represent parking lots or bare lands. After the conversion from raster to vector format in the final step, the user can eliminate irrelevant and false polygons.

In addition to the dilation and erosion, the morphological operators of open and close depend on the type of the *Structuring Element* (SE) which can be of different shapes such as disk, line, square, or diamond. Each parameter of the SE defines the way the information are added or removed. In this paper, several lines, disks, and squares were used to add and remove pixels. There should be some measure to continue or halt the process of morphological operators. Currently this is done manually with several test strategies in order to find a combination of many open and close operators. It is repetitive similar and opposite morphological operators. The research will continue in the future to find an

automated and an optimal process of using morphological operators.

The final step is to use Canny-Deriche to detect edges and to filter out unneeded pixels according to a specific threshold. Figures 5(a) and 5(b) show the canny edge detection and the corresponding vector layer (overlaid on the original image) Figures 5(c) and 5(d) after removing redundant and false information.

The success of this new method depends on the use of a very high-resolution image with little conflict between spectral signatures such as road asphalt, vegetation, and shadow of other urban objects. In other words, it depends on the success of the segmentation approaches.

In general, the method is still in the early phase and the improvements which are planned for the future will make it more successful toward being completely unsupervised.

In addition, to reduce the time required to extract road network features compared to the manual one, another criterion is considered which is the continuity or the discontinuity of the lines representing road networks (percentage of the closed lines). The accuracy can be compared to the proportion of the extracted complete and incomplete road

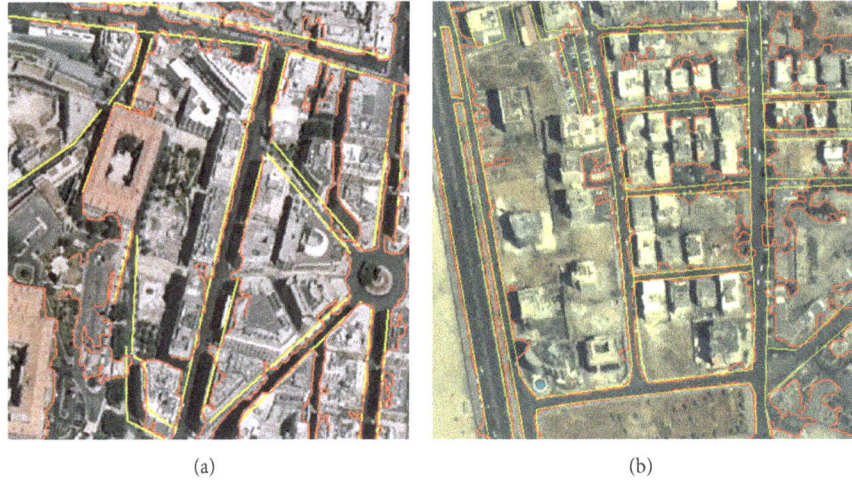

(a)   (b)

FIGURE 6: Road networks: manually digitized (yellow) and extracted by the new model (red).

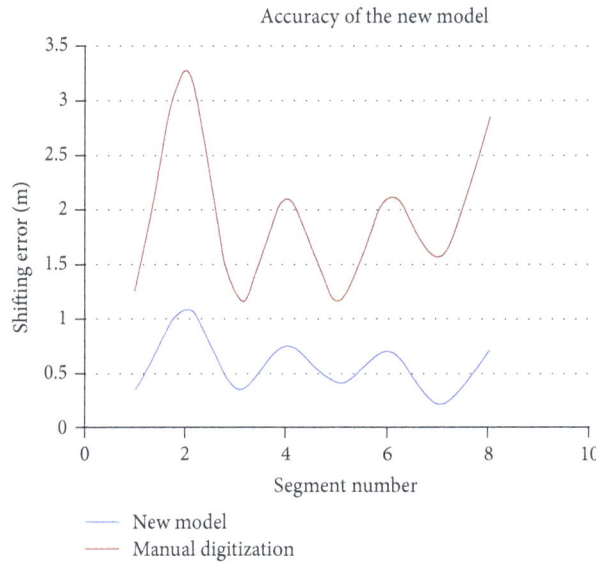

FIGURE 7: Accuracy of the model compared to the manual digitization.

features to the total number of extracted road networks equation:

$$A = \frac{N_C + N_P}{N_T}, \qquad (7)$$

where A stands for accuracy, N_C is the number of complete road segments, N_P is the number of partially complete road segments, and N_T is the total number of road segments.

In the two experiments there is 32 road segments; 30 of them are complete and partially complete. The efficiency of extracting road segments is equal to 93%.

To continue our investigation about the accuracy of the model a computer operator is asked to digitize the road segments according to the normal digitizing process which is used by the GIS experts every day. The digitized layer is overlaid on the extracted road features by our model (Figures 6(a)

and 6(b)). There are several features and characteristics in the road network extraction process which can be compared such as the accurate spatial position of the roads and false detection of the road features. Taking few samples (road segments) and after measuring the distance between these samples and the accurate position of the road edges, we noticed that there are several meter shifting from the correct position of the road edges. This shifting is caused by the limitation of the human visual system caused by low contrast of the images and many other issues related to the quality of these images which in turn cause the operator to create incomplete segments which can be a common issue with the new model. Figure 7 shows the shifting in meters from the real road edges for some selected segments extracted by the manual method and the new model. The accuracy of the new model as indicated by the graph is almost 3 times better than that of the manual digitization.

4. Conclusion

In this paper an attempt is made to improve the process of road network features extraction from high-resolution satellite images. The model consists of many methods and techniques. Each step has specific role such as reducing the efforts incurred on the user by the manual process and increasing the efficiency of road segments extraction. Although, several issues in the model require tuning and adjustments, this method is still much better than the manual one which requires more time and human efforts and in turn requires a larger budget. The accuracy of the position of the line depicting the roads is much higher than the manual one. The continuity of the line is another criterion which indicates the efficiency of the model. The accuracy of the model is almost 3 times better than the manual one with respect to the spatial position of the segments especially where the area is not affected by the shadow of the buildings or covered by vegetation (high-quality satellite image). The human interference in this new model is in the lowest rate. In addition, when the road networks are extracted by the model their direction and positions are respected accurately which is not the same with respect to the manual digitization (supervised or semisupervised). The model can be used not only to extract road network but also to help in identifying problems in the road network (classification phase) such as rehabilitation of some segments due to severe weather or extensive use by heavy machines.

The model can be improved in the future to include a method or a technique which can optimize automatically the quality of all the components of the model in order to become completely unsupervised.

References

[1] J. S. Blundell and D. W. Opitz, *Object Recognition and Feature Extraction from Imagery: the Feature Analyst Approach*, Visual Learning Systems, Missoula, Mont, USA, 2006.

[2] D. Chaudhuri, N. K. Kushwaha, and A. Samal, "Semi-automated road detection from high resolution satellite images by directional morphological enhancement and segmentation techniques," *IEEE Journal of Applied Earth Observations and Remote Sensing*, vol. 5, no. 5, pp. 1538–1544, 2012.

[3] S. Yuanzheng, G. Bingxuan, H. Xiangyun, and D. Liping, "Application of a fast linear feature detector to road extraction from remotely sensed imagery," *IEEE Journal of Applied Earth Observations and Remote Sensing*, vol. 4, no. 3, pp. 626–6631, 2011.

[4] D. A. Lavigne, G. Hong, and Y. Zhang, "Performance assessment of automated feature extraction tools on high resolution imagery," in *Proceedings of the MAPPS/ASPRS 2006 Fall Conference*, San Antonio, Tex, USA, November 2006.

[5] V. Karathanassi, C. Iossifidis, and D. Rokos, "A texture-based classification method for classifying built areas according to their density," *International Journal of Remote Sensing*, vol. 21, no. 9, pp. 1807–1823, 2000.

[6] M. D. Mura, J. A. Benediktsson, F. Bovolo, and L. Bruzzone, "An unsupervised technique based on morphological filters for change detection in very high resolution images," *IEEE Geoscience and Remote Sensing Letters*, vol. 5, no. 3, pp. 433–437, 2008.

[7] R. Bellens, S. Gautama, L. Martinez-Fonte, W. Philips, J. C. W. Chan, and F. Canters, "Improved classification of VHR images of urban areas using directional morphological profiles," *IEEE Transactions on Geoscience and Remote Sensing*, vol. 46, no. 10, pp. 2803–2813, 2008.

[8] G. Akcay and H. Aksoy, "Morphological segmentation of urban structures," in *Proceedings of the 4th IEEE GRSS/ISPRS Joint Workshop on Remote Sensing and Data Fusion over Urban Areas*, Paris, France, April 2007.

[9] J. A. Benediktsson, M. Pesaresi, and K. Arnason, "Classification and feature extraction for remote sensing images from urban areas based on morphological transformations," *IEEE Transactions on Geoscience and Remote Sensing*, vol. 41, no. 9, pp. 1940–1949, 2003.

[10] S. Valero, J. Chanussot, J. A. Benediktsson, H. Talbot, and B. Waske, "Advanced directional mathematical morphology for the detection of the road network in very high resolution remote sensing images," *Pattern Recognition Letters*, vol. 31, no. 10, pp. 1120–1127, 2010.

[11] L. Najman and H. Talbot, *Mathematical Morphology: from Theory to Applications*, ISTE-Wiley, 2010.

[12] J. Bezdek, *Pattern Recognition with Fuzzy Objective Function Algorithm*, Plenum Press, New York, NY, USA, 1981.

[13] J. Tou and R. C. Gonzalez, *Pattern Recognition Principles*, Addison-Wesley, Reading, Mass, USA, 1974.

[14] T. Kohonen, *Self-Organizing Maps*, vol. 30 of *Springer Series in Information Sciences*, Springer, 2001.

[15] M. M. Awad and K. Chehdi, "Satellite image segmentation using hybrid variable genetic algorithm," *International Journal of Imaging Systems and Technology*, vol. 19, no. 3, pp. 199–207, 2009.

[16] M. Awad, K. Chehdi, and A. Nasri, "Multicomponent image segmentation using a genetic algorithm and artificial neural network," *IEEE Geoscience and Remote Sensing Letters*, vol. 4, no. 4, pp. 571–575, 2007.

[17] M. Awad, K. Chehdi, and A. Nasri, "Multi-component image segmentation using a hybrid dynamic genetic algorithm and fuzzy C-means," *IET Image Processing*, vol. 3, no. 2, pp. 52–62, 2009.

[18] M. M. Awad and A. Nasri, "Satellite image segmentation using self-organizing maps and fuzzy C-means," in *Proceedings of the 9th IEEE International Symposium on Signal Processing and Information Technology (ISSPIT '09)*, pp. 398–402, Ajman, UAE, December 2009.

[19] I. Sobel and G. Feldman, "A 3x3 isotropic gradient operator for image processing," in *Pattern Classification and Scene Analysis*, R. Duda and P. Hart, Eds., pp. 271–272, John Wiley and Sons, 1973.

[20] J. Canny, "Computational approach to edge detection," *IEEE Transactions on Pattern Analysis and Machine Intelligence*, vol. 8, no. 6, pp. 679–698, 1986.

[21] J. Shen and S. Castan, "An optimal linear operator for edge detection," in *Proceedings of the IEEE Conference on Computer Vision and Pattern Recognition*, pp. 109–114, Miami Beach, Fla, USA, June 1986.

[22] M. Kass, A. Witkin, and D. Terzopoulos, "Snakes: active contour models," *International Journal of Computer Vision*, vol. 1, no. 4, pp. 321–331, 1988.

[23] R. Deriche, "Using Canny's criteria to derive a recursively implemented optimal edge detector," *International Journal of Computer Vision*, vol. 1, no. 2, pp. 167–187, 1987.

17

Modeling Laterally Loaded Single Piles Accounting for Nonlinear Soil-Pile Interactions

Maryam Mardfekri,[1] Paolo Gardoni,[2] and Jose M. Roesset[1]

[1] *Zachry Department of Civil Engineering, Texas A&M University, College Station,*
  TX 77843-3136, USA
[2] *Department of Civil and Environmental Engineering, University of Illinois at Urbana-Champaign,*
  IL 61801, USA

Correspondence should be addressed to Paolo Gardoni; gardoni@illinois.edu

Academic Editor: Sadhan C. Jana

The nonlinear behavior of a laterally loaded monopile foundation is studied using the finite element method (FEM) to account for soil-pile interactions. Three-dimensional (3D) finite element modeling is a convenient and reliable approach to account for the continuity of the soil mass and the nonlinearity of the soil-pile interactions. Existing simple methods for predicting the deflection of laterally loaded single piles in sand and clay (e.g., beam on elastic foundation, p-y method, and SALLOP) are assessed using linear and nonlinear finite element analyses. The results indicate that for the specific case considered here the p-y method provides a reasonable accuracy, in spite of its simplicity, in predicting the lateral deflection of single piles. A simplified linear finite element (FE) analysis of piles, often used in the literature, is also investigated and the influence of accounting for the pile diameter in the simplified linear FE model is evaluated. It is shown that modeling the pile as a line with beam-column elements results in a reduced contribution of the surrounding soil to the lateral stiffness of the pile and an increase of up to 200% in the predicted maximum lateral displacement of the pile head.

1. Introduction

Pile foundations are widely used to support laterally loaded structures especially offshore. The extensive growth of wind farms around the world has raised new concerns about the accuracy of the analysis and design methods for laterally-loaded large-diameter monopiles (the most popular foundation structure for offshore wind turbines).

Common methods for the analysis of laterally loaded single piles can be generally classified into two categories: (1) Winkler (elastic) foundation models and (2) continuous models accounting for the coupling of forces and displacements in the soil along the pile. In each category the analysis may be static (monotonic or cyclic loading) or dynamic. Also the behavior of the soil, pile, and soil-pile interaction may be considered as linear or nonlinear.

Winkler foundation models are popular because of their simplicity and reasonable accuracy. When the elastic stiffness of the foundation can be considered constant with depth

one can even obtain simple closed form solutions for the pile head stiffness and flexibility [1]. The main difference between the different Winkler foundation models available is in the selection of the foundation stiffness coefficients. For dynamic problems Novak [2] has proposed the use of Winkler foundation coefficients based on Baranov's equations [3] for in plane and out of plane vibrations of a disk. The corresponding horizontal k_x and rotational k_φ springs per unit of length along the pile are functions of a dimensionless frequency $a_0 = \Omega R/C_s$, where Ω = the frequency in radians/second, R = the radius of the pile, and C_s = the shear wave velocity of the soil. Unfortunately the horizontal term tends to zero at a zero frequency representing the static case. As a result it is common to use the values corresponding to a dimensionless frequency of 0.3 for smaller frequencies [1]. In that case, $k_x = 4G$ and $k_\varphi = 2.6666\,GR^2$, where G = the shear modulus of the soil.

For nonlinear analyses the p-y method is the most commonly used in this category. It employs an elastic beam column member to model the pile and nonlinear horizontal springs to represent the soil reactions. The p-y curves describe the nonlinear behavior of the soil springs. They were developed first by Matlock [4] for soft clays under the water table. Reese and Welch [5] and Reese et al. [6] developed p-y curves for hard clays subjected to monotonic and cyclic loading, above and under the water table, respectively. Analyzing the results of the full scale tests conducted by Reese et al. [6], Dewaikar et al. [7] presented a modified approach to construct p-y curves in stiff clay. In another study Kim and Jeong [8] developed a framework based on 3D finite element analysis for determining a p-y curve. The p-y curves for sands were also developed by Reese et al. [9] for monotonic and cyclic loading. Briaud et al. [10] developed an alternative method to obtain the p-y curves directly from pressuremeter tests. The method was reasonably accurate but complicated and time consuming, so Briaud [11] developed a simpler approach called "simple approach for lateral loads on piles" or SALLOP, using the pressuremeter limit pressure and the pressuremeter modulus.

A number of recent studies have been conducted to predict the behavior of laterally loaded piles in different soil conditions. Sanjaya Kumar et al. [12] used ABAQUS and the p-y method to study the behavior of laterally loaded pile foundations in high marine clay with high potential to swell upon wetting and shrink upon drying. Suleiman et al. [13] conducted a test to measure the soil-pile interaction pressure for small diameter piles in loose sand that the results can be used in developing the soil force-displacement relationship (i.e., the soil reaction or the p-y curve). An equivalent model for a laterally loaded linear pile-soil system was presented by Chioui and Chenu [14] using artificial lateral springs.

Continuous modeling of the pile and the surrounding soil are mostly done using finite element or boundary element models. Both methods can provide rigorous solutions accounting for soil-pile interaction under static and dynamic loading. For the linear case an accurate solution was proposed by Blaney et al. [15] using the consistent boundary matrix developed by Kausel [16] to reproduce the soil cavity occupied by the pile and adding then the pile enforcing compatibility of horizontal and vertical displacements between pile and soil along the pile. An extensive number of studies were carried out by Sanchez Salinero [1] comparing the results of this approach to those provided by a variety of other methods and proposing approximate formulas for the pile head stiffness. This approach is only valid however in the linear elastic range. The finite element method is particularly convenient when desiring to account for nonlinear effects including the nonlinear behavior of the soil and of the soil-pile interface.

A 3D nonlinear finite element analysis of a pile foundation in which both the soil and the pile are modeled with 3D finite elements can be quite expensive and time consuming, particularly when incorporating nonlinear behavior. As a result some investigators have used finite element models that represent the pile by an elastic beam column member without transverse dimensions (only the centroidal axis) and only the soil with 3D solid elements. This method takes into account the continuity of the soil mass and is easy to use for linear static and dynamic analysis. However, the most important

limitation of this approach is that it does not take into account the dimension of the pile section.

The work presented in this paper is part of a broader research effort to assess the reliability of foundations for offshore wind turbines. These are normally single large diameter hollow piles. However many different methods have been used and investigated by previous researchers for analysis of laterally loaded piles, most of previous studies are focused on onshore pile foundations with diameters relatively smaller than those in the offshore industry. The first objective of this paper is to compare the results of different methods of analysis of laterally loaded piles and illustrate the possible variability of the results. A second objective is to investigate how the consideration of pile diameter affects the accuracy of simplified models. As a first step the models used for the analyses of these foundations are validated. Then the model selected is implemented in the computer program ABAQUS using 3D brick elements to discretize the soil around the pile and shell elements to model the hollow pile. The results obtained with this model for linear and nonlinear analyses are compared to those provided by a variety of other methods used in practice.

In the following, we first examine four different models used for linear analysis of single pile foundations and evaluate the influence of accounting for the pile diameter in the simplified linear FE analyses. In the next section, we improve the 3D finite element model by accounting for the nonlinearity of the soil and soil-pile interaction. Two common simplified nonlinear models are then assessed using this model for mono-piles in sand and clay.

2. Linear Analyses

Analyses considering linear soil behavior and perfect bonding between the pile and the surrounding soil are conducted first. The pile selected for the study is hollow with a diameter of 4 m and the properties listed in Table 1. Four different models are studied.

(1) The first model is a 3D finite element model of both the soil around the cavity occupied by the pile (solid elements) and for the pile, with shell elements for hollow piles and brick elements for solid piles (shown in Figure 1).

(2) The second, simpler, model reproduces the soil with solid elements filling the space without any cavity. The pile is represented by the centroidal axis of a 1D beam column coinciding with the central axis of the soil model, enforcing only compatibility of horizontal displacements between the nodes of the pile and those of the soil along the axis.

(3) The third model is the one proposed by Blaney et al. [15] with the consistent boundary matrix with the radius of the cavity representing the soil and enforcing compatibility of both horizontal and vertical displacements between the soil and the pile along its sides.

(4) The fourth model is a beam on an elastic (Winkler) foundation with horizontal and rotational springs along the side of the pile. The constants selected for the foundation are $k_x = 4G$ and $k_\varphi = 2.6666GR^2$.

TABLE 1: Properties of the pile.

Parameter	Symbol	Value
Penetration depth (m)	L	21.0
Diameter (m)	D	4.00
Wall thickness (m)	t	0.05
Modulus of elasticity (kPa)	E_p	2.0E8
Unit weight (kN/m^3)	γ_p	87.00
Poisson ratio	ν_p	0.30

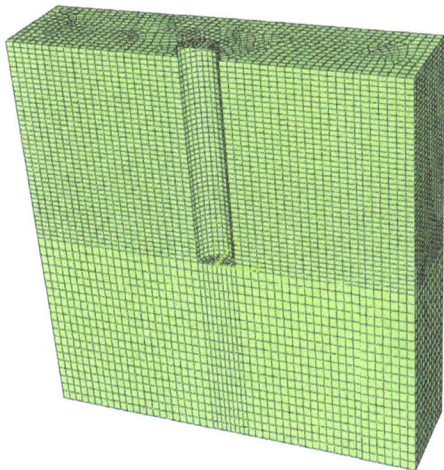

FIGURE 2: Deformation of the soil with the 3D model of the pile.

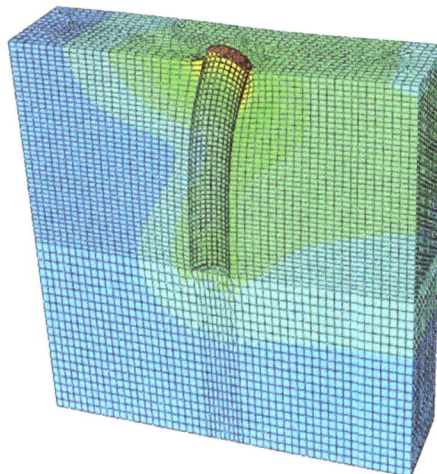

FIGURE 1: 3D finite element model of the pile foundation.

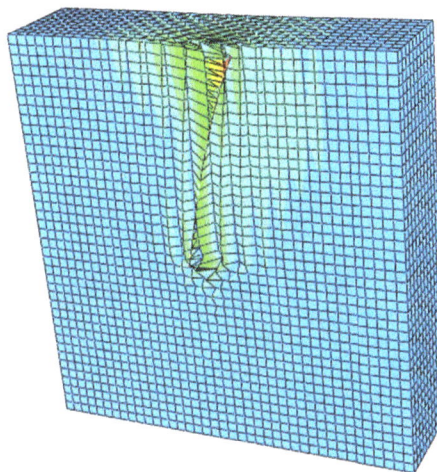

FIGURE 3: Deformation of the soil with the 1D model of the pile.

The pile is subjected at the head to a vertical load of 5,000 kN, a horizontal load of 2,503 kN, and a moment of 84,983 kNm. These are values obtained considering the extreme forces on an offshore wind turbine. For the linear analyses the soil is assumed to have a Young's modulus $E_s = 50,000$ kPa, a Poisson's ratio $\nu_s = 0.3$, and a unit weight $\gamma_s = 20$ kN/m^3.

The predicted deflections at the pile head by the four models are 20.9 mm for the 3D FE pile model, 68.3 mm for the 1D FE pile model, 20.5 mm for the consistent boundary matrix and 24.3 mm for the Winkler foundation. The deformation of the soil with the 3D finite element model is shown in Figure 2 while Figure 3 shows the corresponding deformations with the 1D model of the pile. The results, obtained using the 3D finite element model, are in good agreement with the approach that employs consistent boundary matrix (less than 2% off). The agreement with the results of the Winkler foundation is not quite as good but still acceptable (about 20% off). The model without the cavity and with the pile as a 1D linear element yields deflections that are 200% too large.

To understand better the reasons for this large discrepancy it was decided to conduct studies for other pile sizes. Clearly the results of the 1D model are only a function of the soil properties and of the product EI of the Young's modulus of the pile by the moment of inertia of the cross section but not explicitly of the pile diameter. For a hollow pile the moment of inertia is not uniquely related to the diameter and therefore in this case the actual size of the cavity has no effect

on the results of the model if the moment of inertia is kept constant. This would also be the case for a Winkler foundation model with only horizontal springs.

Table 2 shows the results of the four models for hollow piles with the same EI value of $2.421E + 8$ kN-m^2, but diameters of 1, 2, and 4 m. The agreement between the 3D finite element model and the boundary matrix method is good in all three cases (about 3% off in average). As expected the results for the 1D pile model do not change. The results for the Winkler model vary slightly because of the rotational springs but the variation is still very small and the accuracy deteriorates as the diameter of the pile decreases. To see when the results of the 1D model would become similar to those of the more accurate solutions the boundary matrix model was run for a larger number of diameters going down to 0.02 m. Figure 4 shows the variation of the head displacement with the pile diameter in semi-log scale. The deflection predicted by the boundary matrix model for a pile with a diameter of 0.02 m is 68.5 mm now in good agreement with the prediction of the 1D model. It is interesting to observe that the variation

TABLE 2: Variation of pile head displacement versus pile diameter in different linear analysis methods with constant EI for the pile.

Pile diameter (m)	Pile head deflection (mm)			
	3D pile FEM	1D pile FEM	Consistent boundary matrix	Winkler foundation
1.00	32.5	68.3 (110%)	34.0 (5%)	25.2 (23%)
2.00	27.6	68.3 (148%)	27.5 (1%)	25.0 (10%)
4.00	20.9	68.3 (227%)	20.5 (2%)	24.3 (17%)

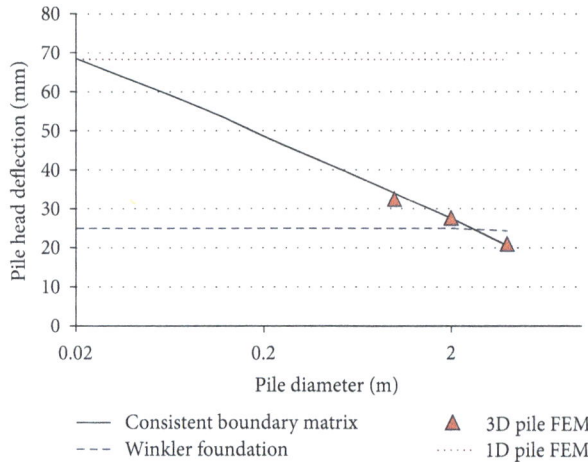

— Consistent boundary matrix ▲ 3D pile FEM
--- Winkler foundation ······ 1D pile FEM

FIGURE 4: Variation of pile head displacement versus pile diameter in linear analyses with constant EI for the pile.

TABLE 3: Elastic-plastic properties of soil.

Parameter	Symbol	Value	
		Sand	Clay
Modulus of elasticity (kPa)	E_s	5.0E4	4.5E4
Unit weight (kN/m^3)	γ_s	20.00	20.00
Poisson ratio	ν_s	0.30	0.30
Angle of internal friction (°)	ϕ	€40.0	—
Undrained shear strength (kPa)	S_u	—	150.0

of the displacement for this hollow pile is approximately inversely proportional to the radius to the power 0.26. It is noteworthy that while clearly the piles with very small diameter and large value of EI are unrealistic, it is desired to cover a wide range of values of the diameter to see more clearly the trend. It must also be remembered that when using a 1D beam-column to represent the pile with a finite element soil mesh that does not include a cylindrical cavity, one is assuming a zero diameter that is not just unrealistic but physically impossible.

3. Nonlinear Analyses

Three different models are used to conduct nonlinear analyses as follows.

(1) The 3D finite element model of the previous runs. In this case however the soil and the soil-pile interface are nonlinear. The finite element model, using ABAQUS, has the capability of taking into account the initial state of stresses in the soil mass. The initial conditions of stress are applied before the pile is installed and as a first step the effective body forces are calculated to account for geostatic equilibrium. The extreme static loads due to the performance of the turbine and wave and wind loading are applied then.

A 22 m long pile with a diameter of 4 m is modeled as a steel pipe using 4-node quadrilateral shell elements with reduced integration. A 1 m long segment of the pile is considered to be above ground level to avoid the soil going over the pile. Linear elastic behavior is assumed for the pile.

For an actual soil profile it would be necessary to select the most appropriate nonlinear constitutive model and to determine the values of the required parameters defining the model from laboratory tests. For the purposes of this work and considering two hypothetical soils, a sand and a clay, a very simple Mohr Coulomb model, as implemented in the program ABAQUS, is used with the properties presented in Table 3. The finite element mesh of the 40 m × 10 m × 41 m soil mass is generated using isoparametric brick elements with reduced integration for the soil.

The nonlinear behavior of the soil-pile contact is modeled using "contact pair" in ABAQUS. Tangential movement between the two parts, pile and surrounding soil, is allowed with a friction coefficient of 0.67. In the radial direction, a "no separation" contact behavior is assumed. The pile outer surface is chosen as the "master surface" and the surface of the soil mass which is in contact with the pile is considered to be the "slave surface." The "small sliding" tracking approach is employed for the contact of the two bodies assuming that even if the two bodies undergo large motions, there is relatively little sliding of one surface along the other. An elastic-plastic Coulomb model is also used to describe the nonlinear behavior of the soil-pile contact. Figure 5 shows the deformation of soil with 3D nonlinear finite element model of pile foundation.

(2) A model using the p-y curves is implemented specifically for this work. As indicated in the introduction section the p-y curves were originally proposed by Matlock [4] for soft clays under the water table and models for hard clays and sands were shortly after introduced by Reese et al. [6]. In this work the sand and the hard clay model are used. The clay model requires the specification of a parameter ε_{50} that has to be determined from experiments. Since the soil considered was not a real one on which experiments could be performed a value of 0.005, as recommended by Reese et al. [6], is used. In the linear elastic range, for very small displacements, the initial stiffness of the springs representing the p-y curves normally varies with depth. In this case

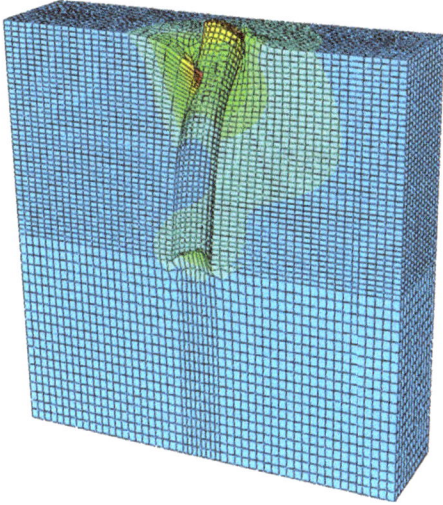

FIGURE 5: Deformed mesh of the pile foundation in 3D nonlinear analysis.

however, to be consistent with the finite element model the initial stiffness value is considered to be constant with the depth and equal to $4G$ as for the linear analyses with the Winkler foundation. Since the p-y curves are in fact a form of the Winkler foundation model with only horizontal springs the solution in the elastic range would be only a function of the EI of the pile and independent of the diameter for a given moment of inertia. The nonlinear variation of the stiffness is on the other hand affected by the pile diameter. It should also be noticed that with the p-y method there are nonlinear springs attached to the side of the pile but not at the bottom. One must decide therefore whether the pile tip is free, hinged or fixed. For long piles the difference between these three cases when considering the pile head displacement is negligible. However, in the present case the parameter $l_0 = (4E_p I_p / K)^{1/4}$, with E_p = modulus of elasticity for the pile (kPa); I_p = moment of inertia for the pile (m4); and K = soil-spring constant (kPa), which is the ratio of the soil resistance P (kN/m) at a depth z to the horizontal pile displacement y (m) at the same depth, is of the order of 10 m so the displacements for a hinged tip may be 25% smaller than for a free tip. For a linear analysis the assumption of a hinged tip, where the displacement of the pile tip is assumed to be constrained and equal to that of the soil mass at the bottom, while the rotation is allowed freely, may be more realistic but for the nonlinear one it is considered that the free end, where both the tip displacement and rotation are allowed, would be more appropriate. For the sake of comparison and to see the effects of such assumption the results are presented for both boundary conditions.

(3) A model implementing the simple approach for lateral loads on piles (SALLOP) proposed by Briaud [11]. It is a semitheoretical or semiempirical method in which the framework is theoretical but the factors in the theoretical equations are adjusted by comparison to some full-scale load tests. SALLOP uses two different theoretical solutions for infinitely long (flexible) piles and for short rigid piles in a

Winkler uniform soil. Defining a transfer length l_0 that is the typical parameter associated with the solution of a beam on elastic foundation, as done earlier, the pile head displacement y_0 for long flexible piles ($L \geq 3l_0$) under a combined loading of a horizontal force and a moment at its head is [17]

$$y_0 = \frac{2H_0}{l_0 K} + \frac{2M_0}{l_0^2 K}, \tag{1}$$

where H_0 = horizontal force applied at the pile head (kN), M_0 = moment applied at the pile head (kNm), and the soil-spring constant K is defined empirically by optimizing the comparison between the predicted deflection and the measured deflections.

Similarly, the pile head displacement for short rigid piles ($L \leq l_0$) is Should we change

$$y_0 = \frac{2(2H_0 L + 3M_0)}{KL^2}. \tag{2}$$

For the SALLOP calculations a linear interpolation between two values will be used if the pile length is between l_0 and $3l_0$. More details on SALLOP are presented at Briaud [17].

For the pile with a diameter of 4 m, the 3D finite element model predicts a displacement of 40 mm in sand and 25.1 mm in clay. The corresponding results with the p-y curves are 38.2 mm and 37.5 mm with a free tip (28.5 mm with a hinge at the bottom); with the SALLOP method 36.0 and 45 mm. The three methods provide results in good agreement for the sand but there are larger differences for the clay particularly for the SALLOP approach and with a free tip for the p-y curves.

The effect of the pile diameter with a constant value of the EI of the pile was again investigated for the nonlinear case. Table 4 and Figure 6 present the results of the three methods for diameters of 1, 2, and 4 m. Again since the SALLOP method is based purely on a Winkler foundation with horizontal springs the results are independent of the diameter for a fixed EI. The p-y curves give results that vary with the diameter but less significantly than the 3D solution. It is interesting to notice that for the sand the best agreement is obtained for a diameter of 4 m. For the 1 m diameter the prediction of the SALLOP method would be about 40% of the FEM result; with the p-y curve it would be about 62%. For the clay on the other hand the best agreement between the three methods is obtained for the diameter of 1 m (almost exactly the same results), whereas the discrepancy increases as the diameter increases. The prediction with the SALLOP method is about 80% too large whereas that with the p-y curves assuming a free tip is about 50% off for the 4 m diameter.

It seems also that given the lack of a spring acting on the bottom face of the pile in the p-y model, for the larger diameter pile the assumption of a hinged tip might be more realistic whereas for the smaller diameters it is better to consider a free tip. Considering the fact that the characteristics of the soils are not actually determined from laboratory tests but some of the parameters are chosen purely as logical values, and that a very simple nonlinear soil model was used, finding an exact agreement between the three methods would have been surprising. The fact that they provide results with the

TABLE 4: Variation of pile head displacement versus pile diameter in different nonlinear analysis methods with constant EI for the pile.

Pile diameter (m)	Pile deflection at the ground level (mm)						
	Sand			Clay			
	3D FEM	p-y	SALLOP	3D FEM	p-y		SALLOP
					Free tip	Hinged tip	
1.00	91.0	57.0 (35%)	36.0 (60%)	45.5	45.3 (1%)	31.2 (32%)	45.0 (1%)
2.00	60.6	43.5 (29%)	36.0 (41%)	33.9	40.5 (20%)	30.0 (12%)	45.0 (33%)
4.00	40.0	38.2 (5%)	36.0 (10%)	25.1	37.5 (50%)	28.5 (14%)	45.0 (80%)

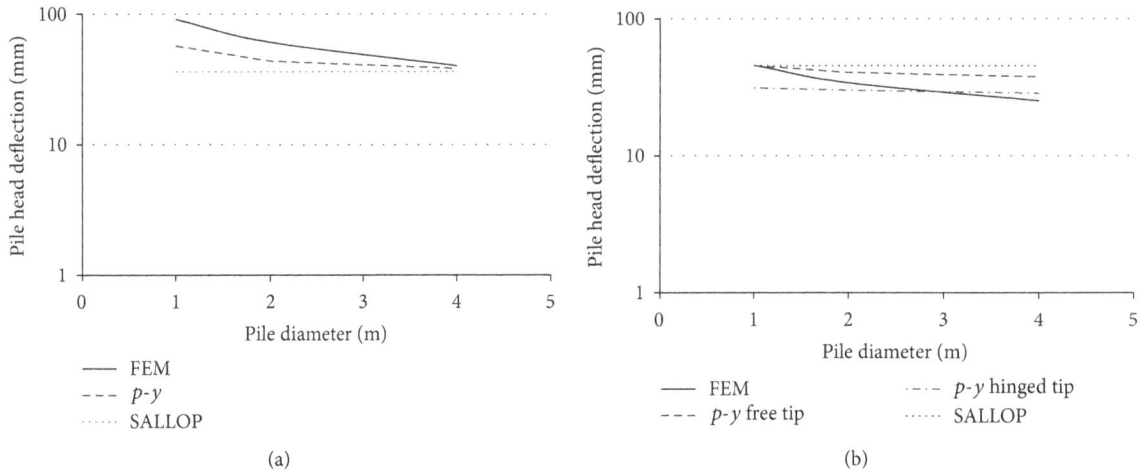

(a)   (b)

FIGURE 6: Variation of pile head displacement versus pile diameter in nonlinear analyses for (a) sand and (b) clay with constant EI for the pile.

same order of magnitude for the range of pile diameters considered is encouraging. On the other hand it is important to notice the effect of the pile diameter on the foundation stiffness beyond the value of the EI, something that would occur irrespective of the constitutive model used. Obtaining a very good agreement for a given pile diameter with a more refined selection of the nonlinear soil model and of the soil parameters will not guarantee similar accuracy for other values of the diameter and the same soil.

4. Summary and Conclusions

This study for the first time provides a comprehensive comparison of common techniques for analysis of laterally loaded single piles in different soil types and for a wide range of pile dimensions. The effect of the pile diameter on its lateral behavior in the linear elastic range was studied using various analysis procedures assuming a constant pile stiffness (EI) and different pile diameters for hollow piles: a 3D ABAQUS finite element model, a model with the soil reproduced with 3D elements but the pile represented by a line, a model using a consistent boundary matrix and a Winkler foundation model. The results show that the pile head lateral deflection is not only a function of EI but also of the pile diameter. It decreases considerably as the pile diameter increases while EI is maintained constant. Modeling a pile as a 1D line with beam-column elements, as done sometimes in the literature, results in a smaller contribution of the surrounding soil to the lateral stiffness of the pile and an increase of up to 200% in the maximum displacement of the pile head.

Nonlinear analyses were next conducted using the three dimensional finite element models of the soil and pile employing ABAQUS for a sand and a clay. The static (monotonic) calculations were conducted for an extreme lateral load and bending moment. A Mohr-Coulomb constitutive model was used for the generic soils. The nonlinear contact between the pile and the soil were accounted for using some of the tools available in ABAQUS. The results were compared to those provided by the use of p-y curves for sand and hard clay and with the SALLOP method suggested by Briaud [11]. Both the p-y model for sand and the SALLOP method provide reasonable answers for the pile with a diameter of 4 m but the accuracy deteriorates for smaller diameters, particularly for the SALLOP method where the results are independent of the diameter for a fixed value of EI. For the clay the p-y curves assuming a free tip and the SALLOP predictions are good for the smaller diameter pile (diameter of 1 m) but deteriorate for larger diameters. It appears that for these cases with the p-y method the assumption of a pile hinged at the bottom would provide better results.

The study conducted uses the 3D nonlinear finite element model as an accurate model for the pile sizes of interest in relation to the foundations of offshore wind turbines to assess other, simpler models. It indicates that when using common simple models and particularly if the pile is modeled as a line, neglecting the size of the soil cavity, the results may be inaccurate.

References

[1] I. Sanchez Salinero, "Static and dynamic stiffnesses of single piles," Geotechnical Engineering Report GR82-31, The University of Texas at Austin, 1982.

[2] M. Novak, "Dynamic stiffness and damping of piles," *Canadian Geotechnical Journal*, vol. 11, no. 5, pp. 574–698, 1975.

[3] V. A. Baranov, "On the calculation of excited vibrations of an embedded foundation," *Voprosy Dynamiki Prochnocti*, vol. 14, pp. 195–209, 1967 (Russian).

[4] H. Matlock, "Correlations for design of laterally loaded piles in soft clay," in *Proceedings of the 2nd Annual Offshore Technology Conference*, vol. 1, pp. 577–588, 1970, Paper 1204.

[5] L. C. Reese and R. C. Welch, "Lateral loading of deep foundations in stiff clay," *Journal of the Geotechnical Engineering Division*, vol. 101, no. 7, pp. 633–649, 1975.

[6] L. C. Reese, W. R. Cox, and F. D. Koop, "Field testing and analysis of laterally loaded pilein stiff clay," in *Proceedings of the 7th Annual Offshore Technology Conference*, vol. 2, pp. 671–690, 1975, Paper 2312.

[7] D. M. Dewaikar, R. S. Salimath, and V. A. Sawant, "A modified P-Y curve for the analysis of a laterally loaded pile in stiff clay," *Australian Geomechanics Journal*, vol. 44, no. 3, pp. 91–100, 2009.

[8] Y. Kim and S. Jeong, "Analysis of soil resistance on laterally loaded piles based on 3D soil-pile interaction," *Computers and Geotechnics*, vol. 38, no. 2, pp. 248–257, 2011.

[9] L. C. Reese, W. R. Cox, and F. D. Koop, "Analysis of laterally loaded piles in sand," in *Proceedings of the 6th Annual Offshore Technology Conference*, vol. 2, pp. 473–483, 1974, Paper 2080.

[10] J. L. Briaud, T. D. Smith, and L. M. Tucker, "A pressuremeter method for laterally loaded piles," in *Proceedings of the 11th International Conference on Soil Mechanics and Foundation Engineering*, vol. 3, pp. 1353–1356, A.A. Balkema, Rotterdam, The Netherlands, 1985.

[11] J. L. Briaud, "SALLOP: simple approach for lateral loads on piles," *Journal of Geotechnical and Geoenvironmental Engineering*, vol. 123, no. 10, pp. 958–964, 1997.

[12] V. Sanjaya Kumar, K. G. Sharma, and A. Varadarajan, "Behaviour of a laterally loaded pile," in *Proceedings of the 10th International Symposium on Numerical Models in Geomechanics (NUMOG '07)*, pp. 447–452, April 2007.

[13] M. T. Suleiman, A. Raich, T. W. Polson, W. J. Kingston, and M. Roth, "Measured soil-pile interaction pressures for small-diameter laterally loaded pile in loose sand," in *GeoFlorida: Advances in Analysis, Modeling and Design Conference*, pp. 1498–1506, February 2010, Geotechnical Special Publication, Paper 199.

[14] J. S. Chioui and C. H. Chenu, "Exact equivalent model for a laterally-loaded linear pile-soil system," *Soils and Foundations*, vol. 47, no. 6, pp. 1053–1061, 2007.

[15] G. W. Blaney, E. Kausel, and J. M. Roesset, "Dynamic stiffness of piles," in *Proceedings of the 2nd International Conference on Numerical Methods in Geomechanics*, pp. 1001–1012, 1976.

[16] E. Kausel, "Forced vibrations of circular foundations on layered media," Research Report R 74-11, Civil Engineering Department, MIT, 1974.

[17] J. L. Briaud, *The Pressuremeter*, A. A. Balkema, Rotterdam, The Netherlands, 1992.

18

Application of Chaos Theory in Trucks' Overloading Enforcement

Abbas Mahmoudabadi,[1] and Arezoo Abolghasem[2]

[1] Department of Industrial Engineering, Payame Noor University (PNU), Shahnaz Alley, Nourian Street, North Dibagi Avenue, Tehran, Iran
[2] Road Maintenance and Transportation Organization, Number 12 Dameshq Street, Vali-e-Asr Avenue, Tehran, Iran

Correspondence should be addressed to Abbas Mahmoudabadi; mahmoudabadi@phd.pnu.ac.ir

Academic Editor: Sang-Min Han

Trucks' overloading is considered as one of the most substantial concerns in road transport due to a possible road surface damage, as well as, are less reliable performance of trucks' braking system. Sufficient human resource and adequate time scheduling are to be planned for surveying trucks' overloading; hence, it seems required to prepare an all-around model to be able to predict the number of overloaded vehicles. In the present research work, the concept of chaos theory has been utilized to predict the ratio of trucks which might be guessed overloaded. The largest Lyapunov exponent is utilized to determine the presence of chaos using experimental data and concluded that the ratio of overloaded trucks reflects chaotic behavior. The prediction based on chaos theory is compared with the results of simple smoothing and moving average methods according to the well-known criterion of mean square errors. The results have also revealed that the chaotic prediction model would act more capably comparing the analogous methods including simple smoothing and moving average to predict the ratio of passing trucks to be possibly overloaded.

1. Introduction

Road transportation is a dominant mode of freight transportation in Iran that accounts for about 80% of the freight movement [1]. The number of heavy vehicles is continuously increasing on the road network. Overloading is growing up, and it is a major cause for significantly accelerating the rate of pavement deterioration [2]. In addition to damage road surface, overloaded trucks endanger other vehicles because of predesigning of vehicles' braking systems particularly in curves and slopes. Having promoted road safety and minimizing road maintenance costs, ordinary road measures imposed on road pavement by axles' load of trucks are made as part of the enforcement process laws [3] is legally controlled by weighing stations.

Relevant studies in the field of pavement designs based on truck axle loads are referred in the literature. Turochy et al. [4] developed truck factors for pavement design and axle load distribution models for mechanistic-empirical pavement design using information from weigh-in-motion sites on arterial roads in Alabama. Till also developed a detailed method of wheel load modeling from overload trucks for bridge decks [5]. Chia-pei calculated average load factors for combined heavy vehicles and axle load ratios for various types of heavy vehicles, to design bridge standard specification [6].

Although some location allocation techniques can be utilized to improve the efficiency of overloading enforcement [7], in order to minimize overloading, scheduling human resources who are employed staff in the process of checking trucks' axles load in weighing stations should be considered in road enforcement. Improving the efficiency, human resources scheduling would require the utilization of prediction methods. Recent techniques of prediction are observed in the literature by applying chaos theory in the field of road traffic. Frazier and Kockelman analyzed traffic flow data and found it chaotic. Their studies showed that predictions based on chaos theory would have greater predictive power than a nonlinear least-squares method [8]. Disbro and Frame demonstrated how the theoretically derived Gazis-Herman-Rothery traffic model [9] is highly chaotic, even though applied to medium or smaller (eight-car) systems [10]. Van Zuylen et al. discussed the implications of human behavior, chaos, and unpredictability for urban and transportation planning as well as forecasting [11].

Safonov et al. showed that chaotic behavior in traffic can be caused by delays in human reaction [12]. Nair et al. analyzed traffic flow data and characterized it as a chaotic factor [13]. Weidlich demonstrated how random-utility-based models of relatively simple social behaviors produced chaotic behavior and offer a detailed application of chaos to traffic flow [14]. Metghalchia et al. examined the profitability of several simple technical trading rules for 16 European stock markets over the 1990 to 2006 period and concluded that increasing trend of moving average rules has predictive power being able to discern recurring price patterns for profitable trading, even after accounting for the effects of data snooping bias [15].

Four methodologies of predicting gas turbine behavior over time have also been utilized by Cavarzere and Venturini [16]. They are used to provide a time prediction when a threshold value will be exceeded in the future. Results presented an aim to select the most suitable methodology that allows both trending and forecasting as a function of data trend over time, in order to predict time evolution of gas turbine characteristic parameters and to provide an estimate of the occurrence of a failure [16].

The main contribution of this paper is to predict the ratio of overloaded trucks based on the concept of chaos theory. Following introduction, some methods of vehicles' axle loads controlling and the well-known prediction methods are briefly presented in the second and third sections, followed by the description of case study. Analytical process includes checking the presence of chaos behaviors using experimental data, as presented in the fifth section together with comparing results consisting of more discussions on the revealed results. A brief discussion of the applied process and some recommendations for further studies are also discussed in the last section.

2. Goods Transportation Controlling and Trucks Axle Load

Pavement management is an important issue, according to its maintenance and repairing roads surface expenses. Heavy vehicles' axle loads are understood as the most important factor of road destruction. Based on observed researches in the literature, the damage on roads surface is related to the vehicles' axle loads by a nonlinear acceleration rate mainly in a polynomial equation of forth degree [2]. Truck weighing control is made by certain equipments, incorporating static, dynamic, and portable scales. Different axle loads and limitations in the scope of case study are shown in Table 1 including three most popular axle group configurations. They include 3 axle-10 wheels, 5 axle-12 wheels, and 5 axle-18 wheels trucks. Maximum axle loads are different from different configurations; however, it does not seem necessary if the total weigh is equal to the sum of axle load limitations. Trucks' axle loads are checked in weighing stations by random and if they exceed the authorized limit, overloaded trucks would

TABLE 1: Axle group configurations and weight limitations of most popular trucks.

Axle group configuration	Weight limitation (Ton)			
	A	B	C	A + B + C
A_2 B_8	6	22	—	28
A_2 B_4 C_6	6	13	24	42
A_2 B_8 C_8	6	22	22	44

be stopped afterwards, requiring the loads adjusted based on regulations.

3. Time Series Prediction

3.1. Moving Average. Moving average method of prediction is obtained by first taking the average of the first subset. The fixed subset size is then shifted forward, creating a new subset of numbers, which is averaged [17]. Equation (1) shows the overall definition of moving average method where P_{n+t+1} is the predicted amount for $n + t + 1$ based on the previous amounts of real demands, and n is the period. Using moving average technique is observed in the literature while Narasimhan et al. proposed new prediction techniques for MPEG-4 encoded variable bit rate video traffic based on the concept of moving average and extension further using the gradient-descent approach [18]:

$$P_{t+n+1} = \frac{(X_{t+1} + X_{t+2} + \cdots + X_{t+n})}{n}. \qquad (1)$$

3.2. Exponential Smoothing. Exponential smoothing is either applied to time series data, or to produce smoothing data, for which the time series are sequence of observations. Raw data sequence is often represented by $\{X_t\}$, and the output of the exponential smoothing algorithm is written as $\{P_t\}$, to be regarded as the best estimate of what the next value of X. When the sequence of observations begins at time $t = 0$, the most simple form of exponential smoothing is given by (2) [17] where α is the smoothing factor, and $0 < \alpha < 1$. Snyder et al. used exponential smoothing method in sales forecasting for inventory control that has always been rationalized in terms of statistical models that possess errors with constant variances. It is shown that exponential smoothing remains appropriate under more general conditions, where the variance is allowed to grow or contract with corresponding movements in the underlying level [19]:

$$P_1 = X_0, \qquad P_{t+1} = \alpha \times X_t + (1 - \alpha) \times P_t, \quad t > 1. \qquad (2)$$

4. Case Study and Experimental Data

Experimental data of trucks' axle load have been stored using weigh in motion (WIM) system installed in highways. WIMs are fully automated systems which record all vehicles

TABLE 2: Coefficients and mean square errors in chaos and simple exponential smoothing.

Axle configuration	Method Number of vehicles	Overload	Simple exponential smoothing		Chaos theory ($\Delta t = 1$)		
			α_{best}	MSE	$\lambda_{\max}$	β_{best}	MSE
3 axle 10 wheel	12898	Axle	0.56	0.0017	0.1769	1.3183	0.0016
		Total	0.52	0.0021	0.2018	1.3528	0.0021
5 axle 12 wheel	4211	Axle	0.24	0.0030	0.2176	1.5065	0.0026
		Total	0.39	0.0022	0.0917	1.1045	0.0024
5 axle 18 wheel	25525	Axle	0.71	≈ 0	0.0289	1.0051	≈ 0
		Total	≈ 1	0.0004	0.0957	1.1360	0.0003
Total	42634	Axle	0.67	0.0004	0.3186	2.9942	0.0003
		Total	0.82	0.0008	0.2600	0.8735	0.0004

and capable of measuring at normal traffic speed, without necessary to stop vehicles or drive at low speed, making them more efficient [20]. Vehicles passing through the sensors connected to a digital video camera system are detected while data are stored simultaneously. Experimental data have been collected by weigh in motion system in Arak-Qom highway and used to analyze in one month period. Three kinds of trucks including 3 axles 10 wheels, 5 axles 12 wheels and 5 axles 18 wheels have been selected for data analyzing because of high frequency of trucks' volume.

5. Analytical Process

After checking the presence of chaos, analytical process has been done based on two methods of time series prediction and the concept of chaos theory. Simple exponential smoothing is corresponding to the chaos theory with time span 1 day, and moving average method corresponding to the chaos theory considering time spans 3, 5, and 7 days as well as data for 42634 vehicles' axle loads that are collected during the time period of case study.

5.1. Presence of Chaos.
A more common technique to determine the presence of chaotic behavior is calculating the largest Lyapunov exponent. Equation (3) determines the largest Lyapunov exponent where $S(t)$ is the observation in t, $S'(t)$ is its nearest neighbor, and N is the number of observations. When the largest Lyapunov exponent exceeds 0, the system is chaotic [21]:

$$\lambda_{\max} = \frac{1}{N\Delta t} \sum_{t=0}^{N-1} \text{Ln} \left(\frac{|S(t + \Delta t) - S'(t - \Delta t)|}{|S(t) - S'(t)|} \right). \quad (3)$$

The ratio of vehicles which are suspicious to be overloaded comparing to passing vehicles is considered as a chaos factor in open interval (0 1). Time span is considered as 24 hours, and data for 30 days were gathered. Eventually the well-known equation of logistic map [22] is utilized to define the ration of overloaded vehicles because of reasonable adaptation to traffic behaviors. Equation (4) defines the logistic map [23], where $P(t)$ is the ratio of trucks to be possibly overloaded:

$$P(t + 1) = \beta \times P(t) \times (1 - P(t)). \quad (4)$$

5.2. Chaos Theory versus Simple Exponential Smoothing.
Analytical process has been done for three types of most popular axle configurations while the numbers of trucks which are checked are different. Because of the same period of analyzing, simple exponential smoothing is compared with chaos theory using ($\Delta t = 1$). In regard to (2) and (3) two coefficients of α and β need to be estimated minimizing mean square errors are notated as α_{best} and β_{best} in Table 2. For each type of trucks, parameters have been estimated separately. Results shown in Table 2 revealed that the mean square error criterion of prediction corresponding to chaos is equal or less than one for simple exponential smoothing.

5.3. Chaos Theory versus Moving Average Method.
Because of considering the same selected periods, moving average method is compared with chaos theory using ($\Delta t = 3$, 5 and 7). Coefficients α and β defined in (2) and (3) are estimated, regarding the minimum mean square errors in similar duration notated as α_{best} and β_{best} in Tables 3 and 4. As it can be observed in Tables 3 and 4, readers are supposed to mention that the concept of chaos theory is more powerful technique for predicting ratio of trucks to be possibly overloaded because of the existing of chaotic behavior in overloading. It must be strictly mentioned that each time span is required to be compared with the same time span. So the results of situation $\Delta t = 3, 5, 7$ must be compared with moving average method of $N = 3, 5$ and 7, respectively.

6. Summary and Conclusion

In this research work, the concept of chaos theory has been applied to predict the ratio of trucks likely to be overloaded in order to make an appropriate time table of human resources in vehicles' axle loads control. Data gathered in one-month period in the case study have been used for analytical process. Validation is checked by the results of two well-known methods of simple exponential smoothing and moving average time series.

Prediction analyses have been presented based on different time intervals including one, three, five, and seven days corresponding to the validation process and method of prediction. The mean square error, most popular criterion of prediction validity, is used to check the validity of proposed method. Discussing results showed that the prediction based

TABLE 3: Coefficients and mean square errors in chaos theory.

Method		Chaos theory					
		$\Delta t = 3$		$\Delta t = 5$		$\Delta t = 7$	
Axle configuration	Overload	β_{best}	MSE	β_{best}	MSE	β_{best}	MSE
3 axle 10 wheel	Axle	1.3259	0.0025	1.3241	0.0002	1.3067	0.0002
	Total	1.3699	0.0028	1.3854	0.0035	1.3472	0.0003
5 axle 12 wheel	Axle	1.5080	0.0038	1.5161	0.0009	1.5005	0.0009
	Total	1.0546	0.0040	1.0663	0.0038	1.1784	≈0
5 axle 18 wheel	Axle	0.9802	≈0	1.0358	≈0	1.0342	≈0
	Total	1.1300	0.0011	1.1386	0.0014	1.1339	≈0
Total	Axle	1.1404	0.0007	1.1481	≈0	1.4119	≈0
	Total	1.2040	0.0016	1.2141	0.0020	1.1951	≈0

TABLE 4: Coefficients and mean square errors in moving average.

Method		Moving average method		
		$N = 3$	$N = 5$	$N = 7$
Axle configuration	Overload	MSE	MSE	MSE
3 axle 10 wheel	Axle	0.0021	0.0022	0.0025
	Total	0.0024	0.0027	0.0030
5 axle 12 wheel	Axle	0.0039	0.0041	0.0044
	Total	0.0029	0.0030	0.0036
5 axle 18 wheel	Axle	≈0	≈0	≈0
	Total	0.0007	0.0008	0.0011
Total	Axle	0.0005	0.0006	0.0006
	Total	0.0010	0.0013	0.0016

on the concept of chaos theory using logistic map equation is more capable than simple exponential smoothing and moving average time series methods. It has been shown, of course, more capable than simple exponential method of prediction when time interval is 1 day (next day), in respect, more powerful than moving average method if time intervals are 3, 5, and 7 days. In brief, chaos theory performs deeply better to predict the ratio of vehicles which are possibly overloaded rather than the other methods of prediction. Having interested to study in due field, it is recommended to focus on the application of chaos theory in the other road traffic measures to disclose which measures are able to be defined based on chaotic behavior.

References

[1] Road Maintenance and Transportation Organization, "Annual survey of road transport in Iran," 1388 local calender, 2009, http://www.rmto.ir/NewTTO/MainF.asp/.

[2] A. Mahmoudabadi and A. Abolghasem, "A cluster-based method for evaluation of truck's weighing control stations," in *MATLAB—A Ubiquitous Tool for the Practical Engineer*, C. M. Ionescu, Ed., 2011.

[3] A. M. Asa, "Laws and regulations in rural transport," Road maintenance and transportation organization, 1378 local calendar, 1999.

[4] R. E. Turochy, D. H. Timm, and S. M. Tisdale, "Truck equivalency factors, load spectra modeling and effects on pavement design," Tech. Rep. 930-564, Harbert Engineering Center, 2005.

[5] R. D. Till, "Overload truck wheel load distribution on bridge decks," Research Report R-1529, Michigan Transportation Commission, Structural Section Construction and Technology Division, 2009.

[6] C. J. Chia-pei, "Effect of overloaded heavy vehicles on pavement and bridge design," *Transportation Research Board of the National Academies*, no. 1539, pp. 58–65, 2007.

[7] A. Mahmoudabadi and S. M. Seyedhosseini, "Improving the efficiency of weigh in motion systems through optimized allocating truck checking oriented procedure," *IATSS Research*. In Press.

[8] C. Frazier and K. M. Kockelman, "Chaos theory and transportation systems: instructive example," in *Proceedings of the 83rd Annual Meeting of the Transportation Research Board*, Washington, DC, USA, January 2004.

[9] D. C. Gazis, R. Herman, and R. W. Rothery, "Nonlinear follow-the-leader models for trafficflow," *Operations Research*, vol. 9, no. 4, pp. 545–567, 1961.

[10] J.E. Disbro and M. Frame, "Traffic flow theory and chaotic behavior," Transportation Research Record 1225, National Research Council, Washington, DC, USA, 1989.

[11] H. J. van Zuylen, M. S. van Geenhuizen, and P. Nijkamp, "(Un)predictability in traffic and transport decision making," Transportation Research Record 1685, National Research Council, Washington, DC, USA, 1999.

[12] L. A. Safonov, E. Tomer, V. V. Strygin, Y. Ashkenazy, and S. Havlin, "Delay-induced chaos with multifractal attractor in a traffic flow model," *Europhysics Letters*, vol. 57, no. 2, pp. 151–157, 2002.

[13] A. S. Nair, Liu Jyh-Charn, L. Rilett, and S. Gupta, "Nonlinear analysis of traffic flow," 2001, http://translink.tamu.edu/docs/Research/LinearAnalysisTrafficFlow/chaos1.pdf/.

[14] W. Weidlich, *Socio Dynamics: A Systematic Approach to Mathematical Modeling in the Social Sciences*, Harwood Academic, Amsterdam, The Netherlands, 2000.

[15] M. Metghalchia, J. Marcucci, and Y.-H. Chang, "Are moving average trading rules profitable? evidence from the European stock markets," *Applied Economics*, vol. 44, no. 12, pp. 1539–1559, 2012.

[16] A. Cavarzere and M. Venturini, "Application of forecasting methodologies to predict gas turbine behavior over time,"

Journal of Engineering for Gas Turbines and Power, vol. 134, no. 1, Article ID 012401, 2012.

[17] B. R. Goodell, *Smoothing Forecasting and Prediction of Discrete Time Series*, Prentice-Hall, Englewood Cliffs, NJ, USA, 1963.

[18] H. Narasimhan, R. Tripuraribhatla, and K. S. Easwarakumar, "Moving average based predictors for MPEG-4 VBR traffic sources," in *Proceedings of the IEEE Globecom Workshops (GC '10)*, pp. 924–928, Miami, Fla, USA, December 2010.

[19] R. D. Snyder, A. B. Koehler, and J. K. Ord, "Forecasting for inventory control with exponential smoothing," *International Journal of Forecasting*, vol. 18, no. 1, pp. 5–18, 2002.

[20] B. Jacob and V. F. L. Beaumelle, "Improving truck safety: potential of weigh-in-motion technology," *IATSS Research*, vol. 34, no. 1, pp. 9–15, 2010.

[21] C. Frazier and K. M. Kockelman, "Chaos theory and transportation systems: instructive example," *Journal of the Transportation Research Record*, no. 1897, pp. 9–17, TRB, National Research Council, Washington, DC, USA, 2004.

[22] S. C. Lo and H. J. Cho, "Chaos and control of discrete dynamic traffic model," *Journal of the Franklin Institute*, vol. 342, no. 7, pp. 839–851, 2005.

[23] J. Mingjun and T. Huanwen, "Application of chaos in simulated annealing," *Chaos, Solitons and Fractals*, vol. 21, no. 4, pp. 933–941, 2004.

19

Estimation of Optimum Dilution in the GMAW Process Using Integrated ANN-GA

P. Sreeraj,[1] T. Kannan,[2] and Subhashis Maji[3]

[1] Department of Mechanical Engineering, Valia Koonambaikulathamma College of Engineering and Technology, Trivandrum, Kerala 692574, India
[2] SVS College of Engineering, Coimbatore, Tamilnadu 642109, India
[3] Department of Mechanical Engineering, IGNOU, New Delhi 110068, India

Correspondence should be addressed to P. Sreeraj; pathiyasseril@yahoo.com

Academic Editor: Keat Teong Lee

To improve the corrosion resistant properties of carbon steel, usually cladding process is used. It is a process of depositing a thick layer of corrosion resistant material over carbon steel plate. Most of the engineering applications require high strength and corrosion resistant materials for long-term reliability and performance. By cladding these properties can be achieved with minimum cost. The main problem faced on cladding is the selection of optimum combinations of process parameters for achieving quality clad and hence good clad bead geometry. This paper highlights an experimental study to optimize various input process parameters (welding current, welding speed, gun angle, and contact tip to work distance and pinch) to get optimum dilution in stainless steel cladding of low carbon structural steel plates using gas metal arc welding (GMAW). Experiments were conducted based on central composite rotatable design with full replication technique, and mathematical models were developed using multiple regression method. The developed models have been checked for adequacy and significance. In this study, artificial neural network (ANN) and genetic algorithm (GA) techniques were integrated and labeled as integrated ANN-GA to estimate optimal process parameters in GMAW to get optimum dilution.

1. Introduction

Prevention of corrosion is a major problem in industries. Even though it cannot be eliminated completely, it can be reduced to some extent. A corrosion resistant protective layer is made over the less corrosion resistant substrate by a process called cladding. This technique not only used to improve life of engineering components but also to reduce their cost. This process is mainly now used in industries such as chemical, textiles, nuclear, steam power plants, food processing, and petrochemical industries [1].

The most accepted method employed in weld cladding is GMAW. It has got the following advantages [2]:

(i) high reliability,

(ii) all-position capability,

(iii) ease to use,

(iv) low cost,

(v) high Productivity,

(vi) suitable for both ferrous and nonferrous metals,

(vii) high deposition rate,

(viii) absences of fluxes,

(ix) cleanliness and ease of mechanization.

The mechanical strength of clad metal is not only highly influenced by the composition of metal but also by clad bead shape. This is an indication of bead geometry. Figure 1 shows the clad bead geometry. It mainly depends on wire feed rate, welding speed, arc voltage, and so forth. Therefore it is necessary to study the relationship between in-process parameters and bead parameters to study clad bead geometry. Using mathematical models, it can be achieved. This paper highlights the study carried out to develop mathematical and ANN-GA models to optimize clad bead geometry, in stainless steel cladding deposited by GMAW. The experiments were

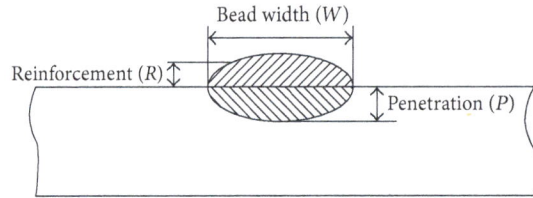

FIGURE 1: Clad bead geometry.

conducted based on four-factor-five-level central composite rotatable designs with full replication technique [3]. The developed models have been checked for their adequacy and significance. Again using ANN-GA, the bead parameters were optimized.

This study can be divided into three parts that are experimental analysis, analytical model, and artificial intelligence model. Experimental and analytical models can be developed using conventional and statistical models. Artificial intelligent models such as ANN and GA can be used for prediction. There are many advantages using ANN for prediction such as ANN's ability to handle nonlinear form of modeling that learn the mapping of input and output. ANN is more successful than conventional methods in terms of speed and simplicity and its capacity to learn from examples. Moreover it does not need any preliminary assumptions. Simple MATLAB Toolbox can be used for prediction.

In this study an integration system is opted for study to get an improvement in GMAW process. In this study two techniques were used: ANN is used for prediction and GA is used for optimization. These two techniques were integrated to form a new system integrated ANN-GA. It is proved that the proposed hybrid system ANN-GA can produce more significant results than conventional techniques.

2. Experimental Procedure

The following machines and consumables were used for the purpose of conducting experiments:

(1) a constant current gas metal arc welding machine (Invrtee V 350-PRO advanced processor with 5–425 amp output range),

(2) welding manipulator,

(3) wire feeder (LF-74 Model),

(4) filler material Stainless Steel wire of 1.2 mm diameter (ER-308 L),

(5) gas cylinder containing a mixture of 98% argon and 2% oxygen,

(6) mild steel plate (grade IS-2062).

Test plates of size $300 \times 200 \times 20$ mm were cut from mild steel plate of grade IS-2062, and one of the surfaces is cleaned to remove oxide and dirt before cladding. ER-308 L stainless steel wire of 1.2 mm diameter was used for depositing the clad beads through the feeder. Argon gas at a constant flow rate of 16 litres per minute was used for shielding. The properties of

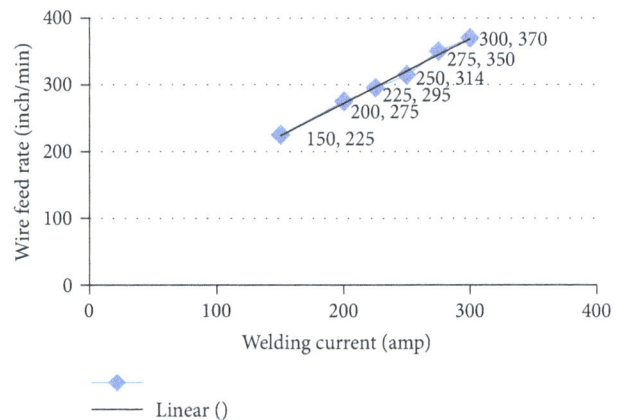

FIGURE 2: Relationship between Current and Wire Feed Rate.

base metal and filler wire are shown in Table 1. The important and most difficult parameter found from trial run is wire feed rate. The wire feed rate is proportional to current. Wire feed rate must be greater than critical wire feed rate to achieve pulsed metal transfer. The relationship found from trial run is shown in (1). The formula derived is shown in Figure 2:

$$\text{wire feed rate} = 0.96742857 * \text{current} + 79.1. \quad (1)$$

The selection of the welding electrode wire is based on the matching of the mechanical properties and physical characteristics of the base metal, weld size, and existing electrode inventory [4]. The candidate material for cladding which has excellent corrosion resistance and weldability is stainless steel. It has chloride stress corrosion cracking resistance and strength significantly greater than other materials. It has good surface appearance, good radiographic standard quality, and minimum electrode wastage. Experimental design procedure used for this study is shown in Figure 3 and important steps are briefly explained.

3. Plan of Investigation

The research work was planned to be carried out in the following steps [5]:

(1) identification of factors and responses,

(2) finding limits of process variables,

(3) development of design matrix,

(4) conducting experiments as per design matrix,

TABLE 1: Chemical composition of base metal and filler wire.

Materials	Elements wt%								
	C	SI	Mn	P	S	Al	Cr	Mo	Ni
IS 2062	0.150	0.160	0.870	0.015	0.016	0.031	—	—	—
ER 308L	0.03	0.57	1.76	0.021	0.008	—	19.52	0.75	10.02

TABLE 2: Welding parameters and their levels.

Parameters	Unit	Notation	Factor levels				
			−2	−1	0	1	2
Welding current	A	I	200	225	250	275	300
Welding speed	mm/min	S	150	158	166	174	182
Contact tip to work distance	mm	N	10	14	18	22	26
Welding gun angle	Degree	T	70	80	90	100	110
Pinch	—	Ac	−10	−5	0	5	10

(5) recording the responses,

(6) development of mathematical models,

(7) checking the adequacy of developed models,

(8) conducting conformity tests.

3.1. Identification of Factors and Responses.

The following independently controllable process parameters were found to be affecting output parameters. These are wire feed rate (W), welding speed (S), welding gun angle (T), contact tip to work to distance (N), and pinch (Ac). The responses chosen were clad bead width (W), height of reinforcement (R), depth of penetration (P), and percentage of dilution (D). The responses were chosen based on the impact of parameters on final composite model.

The basic difference between welding and cladding is the percentage of dilution. The properties of the cladding is significantly influenced by the dilution obtained. Hence control of dilution is important in cladding where a low dilution is highly desirable. When dilution is quite low, the final deposit composition will be closer to that of filler material, and hence corrosion resistant properties of cladding will be greatly improved. The chosen factors have been selected on the basis of getting minimal dilution and optimal clad bead geometry.

Few significant research works have been conducted in these areas using these process parameters, and so these parameters were used for an experimental study.

3.2. Finding the Limits of Process Variables.

Working ranges of all selected factors are fixed by conducting trial runs. This was carried out by varying one of the factors while keeping the rest of them as constant values. Working range of each process parameter was decided upon by inspecting the bead for smooth appearance without any visible defects. The upper limit of a given factor was coded as −2. The coded value of intermediate values were calculated using (2):

$$X_i = \frac{2\left[2X - \left(X_{max} + X_{min}\right)\right]}{\left[\left(X_{max} - X_{min}\right)\right]}, \tag{2}$$

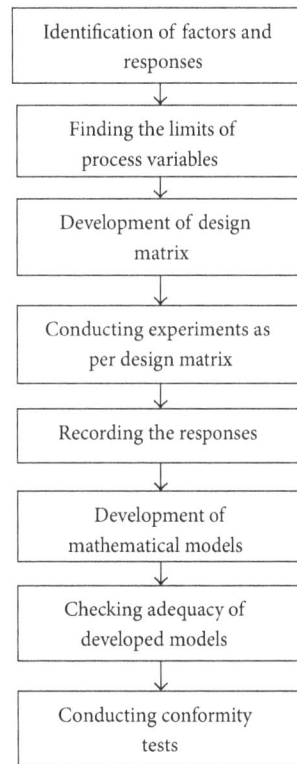

FIGURE 3: Experimental design procedure.

where X_i is the required coded value of parameter and X is any value of parameter from $X_{min} - X_{max}$. X_{min} is the lower limit of parameters and X_{max} is the upper limit of parameters [4].

The chosen level of the parameters with their units and notations are given in Table 2.

3.3. Development of Design Matrix.

Design matrix chosen to conduct the experiments was central composite rotatable design. The design matrix comprises full replication of 2^5 (=32), factorial designs. All welding parameters in

TABLE 3: Design matrix.

Trial no.	Design matrix				
	I	S	N	T	Ac
1	−1	−1	−1	−1	1
2	1	−1	−1	−1	−1
3	−1	1	−1	−1	−1
4	1	1	−1	−1	1
5	−1	−1	1	−1	−1
6	1	−1	1	−1	1
7	−1	1	1	−1	1
8	1	1	1	−1	−1
9	−1	−1	−1	1	−1
10	1	−1	−1	1	1
11	−1	1	−1	1	1
12	1	1	−1	1	−1
13	−1	−1	1	1	1
14	1	−1	1	1	−1
15	−1	1	1	1	−1
16	1	1	1	1	1
17	−2	0	0	0	0
18	2	0	0	0	0
19	0	−2	0	0	0
20	0	2	0	0	0
21	0	0	−2	0	0
22	0	0	2	0	0
23	0	0	0	−2	0
24	0	0	0	2	0
25	0	0	0	0	−2
26	0	0	0	0	2
27	0	0	0	0	0
28	0	0	0	0	0
29	0	0	0	0	0
30	0	0	0	0	0
31	0	0	0	0	0
32	0	0	0	0	0

I: welding current; S: welding speed; N: contact tip to work distance; T: welding gun angle; Ac: pinch.

the intermediate levels (0) constitute the central points and combination of each welding parameters at either the highest value (+2) or the lowest value (−2) with other parameters of intermediate levels (0) constitutes star points. 32 experimental trails were conducted to make the estimation of linear quadratic and two-way interactive effects of process parameters on clad geometry [5].

3.4. Conducting Experiments as per Design Matrix. The experiments were conducted at SVS College of Engineering, Coimbatore, India. In this work thirty-two experimental runs were allowed for the estimation of linear quadratic and two-way interactive effects correspond to each treatment combination of parameters on bead geometry as shown in Table 3 at random. At each run settings for all parameters were

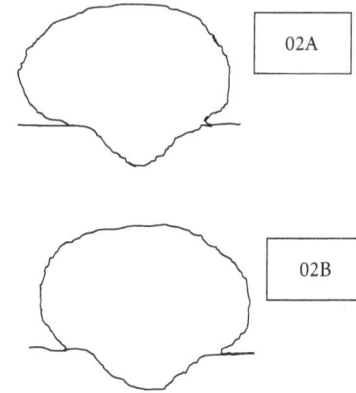

FIGURE 4: Traced Profiles (Specimen No. 2). 02A represents profile of the specimen (front side) and 02B represents profile of the specimen (rear side). The measured clad bead dimension and percentage of dilution is shown in Table 4.

disturbed and reset for next deposit. This is very essential to introduce variability caused by errors in experimental setup.

3.5. Recording of Responses. In order to measure clad bead geometry of transverse section of each weld, overlays were cut using band saw from mid length. Positions of the weld and end faces were machined and grinded. The specimen and faces were polished and etched using a 5% nital solution to display bead dimensions. The clad bead profiles were traced using a reflective type optical profile projector at a magnification of ×10. Then the bead dimensions such as depth of penetration, height of reinforcement, and clad bead width were measured [6]. The traced bead profiles were scanned in order to find various clad parameters and the percentage of dilution with the help of AUTO CAD software. This is shown in Figure 4.

3.6. Development of Mathematical Models. The response function representing any of the clad bead geometry can be expressed as [7–9]

$$Y = f(A, B, C, D, E), \tag{3}$$

where Y = response variable, A = welding current (I) in amp, B = welding speed (S) in mm/min, C = contact tip to work distance (N) in mm, D = welding gun angle (T) in degrees, and E = pinch (Ac).

The second-order surface response model can be expressed as below:

$$\begin{aligned} Y = {} & \beta_0 + \beta_1 A + \beta_2 B + \beta_3 C + \beta_4 D + \beta_5 E + \beta_{11} A^2 \\ & + \beta_{22} B^2 + \beta_{33} C^2 + \beta_{44} D^2 + \beta_{55} E^2 + \beta_{12} AB \\ & + \beta_{13} AC + \beta_{14} AD + \beta_{15} AE + \beta_{23} BC + \beta_{24} BD \\ & + \beta_{25} BE + \beta_{34} CD + \beta_{35} CE + \beta_{45} DE, \end{aligned} \tag{4}$$

where β_0 is the free term of the regression equation, the coefficient β_1, β_2, β_3, β_4 and β_5 are linear terms, the coefficients, $\beta_{11}, \beta_{22}, \beta_{33}, \beta_{44}$, and β_{55} quadratic terms, and the coefficients

TABLE 4: Design matrix and observed values of clad bead geometry.

Trial no.	Design matrix					Bead parameters			
	I	S	N	T	Ac	W (mm)	P (mm)	R (mm)	D (%)
1	−1	−1	−1	−1	1	6.9743	1.67345	6.0262	10.72091
2	1	−1	−1	−1	−1	7.6549	1.9715	5.88735	12.16746
3	−1	1	−1	−1	−1	6.3456	1.6986	5.4519	12.74552
4	1	1	−1	−1	1	7.7635	1.739615	6.0684	10.61078
5	−1	−1	1	−1	−1	7.2683	2.443	5.72055	16.67303
6	1	−1	1	−1	1	9.4383	2.4905	5.9169	15.96692
7	−1	1	1	−1	−1	6.0823	2.4672	5.49205	16.5894
8	1	1	1	−1	−1	8.4666	2.07365	5.9467	14.98494
9	−1	−1	−1	1	−1	6.3029	1.5809	5.9059	10.2749
10	1	−1	−1	1	1	7.0136	1.5662	5.9833	9.707297
11	−1	1	−1	1	1	6.2956	1.58605	5.5105	11.11693
12	1	1	−1	1	−1	7.741	1.8466	5.8752	11.4273
13	−1	−1	1	1	1	7.3231	2.16475	5.72095	15.29097
14	1	−1	1	1	−1	9.6171	2.69495	6.37445	18.54077
15	−1	1	1	1	−1	6.6335	2.3089	5.554	17.23138
16	1	1	1	1	1	10.514	2.7298	5.4645	20.8755
17	−2	0	0	0	0	6.5557	1.99045	5.80585	13.65762
18	2	0	0	0	0	7.4772	2.5737	6.65505	15.74121
19	0	−2	0	0	0	7.5886	2.50455	6.4069	15.77816
20	0	2	0	0	0	7.5014	2.1842	5.6782	16.82349
21	0	0	−2	0	0	6.1421	1.3752	6.0976	8.941799
22	0	0	2	0	0	8.5647	3.18536	5.63655	22.94721
23	0	0	0	−2	0	7.9575	2.2018	5.8281	15.74941
24	0	0	0	2	0	7.7085	1.85885	6.07515	13.27285
25	0	0	0	0	−2	7.8365	2.3577	5.74915	16.63287
26	0	0	0	0	2	8.2082	2.3658	5.99005	16.38043
27	0	0	0	0	0	7.9371	2.1362	6.0153	15.18374
28	0	0	0	0	0	8.4371	2.17145	5.69895	14.82758
29	0	0	0	0	0	9.323	3.1425	5.57595	22.8432
30	0	0	0	0	0	9.2205	3.2872	5.61485	23.6334
31	0	0	0	0	0	10.059	2.86605	5.62095	21.55264
32.	0	0	0	0	0	8.9953	2.72068	5.7052	19.60811

W: width; P: penetration; R: reinforcement; D: dilution %.

$\beta_{12}, \beta_{13}, \beta_{14}, \beta_{15}$, and so forth are the interaction terms. The coefficients were calculated using Quality America Six Sigma Software (DOE-PC IV). After determining the coefficients, the mathematical models were developed. The developed mathematical models are given as follows:

clad bead width (W), mm

$$= 8.923 + 0.701A + 0.388B$$

$$+ 0.587C + 0.040D + 0.088E$$

$$- 0.423A^2 - 0.291B^2 - 0.338C^2$$

$$- 0.219D^2 - 0.171E^2 + 0.205AB$$

$$+ 0.405AC + 0.105AD + 0.070AE$$

$$- 0.134BC + 0.225BD + 0.098BE$$

$$+ 0.26CD + 0.086CE + 0.012DE, \tag{5}$$

depth of penetration (P), mm

$$= 2.735 + 0.098A - 0.032B + 0.389C - 0.032D$$

$$- 0.008E - 0.124A^2 - 0.109B^2 - 0.125C^2$$

$$- 0.187D^2 - 0.104E^2 - 0.33AB + 0.001AC \tag{6}$$

$$+ 0.075AD + 0.005AE - 0.018BC + 0.066BD$$

$$+ 0.087BE + 0.058CD + 0.054CE - 0.036DE,$$

height of reinforcement (R), mm

$$= 5.752 + 0.160A - 0.151B - 0.060C + 0.016D$$

$$-0.002E + 0.084A^2 + 0.037B^2 - 0.0006C^2$$
$$+ 0.015D^2 - 0.006E^2 + 0.035AB + 0.018AC$$
$$- 0.008AD - 0.048AE - 0.024BC - 0.062BD$$
$$- 0.003BE + 0.012CD - 0.092CE - 0.095DE, \qquad (7)$$

percentage dilution (D), %

$$=19.705 + 0.325A + 0.347B + 3.141C - 0.039D$$
$$- 0.153E - 1.324A^2 - 0.923B^2 - 1.012C^2$$
$$- 1.371D^2 - 0.872E^2 - 0.200AB + 0.346AC \qquad (8)$$
$$+ 0.602AD + 0.203AE + 0.011BC + 0.465BD$$
$$+ 0.548BE + 0.715CD + 0.360CE + 0.137DE.$$

3.7. Checking the Adequacy of the Developed Models. The adequacy of the developed model was tested using the analysis of variance (ANOVA) technique. As per this technique, if the F-ratio values of the developed models do not exceed the standard tabulated values for a desired level of confidence (95%) and the calculated R-ratio values of the developed model exceed the standard values for a desired level of confidence (95%), then the models are said to be adequate within the confidence limit [10]. These conditions were satisfied for the developed models. The values are shown in Table 5.

4. Artificial Neural Networks

Artificial neural network (ANN) is biologically inspired by intelligent techniques [11]. Neural network consists of many nonlinear computational elements operating in parallel. Basically it consists of neurons; it represents our biological nervous system. The basic unit of ANN is the neuron. The neurons are connected to each other by link and are known as synapses which are associated to a weight factor. An artificial neuron receives signals from other neurons through the connection between them. Each of the connection strengths has a synaptic connection strength which is represented by a weight of that connection strength. This artificial neuron receives a weighted sum of outputs of all neurons to which it is connected. The weighted sum is then compared with the threshold for an ANN and if it exceeds this threshold ANN fires. When it fires it goes to higher excitation state and a signal is sent down to other connected neurons. The output of a typical neuron is obtained as a result of nonlinear function of weighted sum. This process is clearly shown in Figure 5.

Neural networks have many usages in the present decade. It has been successfully applied across an entire ordinary range of problems domains such as finance, medicine, energy, geology and physics where there is a problem of prediction neural network can be used successfully [12]. It is an adaptable system that can learn relationship through repeated presentation of data and is capable of generalizing a new previously

unseen data. One of the most popular learning algorithms is the back propagation algorithm [13, 14]. In this study, feedback propagation algorithm was used with a single hidden layer improved with numerical optimization technique called Levenberg-Marquard approximation algorithm (LM) [15]. The topology of architecture three-layer of feedforward back propagations network is illustrated in Figure 6.

MATLAB 7 was used for tracing the network for the prediction of clad bead geometry. Statistical mathematical model was used to compare results produced by the work. For normalizing the data, the goal is to examine the statistical distribution of values of each net input, and outputs are roughly uniform; in addition the value should be scaled to match range of input neurons [16].

This is basically in the range of 0 to 1 in practice it is found to between 01 and 9 [17]. In this paper database is normalized using (9):

$$X_{\text{norm}} = 0.1 + \frac{(X - X_{\min})}{1.25\left(X_{\max} - (X_{\min})\right)}, \qquad (9)$$

where X_{norm} = normalized value between 0 and 1, X = value to be normalized, and $X_{\min}$ = minimum value in the dataset range the particular data set rage which is to be normalized. $X_{\max}$ = maximum value in the particular data set range which is to be normalized.

In this study five welding process parameters were employed as input to the network. The Levenberg-Marquardt approximation algorithm was found to be the best fit for application because it can reduce the MSE to a significantly small value and can provide better accuracy of prediction. So neural network model with feed forward back propagation algorithm and Levenberg-Marquardt approximation algorithm was trained with data collected for the experiment. Error was calculated using (10):

$$\text{error} = \frac{(\text{actual value} - \text{predicted value}) \times 100}{\text{predicted value}} \qquad (10)$$

The difficulty in using the regression equation is the possibility of overfitting the data. To avoid this, the experimental data is divided into two sets: one training set and other test data set. The ANN model is created using only training data; the other test data is used to check the of behavior the ANN model created. All variables are normalized using (9). The data was randomized and portioned into two:one training and other test data,

$$y = \sum_i w_{ij}\, h_i + \theta,$$
$$h_i = \tanh\left(\sum_j w_{ij}\, x_j + \theta_i\right). \qquad (11)$$

Neural network general form can be defined as a model shown above y representing the output variables and x_j the set of inputs, shown in (11) and (12). The subscript i represents the hidden units shown in Figure 6, θ represents bias, and w_j represents the weights. The above equation defines the function giving output as a function of input.

TABLE 5: Analysis of variance for testing adequacy of the model.

Parameter	1st-order terms		2nd-order terms		Lack of fit		Error terms		F-ratio	R-ratio	Whether model is adequate
	SS	DF	SS	DF	SS	DF	SS	DF			
W	36.889	20	6.233	11	3.513	6	2.721	5	1.076	3.390	Adequate
P	7.810	20	0.404	11	0.142	6	0.261	5	0.454	7.472	Adequate
R	1.921	20	0.572	11	0.444	6	0.128	5	2.885	3.747	Adequate
D	506.074	20	21.739	11	6.289	6	15.45	5	0.339	8.189	Adequate

SS: sum of squares; DF: degree of freedom; F-ratio $(6, 5, 0.5) = 3.40451$; R-ratio $(20, 5, 0.05) = 3.20665$.

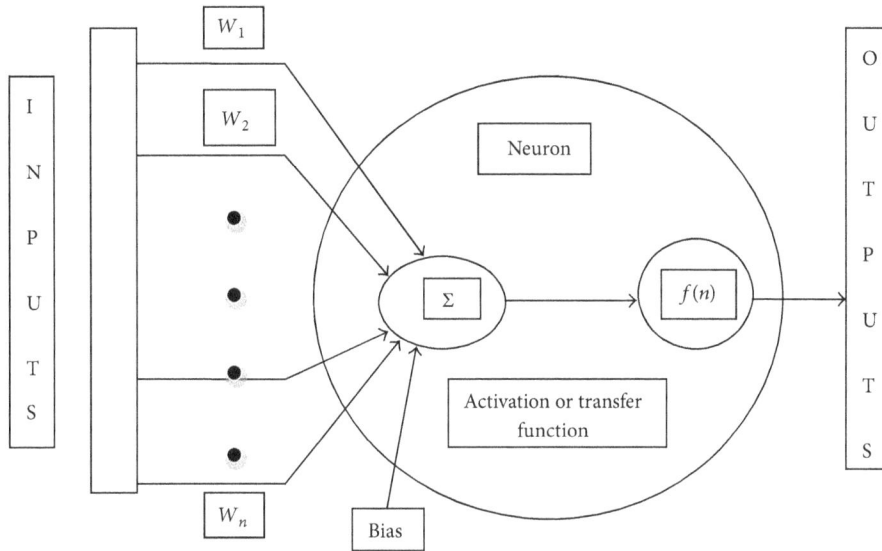

FIGURE 5: Model of an Artificial Neuron.

The training process involves the derivation of weights by minimization of the regularized sum of squared error.

The complexity of model is controlled by the number of hidden level values of regularization constants and are associated with each input one for biases and one for all weights connected to output.

4.1. Procedure for Prediction. The effectiveness of ANN model fully depends on the trial and error process. This study considers the factors that could be influencing the effectiveness of the model developed. In the MATLAB 7 Toolbox there are five Influencing factors listed below:

(1) network algorithm,

(2) transfer function,

(3) training function,

(4) learning function,

(5) performance function.

The ANN structure consists of three layers which are input, hidden, and output layers. It is known that ANN model is designed on trial and error basis. The trial and error is carried out by adjusting the number of layers and the number of neurons in the hidden structure. Too many neurons in hidden layer result in a waste of computer memory and computation time, while too few neurons may not provide the desired data control effect. The process is conducted using 28 randomly selected samples. Seventeen data are used for training and eleven used for testing the data. Table 4 shows randomized data with 1–11 for test data and 12–28 for training data. It is suggested that guide lines should be followed for selecting training and testing of data such as 90% : 10%, 85% : 15%, and 80% : 20% with a total of 100% combined ratio. To fit the randomized sample of 28, preferred ratio selected is 70% : 30%:

(1) $(70/100) \times 28 = 19 - 20$ training samples,

(2) $(30/100) \times 28 = 8 - 9$ data testing samples.

It is necessary to normalize the quantitative variable to some standard range from 0 to 1. The number of neurons' hidden layers should be approximately equal to $n/2, n, 2n$, and $2n+1$ where n is the number of input neurons. Many different ANN network algorithms have been proposed by researchers, but back propagation (BP) algorithm was found to be the best for prediction. Researchers developed the model by using feed forward BP and radial basis network algorithm, and it was found that feed forward BP gives more accurate results. Basically a feed forward network based on BP is a multilayered architecture made up of one or more hidden layers placed between input and output layers, as shown in Figure 6. Transfer function, training function, learning

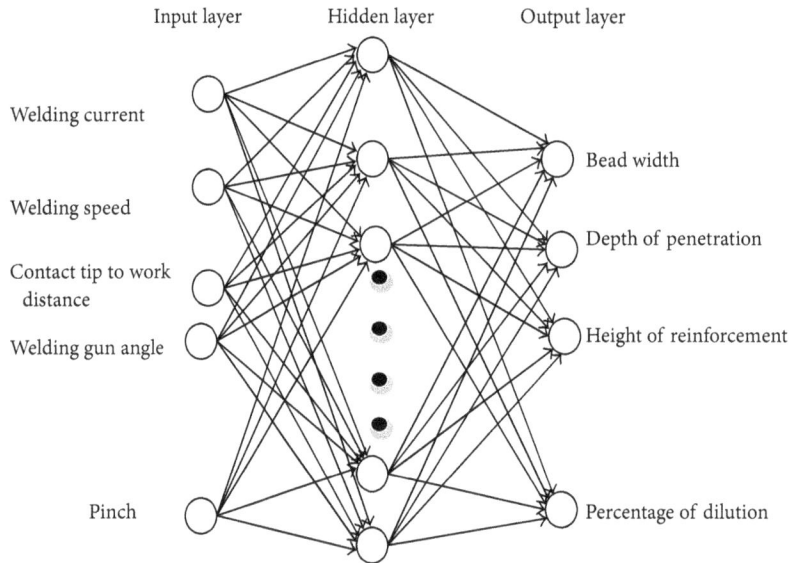

FIGURE 6: Neural network architecture.

function, and performance function used in this study are logsig, traindgm, traingdx, and MSE.

4.2. Determination of the Best ANN Model. In order to determine the best network structure of ANN prediction model, the randomized data set is divided into two sets: one training data set which is used for the prediction of the model, and the other test data which is used to validate the model. Seventeen data were used for training set and eleven data for test set. This to avoid the over fitting of the data. An ANN model is created and trained using MATLAB 7 ANN Toolbox. The lowest MSE obtained is for twelve hidden neurons. So a network structure is 5-12-4: five input neurons, twelve hidden neurons, and four output neurons. Then the test data is validated against the ANN model created; the results are shown in Table 6. Training data is shown in Table 7.

5. Genetic Algorithm Optimization

Genetic algorithm is meta heuristic searching techniques, which mimics the principles of evaluation and natural genetics. These are guided by the random search which scans through entire sample space and therefore provide a reasonable solution. It was introduced by Holland (1975). It is also considered as a heuristic technique inspired by natural biological evolution process comprising selection, cross over mutation, and so forth.

In biological population genetic information is stored in the form of binary strings. The basic processes which affect the binary strings makeup in natural evolution are a selection, a crossover of genetic information between reproducing parents, a mutation of genetic information, and an elitist strategy that keeps the best individual to next generation. The main operations of GA are characterized as follows.

Selection is a method that randomly picks up chromosomes out of the population according to the evolution

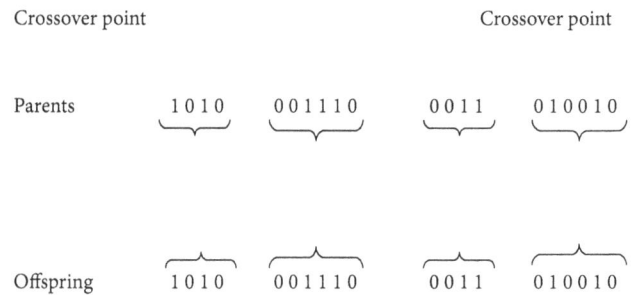

FIGURE 7: Single—Point cross over.

function. The higher the fitness function is, the more chance of an individual has to be selected. The selection pressure is defined as the degree to which the better individuals are favoured.

It takes two individuals and puts their chromosome strings at some randomly chosen position to produce two head segments and two tail segments. The tail segments are then supposed to produce two new full length chromosomes as shown in Figure 7. Two off spring, each inherits some genes from each parent. This is known as single-point cross over. Cross over is not usually applied to all pairs of individuals selected or mating. A random choice is made where likelihood of cross over appeared typically between 6 and 10. If cross over is not applied, off spring are produced simply by duplicating the parents. This gives each individual a chance of passing on its genes without duplication of cross over.

Mutation is applied to each child individually after cross over. It randomly alters each gene with small probability (typically .001) Table 8 shows the fifth gene of a chromosome being mutated. Traditional view is that cross over is more important than the two techniques for rapidly exploring a search space. Mutation provides a small amount of random

TABLE 6: Comparison of actual and predicted values of the clad bead parameters using neural network data (test).

Trial no.	Actual bead parameters				Predicted bead parameters				Error			
	W (mm)	P (mm)	R (mm)	D (%)	W (mm)	P (mm)	R (mm)	D (%)	W (mm)	P (mm)	R (mm)	D (%)
1	6.9743	1.6735	6.0262	10.721	6.1945	1.85	5.9611	12.367	0.7798	−0.177	0.0651	−1.646
2	7.6549	1.9715	5.8873	12.167	7.1815	2.1507	6.5553	10.268	0.4734	−0.179	−0.668	1.899
3	6.3456	1.6986	5.4519	12.746	7.4954	1.5339	5.4923	9.3808	−1.15	0.1647	−0.04	3.3652
4	7.7635	1.7396	6.0684	10.611	6.4936	1.854	6.5573	9.4799	1.2699	−0.114	−0.489	1.1311
5	7.2683	2.443	5.7206	16.673	7.3354	2.6576	5.5657	19.104	−0.067	−0.215	0.1549	−2.431
6	9.4383	2.4905	5.9169	15.967	7.6066	2.1045	6.4342	18.49	1.8317	0.386	−0.517	−2.523
7	6.0823	2.4672	5.492	16.589	8.0417	2.1722	5.5126	16.874	−1.959	0.295	−0.021	−0.285
8	8.4666	2.0737	5.9467	14.985	8.3236	2.2349	5.9031	16.972	0.143	−0.161	0.0436	−1.987
9	6.3029	1.5809	5.9059	10.275	8.2381	1.7955	5.6022	11.219	−1.935	−0.215	0.3037	−0.944
10	7.0136	1.5662	5.9833	9.7073	7.5899	2.4579	6.542	13.415	−0.576	−0.892	−0.559	−3.708
11	6.2956	1.586	5.5105	11.117	7.7318	1.7647	5.8676	10.71	−1.436	−0.179	−0.357	0.407

TABLE 7: Comparison of actual and predicted values of the clad bead parameters using neural network data (training).

Trial no.	Actual bead parameters				Predicted bead parameters				Error			
	W (mm)	P (mm)	R (mm)	D (%)	W (mm)	P (mm)	R (mm)	D (%)	W (mm)	P (mm)	R (mm)	D (%)
1	7.741	1.8466	5.8752	11.4273	7.335	2.0986	6.0792	10.8222	0.406	−0.252	−0.204	0.6051
2	7.3231	2.16475	5.72095	15.29097	6.8214	2.0617	5.6946	14.9379	0.5017	0.10305	0.02635	0.35307
3	9.6171	2.69495	6.37445	18.54077	9.3713	2.8982	6.4084	17.4578	0.2458	0.20325	−0.0339	1.08297
4	6.6335	2.3089	5.554	17.23138	7.4306	2.2927	5.6232	15.7908	−0.7971	0.0162	−0.0692	1.44058
5	10.514	2.7298	5.4645	20.8755	7.8991	2.5154	5.8078	18.0664	2.6149	0.2144	−0.3433	2.8091
6	6.5557	1.99045	5.80585	13.65762	6.5761	1.9158	5.7867	14.2039	−0.0204	0.07465	0.01915	−0.5462
7	7.4772	2.5737	6.65505	15.74121	7.393	2.7191	6.7112	14.7525	0.0842	−0.1454	−0.0561	0.98871
8	7.5886	2.50455	6.4069	15.77816	7.5943	2.4317	6.3834	15.9881	−0.0057	0.07285	0.0235	−0.2099
9	7.5014	2.1842	5.6782	16.82349	7.4652	2.2814	5.7674	16.5744	0.0362	−0.0972	−0.0892	0.24909
10	6.1421	1.3752	6.0976	8.941799	5.6583	1.44	6.2054	9.3753	0.4838	−0.0648	−0.1078	−0.4335
11	8.5647	3.18536	5.63655	22.94721	9.9724	2.962	5.5227	18.9566	−1.4077	0.22336	0.11385	3.99061
12	7.9575	2.2018	5.8281	15.74941	9.0693	2.6919	6.2337	17.5548	−1.1118	−0.4901	−0.4056	−1.8053
13	7.7085	1.85885	6.07515	13.27285	6.7699	1.7807	6.109	12.8584	0.9386	0.07815	−0.0338	0.41445
14	7.8365	2.3577	5.74915	16.63287	8.5364	2.9431	6.6735	15.9653	−0.6999	−0.5854	−0.9243	0.66757
15	8.2082	2.3658	5.99005	16.38043	8.0083	2.371	6.0186	16.3701	0.1999	−0.0052	−0.0285	0.01033
16	7.9371	2.1362	6.0153	15.18374	7.9441	2.1197	6.01	15.3735	−0.007	0.0165	0.0053	−0.1897
17	8.4731	2.17145	5.69895	14.82758	8.6735	2.5165	5.4985	15.2875	−0.2001	−0.3450	0.2031	−0.4599

TABLE 8: Single mutations.

Offspring	1	0	1	0	0	1	0	0	1	0
Mutual offspring	1	0	1	0	1	1	0	0	1	0

TABLE 9: Details of individual.

Individual	X	Fitness	Chromosome
Parent I	0.08	0.05	0001010010
Parent II	0.73	0.000002	1011101011
Offspring I	0.23	0.47	0011101011
Offspring II	0.58	0.00007	1001010010

search and help to ensure that no point in the space has zero probability of being examined. Two individuals can reproduce to give two off spring as shown in Figure 7.

The fitness function is an exponential function of one variable with a maximum $x = .2$ which is coded as a 10-bit binary number in Table 9. Table 9 shows two parents as off spring they produce crossed over after second bit.

If genetic algorithm has been correctly implemented, the population will evolve over successive generation so that fitness of the best and the average individual in each generation increases towards the global optimum. Convergence is the progression towards increasing uniformity. A gene is said to

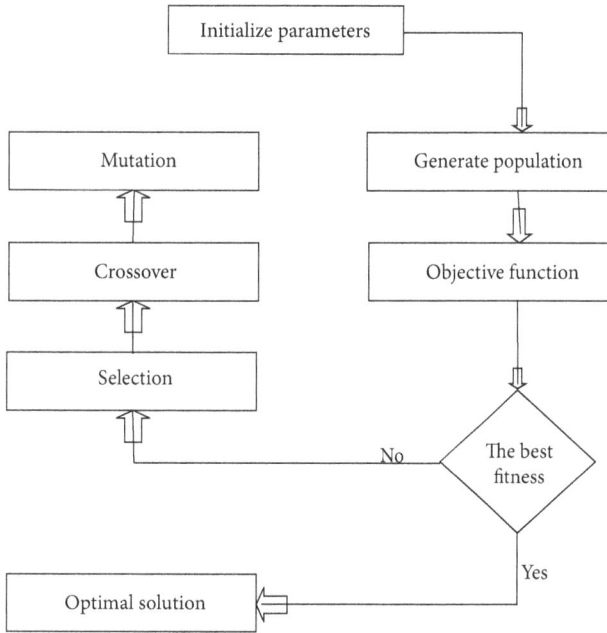

FIGURE 8: Traditional Genetic Algorithm.

TABLE 10: GA search ranges.

Parameters	Range
Welding current (I)	200–300 amp
Welding speed (S)	150–182 mm/min
Contact tip to work distance (N)	10–26 mm
Welding gun angle (T)	70–110 deg
Pinch (Ac)	−10–10

TABLE 11: For GA computation.

Population type	Double vector
Population size	30
Fitness scaling function	Rank
Selection function	Roulette
Reproduction elite count	2
Cross over rate	100
Cross over function	Intermediate
Mutation function	Uniform
Mutation rate	1%
Number of generation	52
Migration	Forward

have converged when 95% of the population shares the same value (Dejony 1975). The population is said to have converged when all of the genes have converged. Traditional genetic algorithm is shown in Figure 8.

6. Optimization of Clad Bead Geometry Using GA

The experimental data related to welding current (I), welding speed (S), welding gun angle (T), contact tip to work distance (N), and pinch (Ac) is used in the experiments conducted.

The aim of the study is to find for optimum adjustments for welding current (I), welding speed (S), welding gun angle (T), contact tip to work distance (N), and pinch (Ac) in a GMAW cladding process. Table 10 shows the options used for study. Table 11 shows GA computation.

The objective function selected for optimizing was the percentage of dilution. The response variables bead width (W), Penetration (P), reinforcement (R), and dilution (D) were given as constraints in their equation. The constrained nonlinear optimiation is mathematically stated as follows:

minimize $f(x)$

subject to $f(X(1), X(2), X(3), X(4), X(5)) < 0$.

Genetic algorithms are nowadays a popular tool in optimizing because GA uses only the values of objective function. The derivatives are not used in the procedure. Secondly the objective function value corresponding to a design vector plays the role of fitness in natural genetics. The aim of the study is to find the optimum adjustments for welding current, welding speed, pinch, welding angle, and contact to tip distance. Objective function selected for optimization was the percentage of dilution. The process parameters and their notations used in writing the program in MATLAB 7 software are given below:

$X(1) = A$ = welding current (I) in amp,

$X(2) = B$ = welding Speed (S) in mm/min,

$X(3) = C$ = contact to work piece distance (N) in mm,

$X(4) = D$ = Welding gun angle (T) in degree,

$X(5) = E$ = pinch (Ac),

Objective function for percentage of dilution which must be minimized was derived from (5)–(8). The constants of welding parameters are given in Table 2

subjected to bounds

$200 \leq X(1) \leq 300$,

$150 \leq X(2) \leq 182$,

$10 \leq X(3) \leq 26$,

$70 \leq X(4) \leq 110$,

$-10 \leq X(5) \leq 10$.

6.1. Objective Function. Consider the following:

$$f(x) = 19.75 + 0.325 * x(1) + 0.347 * x(2)$$
$$+ 3.141 * x(3) - 0.039 * x(4) - 0.153 * x(5)$$

$$-1.324 * x(1)^2 - 0.923 * x(2)^2 - 1.012 * x(3)^2$$
$$-1.371 * x(4)^2 - 0.872 * x(5)^2 \, 0.200 * x(1)$$
$$* x(2) + 0.346 * x(1) * x(3) + 0.602 * x(1)$$
$$* x(4) + 0.203 * x(1) * x(5) + 0.011 * x(2)$$
$$* x(3) + 0.465 * x(2) * x(4) + 0.548 * x(2)$$
$$* x(5) + 0.715 * x(3) * x(4) + 0.360 * x(3)$$
$$* x(5) + 0.137 * x(4) * x(5).$$

(12)

6.2. Constraint Equations. Consider

$$W = (8.923 + 0.701 * x(1) + 0.388 * x(2)$$
$$+ 0.587 * x(3) + 0.040 * x(4) + 0.088 * x(5)$$
$$- 0.423 * x(1)^2 - 0.291 * x(2)^2 - 0.338 * x(3)^2$$
$$- 0.219 * x(4)^2 \, 0.171 * x(5)^2 + 0.205 * x(1) * x(2)$$
$$+ 0.405 * x(1) * x(3) + 0.105 * x(1) * x(4)$$
$$+ 0.070 * x(1) * x(5) \, 0.134 * x(2) * x(3)$$
$$+ 0.2225 * x(2) * x(4) + 0.098 * x(2) * x(5)$$
$$+ 0.26 * x(3) * x(4) + 0.086 * x(3) * x(5)$$
$$+0.12 * x(4) * x(5)) - 3$$

(13)

(clad bead width (W) mm lower limit). Consider

$$P = (2.735 + 0.098 * x(1) - 0.032 * x(2)$$
$$+ 0.389 * x(3) - 0.032 * x(4) - 0.008 * x(5)$$
$$- 0.124 * x(1)^2 - 0.109 * x(2)^2 - 0.125 * x(3)^2$$
$$- 0.187 * x(4)^2 - 0.104 * x(5)^2 - 0.33 * x(1) * x(2)$$
$$+ 0.001 * x(1) * x(3) + 0.075 * x(1) * x(4)$$
$$+ 0.005 * x(1) * x(5) - 0.018 * x(2) * x(3)$$
$$+ 0.066 * x(2) * x(4) + 0.087 * x(2) * x(5)$$
$$+ 0.058 * x(3) * x(4) + 0.054 * x(3) * x(5)$$
$$-0.036 * x(4) * x(5)) - 3$$

(14)

(depth of penetration (P) upper limit). Consider

$$P = (2.735 + 0.098 * x(1) - 0.032 * x(2) + 0.389 * x(3)$$
$$- 0.032 * x(4) - 0.008 * x(5) - 0.124 * x(1)^2$$
$$- 0.109 * x(2)^2 - 0.125 * x(3)^2 - 0.187 * x(4)^2$$
$$- 0.104 * x(5)^2 - 0.33 * x(1) * x(2) + 0.001 * x(1)$$
$$* x(3) + 0.075 * x(1) * x(4) + 0.005 * x(1)$$
$$* x(5) - 0.018 * x(2) * x(3) + 0.066 * x(2)$$
$$* x(4) + 0.087 * x(2) * x(5) + 0.058 * x(3) * x(4)$$
$$+0.054 * x(3) * x(5) - 0.036 * x(4) * x(5)) + 2$$

(15)

(depth of penetration (P) lower limit). Consider

$$W = (8.923 + 0.701 * x(1) + 0.388 * x(2)$$
$$+ 0.587 * x(3) + 0.040 * x(4) + 0.088 * x(5)$$
$$- 0.423 * x(1)^2 - 0.291 * x(2)^2 - 0.338 * x(3)^2 \, 0.219$$
$$* x(4)^2 \, 0.171 * x(5)^2 + 0.205 * x(1) * x(2) + 0.405$$
$$* x(1) * x(3) + 0.105 * x(1) * x(4) + 0.070 * x(1)$$
$$* x(5) \, 0.134 * x(2) * x(3) + 0.225 * x(2) * x(4)$$
$$+ 0.098 * x(2) * x(5) + 0.26 * x(3) * x(4)$$
$$+0.086 * x(3) * x(5) + 0.012 * x(4) * x(5)) - 10$$

(16)

(clad bead width (W) upper limit). Consider

$$R = (5.752 + 0.160 * x(1) - 0.151 * x(2) - 0.060$$
$$* x(3) + 0.016 * x(4) - 0.002 * x(5) + 0.084$$
$$* x(1)^2 + 0.037 * x(2)^2 - 0.0006 * x(3)^2 + 0.015$$
$$* x(4)^2 - 0.006 * x(5)^2 + 0.035 * x(1)$$
$$* x(2) + 0.018 * x(1) * x(3) - 0.008 * x(1)$$
$$* x(4) - 0.048 * x(1) * x(5) - 0.024 * x(2)$$
$$* x(3) - 0.062 * x(2) * x(4) - 0.003 * x(2)$$
$$* x(5) + 0.012 * x(3) * x(4) - 0.092 * x(3)$$
$$*x(5) - 0.095 * x(4) * x(5)) - 6$$

(17)

(height of reinforcement (R) lower limit). Consider

$$R = (5.752 + 0.160 * x(1) - 0.151 * x(2) - 0.060$$
$$* x(3) + 0.016 * x(4) - 0.002 * x(5) + 0.084$$
$$* x(1)^2 + 0.037 * x(2)^2 - 0.0006 * x(3)^2 + 0.015$$
$$* x(4)^2 - 0.006 * x(5)^2 + 0.035 * x(1) * x(2)$$
$$+ 0.018 * x(1) * x(3) - 0.008 * x(1) * x(4)$$
$$- 0.048 * x(1) * x(5) - 0.024 * x(2) * x(3)$$
$$- 0.062 * x(2) * x(4) - 0.003 * x(2) * x(5)$$
$$+ 0.012 * x(3) * x(4) - 0.092 * x(3) * x(5)$$
$$-0.095 * x(4) * x(5)) + 6$$

$$\tag{18}$$

(heights of reinforcement (R) upper limit),

$$f(x) - 23.6334$$
$$8.94441799 - f(x).$$

$$\tag{19}$$

(dilution upper and lower limits),

$$x(1), x(2), x(3), x(4), x(5) \leq 2,$$
$$x(1), x(2), x(3), x(4), x(5) \geq -2.$$

$$\tag{20}$$

MATLAB program in GA and GA function were used for optimizing the problem. The program was written in GA and constraint bounds were applied. The minimum percentage of dilution obtained from the results was while obtained running the GA tool. The minimum percentage of dilution obtained is 10.396. the value of process parameters obtained is $I = 240$ amp, $S = 157$ mm/min, $N = 13$ mm, $T = 87$ degrees, and Ac = 4.5. The fitness function is shown in Figure 9.

7. Methodology of Integrated ANN-GA

The methodology applied in this study involves six cases [18]. They are experimental data, regression modeling, ANN single-based modeling, GA-single based optimization, integrated ANN-GA-type-A-based optimization and integrated ANN-GA-type-B-based optimization. The objectives of type A and type B are as follows.

(1) To estimate the minimum values of cladding parameters compared to the performance values of the experimental data, regression modeling, and ANN single-based modeling.

(2) To estimate the optimal process parameters values that have been within the range of minimum and maximum coded values for process parameters of experimental design that are used for experimental trial.

(3) To estimate the optimal solution of the process parameters within the small number of iterations compared to the optimal solution of the process parameters with single-based GA optimization.

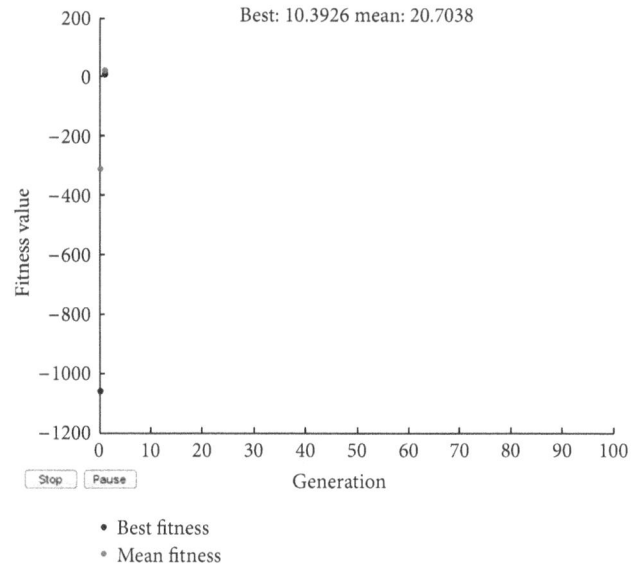

FIGURE 9: Fitness value of GA function.

The steps in order to implement the integrated ANN-GA type A and integrated ANN-GA type-B in fulfilling the three objectives are as follows.

(1) In the experimental data module the values of dilution for different combinations of process parameters are used for modeling.

(2) In the regression modeling schedule model was developed using cladding process parameters. A multilinear regression analysis was performed to predict dilution, and a governing equation was constructed [19].

(3) A predicted model was developed using ANN. The percentage of error was calculated between predicted and actual values.

(4) In the single-based GA optimization, the predicted equation of the regression model would become the objective function. The minimum and maximum coded values of the process parameters of the experimental design would define the boundaries for minimum and maximum values of the optimal solution.

(5) In the integrated ANN-GA type A module which was the first integration system proposed in this study. Similar to GA single-based optimization process, the predicted equation of the regression would become the objective function. The integration system possesses the optimal process parameters value of the single based GA optimization process combined with the process parameters of ANN system would define the boundaries for minimum and maximum values for optimal solution.

(6) In the integrated ANN-GA type B system which is the second integration system proposed in this study. Similar to type A, the predicted regression

TABLE 12: Conditions to define limitation constraint bounds of integrated ANN-GA.

Condition	Decision	
	Lower bound	Upper bound
(1) $(\text{Opt}_{\text{ANN}}) < (\text{Opt}_{\text{GA}})$	Opt_{ANN}	Opt_{GA}
(2) $(\text{Opt}_{\text{ANN}}) > (\text{Opt}_{\text{GA}})$	Opt_{GA}	Opt_{ANN}
(3) $(\text{Opt}_{\text{ANN}}) = (\text{Opt}_{\text{GA}})$	Nearest lower bound of the coded value of experimental design	Nearest upper bound of the coded value of experimental design

equation would become the objective function and the optimal process parameters value of single-based GA optimization combined with process parameters value of the ANN model would define the boundaries for the minimum and maximum values of the optimal solution. This integration system proposes the process parameters values of the best ANN model to define the initial point for optimization solution. This process is shown in Figure 10 [20].

7.1. Integrated ANN-GA Type A optimization Solution. The strategy of this study in implementing integrated ANN-GA type A is implemented by proposing the optimal process parameters value of the GA combined with the nonoptimal process parameters value of the ANN model to define the boundaries for the minimum and maximum value for optimization solution [21]. As given in Table 4, the nonoptimal process parameters values that yield to the minimum prediction value of the ANN model for dilution are $D\% = 9.353$, $I = 250$ amp, $S = 166$ mm/min, $N = 10$ mm, $T = 10$ degree, and $\text{Ac} = 0$. As given in Figure 9, the values of optimal process parameters from GA are $I = 240$ amp, $S = 157$ mm/min, $N = 13$ mm. $T = 87$ degrees, and $\text{Ac} = 4.5$.

Three conditions could be stated for the nonoptimal process parameters values of the ANN model (OptANN) and optimal process parameters values of the GA (OptGA) as classified in Table 12.

By using the conditions stated in Table 12, the decision to define the limitation constraint bound values of the optimization solution is given. Fulfilling the above conditions, (21)–(28) are formulated to define the limitation constraint bounds for welding current, welding speed, contact tip to work distance, welding angle, and pinch as process parameters, respectively. The limitation constraint bound for each process parameter is stated as follows [22]:

$$240 \leq I \leq 250,$$

$$157 \leq S \leq 166,$$

$$10 \leq N \leq 13, \quad (21)$$

$$87 \leq T \leq 90,$$

$$0 \leq \text{Ac} \leq 4.5.$$

As per Figure 10 the optimal process parameters values that lead to the minimum cladding performance of the GA are proposed to define initial points for the integrated ANN-GA-type A to search for the optimal solution [23]. By considering the optimal solution of process parameters of the GA shown in Figure 9, the equations for the initial point of optimization solution for integrated ANN-GA type A are given in (22), (30), (31), (32), and (33) as follows:

$$\text{initial point of } I = 240,$$

$$\text{initial point of } S = 157,$$

$$\text{initial point of } N = 10, \quad (22)$$

$$\text{initial point of } T = 87,$$

$$\text{initial point of Ac} = 0.$$

By using the objective function formulated in (12), the limitation constraint bounds of process parameters formulated in (21), (25), (26), (27), and (28), the initial points formulated in (22), (30), (31), (32), and (33), and the same setting of parameters applied in GA given in Table 12, results of the integrated ANN-GA type A by using MATLAB 7 Optimization Toolbox fitness function are obtained as shown in Figure 11 [19].

It can be seen that the set values of optimal process parameters that lead to the minimum dilution $D = 9.7467\%$ are $I = 274$ amp, $S = 152$ mm/min, $N = 12$ mm, $T = 73$ degrees, and $\text{Ac} = 1.4$.

7.2. Integrated ANN-GA Type B Optimization Solution. Similar to integrated ANN-GA type A approach, the objective function formulated in (12), The basic difference to the integrated ANN-GA type A approach, the nonoptimal process parameters values that led to minimum WELDING performance of the best ANN model shown in Figure 11 is proposed to define initial points for the integrated ANN-GA Type B is to search the optimal solution. Therefore, the equation for the initial point for integrated ANN-GA type B could be given in (23) to (25) as follows:

$$\text{initial point of } I = 250, \quad (23)$$

$$\text{initial point of } S = 166, \quad (24)$$

$$\text{initial point of } N = 10, \quad (25)$$

$$\text{initial point of } T = 90, \quad (26)$$

$$\text{initial point of Ac} = 0. \quad (27)$$

The results of the integrated ANN-GA type B by using MATLAB Optimization Toolbox fitness function are shown in Figure 12.

From Figure 8 it can be seen that the set values of optimal process parameters that lead to the minimum value of dilution $D = 9.7467$ are $I = 212$ amp $S = 138$ mm/min, $N = 16$ mm, $T = 78$ degrees, and $\text{Ac} = 1.5$.

Experimental with several trials modeled points for teaching the prediction values of the regression and ANN.

The minimum and maximum process parameters coded value experimental design maximum and minimum points of the boundaries value for optimization solution for conventional GA

GA optimization

State the optimal process parameters that lead to the minimum cladding performance → one of the maximum/minimum points of the boundaries value for optimization solution for integrated ANN-GA type A and B

State the optimal process parameters that lead to the minimum cladding performance initial point for optimization solution for integrated ANN-GA type A.

Estimate the minimum value of cladding performance

Integrated ANN-GA type A optimization

Estimate the optional process parameters

Estimate the minimum value of cladding performance

Regression modeling

State the optimal process parameters that lead to the minimum cladding performance → initial points of the boundaries value for optimization solution for conventional GA

Estimate the minimum value of cladding performance

Predicted equation of the best regression → objective function for conventional SA and integrated ANN-GA type A and B.

Integrated ANN-GA type B optimization

Estimate the optional process parameters

Estimate the minimum value of cladding performance

ANN modeling

State the optimal process parameters that lead to the minimum cladding performance → initial points of the boundaries value for optimization solution for conventional GA

State the optimal process parameters that lead to the minimum cladding performance → one of the maximum/minimum points of the boundaries value for optimization solution for integrated ANN-GA type A and B

Estimate the minimum value of cladding performance

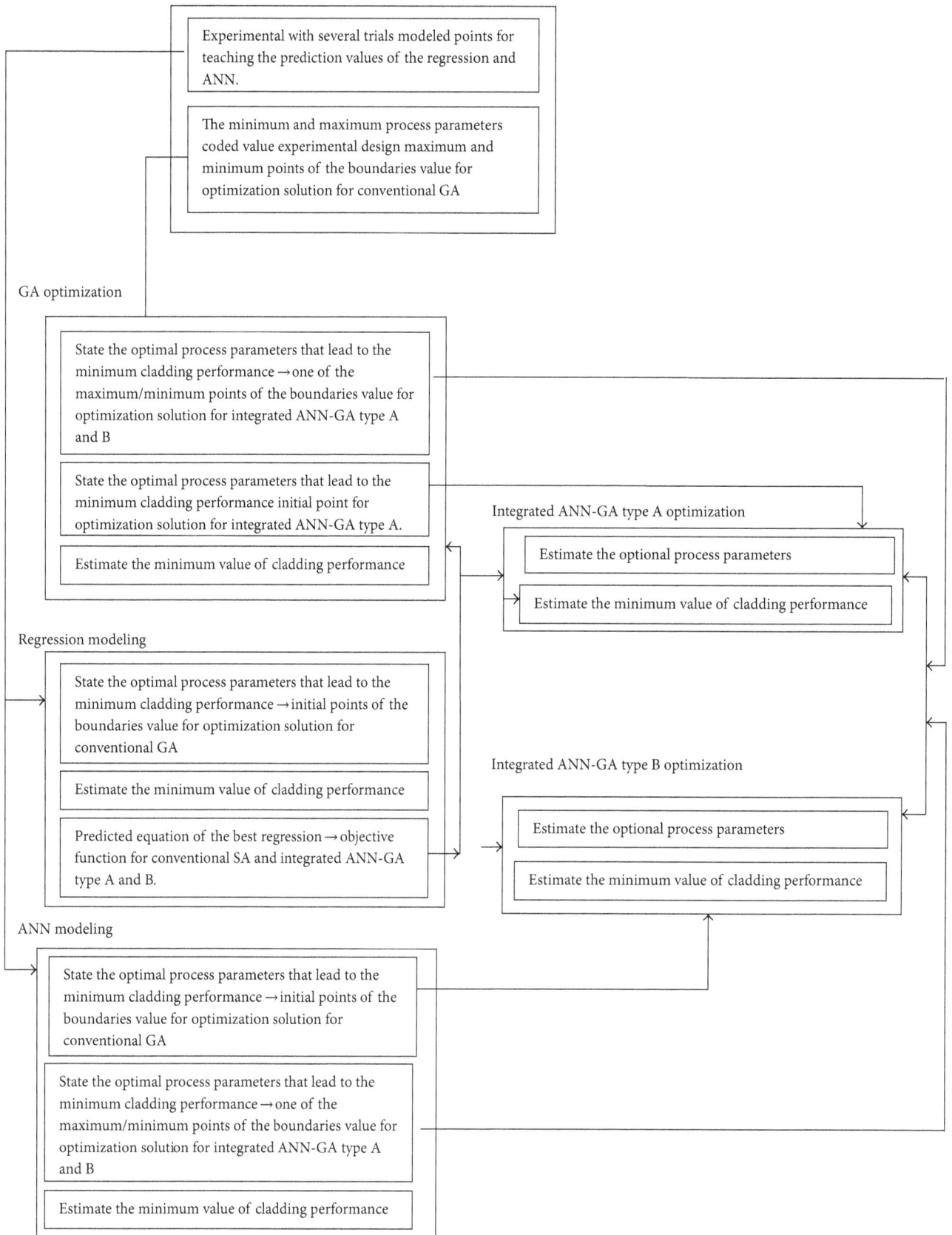

FIGURE 10: Integrated ANN-GA optimization procedure.

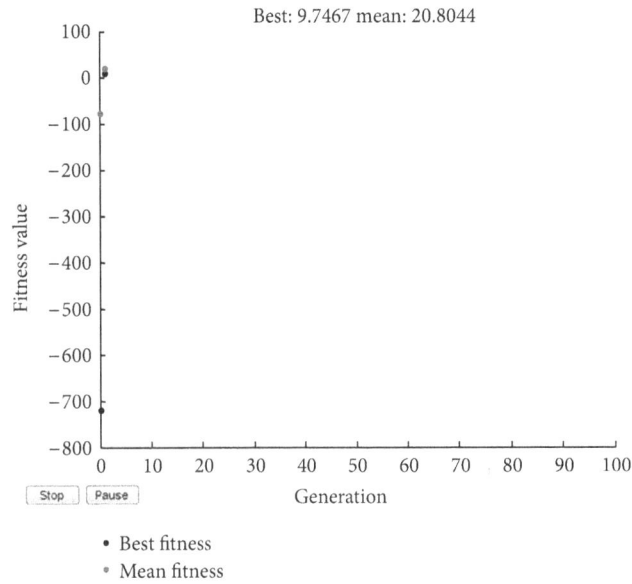

FIGURE 11: Fitness value of ANN-GA Type-A.

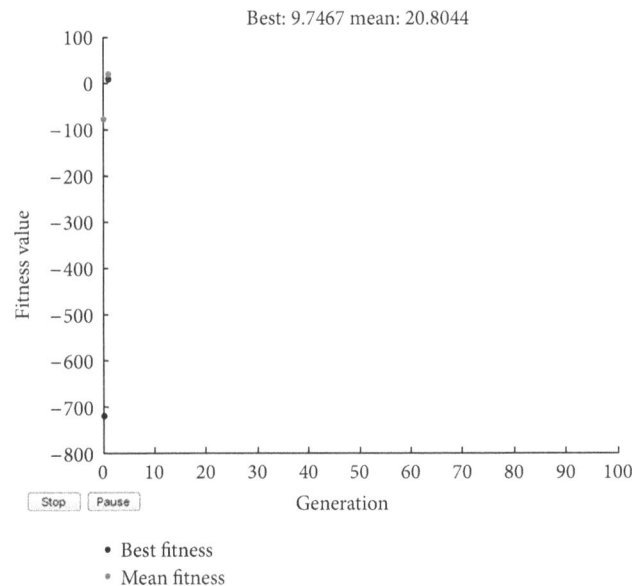

FIGURE 12: Fitness value of ANN-GA Type-B.

7.3. ANN-GA Type A. Consider

$$
\begin{aligned}
D = {} & 19.75 + 0.325 * x\,(1) + 0.347 * x\,(2) + 3.141 \\
& * x\,(3) - 0.039 * x\,(4) - 0.153 * x\,(5) - 1.324 \\
& * x(1)^2 - 0.923 * x(2)^2 - 1.012 * x(3)^2 - 1.371 \\
& * x(4)^2 - 0.872 * x(5)^2 - 0.200 * x\,(1) * x\,(2) + 0.346 \\
& * x\,(1) * x\,(3) + 0.602 * x\,(1) * x\,(4) + 0.203 * x\,(1) \\
& * x\,(5) + 0.011 * x\,(2) * x\,(3) + 0.465 * x\,(2) * x\,(4)
\end{aligned}
$$

$$
\begin{aligned}
& + 0.548 * x\,(2) * x\,(5) + 0.715 * x\,(3) * x\,(4) \\
& + 0.360 * x\,(3) * x\,(5) + 0.137 * x\,(4) * x\,(5),
\end{aligned}
\tag{28}
$$

optimum dilution obtained is 9.9532.

7.4. ANN-GA Type B

$$
\begin{aligned}
D = {} & 19.75 + 0.325 * x\,(1) + 0.347 * x\,(2) + 3.141 \\
& * x\,(3) - 0.039 * x\,(4) - 0.153 * x\,(5) - 1.324 \\
& * x\,(1)^2 - 0.923 * x\,(2)^2 - 1.012 * x\,(3)^2 \\
& - 1.371 * x\,(4)^2 - 0.872 * x\,(5)^2 - 0.200 \\
& * x\,(1) * x\,(2) + 0.346 * x\,(1) * x\,(3) + 0.602 * x\,(1) \\
& * x\,(4) + 0.203 * x\,(1) * x\,(5) + 0.011 * x\,(2) \\
& * x\,(3) + 0.465 * x\,(2) * x\,(4) + 0.548 * x\,(2) \\
& * x\,(5) + 0.715 * x\,(3) * x\,(4) + 0.360 * x\,(3) \\
& * x\,(5) + 0.137 * x\,(4) * x\,(5),
\end{aligned}
\tag{29}
$$

optimum dilution obtained is 9.7895.

8. Validation of the Integrated ANN-GA Results

Theoretically, to validate the results of the proposed approach of this study, the optimal process parameters values of the integrated ANN-GA will be transferred into the regression model equation. Equation (12) is taken as the objective function of the optimization solution. With I being an optimal solution of the welding current. S is an optimal solution of the welding speed; N is an optimal solution of contact to work distance. T is an optimal solution of the welding gun angle, and Ac is an optimal solution of the pinch of integrated ANN-GA. This study discusses the calculation for validating the result of integrated ANN-GA type A and integrated ANN-GA type B.

9. Evaluation of the Integrated ANN-GA Results

In this study, discussion is carried out to highlight all the objectives of the study and is separated into three parts which are evaluation of the minimum dilution value, evaluation of the optimal process parameters, and evaluation of the number of iteration of the integrated ANN-GA results.

9.1. First Objective: Evaluation of the Minimum Dilution Value. Figures 11 and 12 show the minimum value of dilution for both integration systems, integrated ANN-GA type A and ANN-GA type B, is 9.7467. For fulfilling the first objective of this study, evaluation against the minimum dilution is 10.3926 by GA.

9.2. Experimental Data versus Integrated ANN-GA. Optimum dilution should be between 8 and 15% from a previous study. This objective is fulfilled in this case.

9.3. Regression versus Integrated ANN. The minimum predicted dilution value of the regression model is 9.875. Therefore, with dilution is 9.7467, It can be concluded that both integration systems have given the more minimum result of the dilution compared to regression model. Consequently, integrated ANN-GA type A and integrated ANN-GA type B have reduced the value of dilution and optimized it between 8 to 15.

9.4. ANN versus Integrated ANN-GA. The minimum predicted Dilution value of the ANN model is 9.353; it can be concluded that both integration systems have given the more optimum result of the dilution compared to ANN model. Consequently, integrated ANNSA type A and integrated ANN-GA type A are well within the limits of standard dilution.

9.5. GA versus Integrated ANN-GA. The minimum predicted dilution value of the GA was 10.375. Therefore, with dilution 9.745, it can be concluded that both integrated systems have given the more minimum of the Dilution compared to GA technique.

9.6. Second Objective: Evaluation of the Optimal Process Parameters. For the second objective of this study, the optimal values of the integrated ANN-GA type A and integrated ANN-SA type B for each process parameter are within the range of minimum and maximum values of experimental design; thus, this study concludes that the second objective of this study is fulfilled.

9.7. Third Objective: Evaluation of the Number of Iteration. Integrated ANN-GA type A and integrated ANN-GA type B are approximately the same as or lower than the number of iterations by GA; thus, this study concludes that the third objective of this study is fulfilled.

10. Conclusion

(1) A five-level-five-factor full factorial design matrix based on central composite routable design technique was used for building the mathematical model.

(2) ANN tool box available in MATLAB was used for the prediction of bead geometry.

(3) In this study two models, ANN and GA are used for prediction and optimization of bead geometry.

(4) In GMAW process bead geometry plays an important role in economising the material. This study effectively used the integration method for optimization.

(5) This study proposed two integration systems, integrated ANNGA type A and integrated ANN-GA type B, in order to estimate the optimal solutions of process parameters that lead to minimum dilution found to satisfy three conditions. Overall, integrated ANN-GA type A and integrated ANN-GA type B have been very effective in estimating optimum values of dilution compared to experimental data and regression method. In the second process the optimal value of process parameters recommended by integrated type A and integrated type B are well within the range of minimum and maximum value of process parameters of experimental design. In the third issue it was found that the proposed integration system satisfies the number of iterations than the single system of optimization.

References

[1] P. K. Palani and N. Murugan, "Prediction of delta ferrite content and effect of welding process parameters in claddings by FCAW," *Materials and Manufacturing Processes*, vol. 21, no. 5, pp. 431–438, 2006.

[2] T. Kannan and N. Murugan, "Prediction of Ferrite Number of duplex stainless steel clad metals using RSM," *Welding Journal*, vol. 85, no. 5, pp. 91-s–99-s, 2006.

[3] N. Murugan and V. Gunaraj, "Prediction and control of weld bead geometry and shape relationships in submerged arc welding of pipes," *Journal of Materials Processing Technology*, vol. 168, no. 3, pp. 478–487, 2005.

[4] I. S. Kim, K. J. Son, Y. S. Yang, and P. K. D. V. Yaragada, "Sensitivity analysis for process parameters in GMA welding processes using a factorial design method," *International Journal of Machine Tools and Manufacture*, vol. 43, no. 8, pp. 763–769, 2003.

[5] W. G. Cochran and G. M. Coxz, *Experimental Design*, John Wiley & Sons, New York, NY, USA, 1987.

[6] S. Karaoğlu and A. Seçgin, "Sensitivity analysis of submerged arc welding process parameters," *Journal of Materials Processing Technology*, vol. 202, no. 1–3, pp. 500–507, 2008.

[7] P. K. Ghosh, P. C. Gupta, and V. K. Goyal, "Stainless steel cladding of structural steel plate using the pulsed current GMAW process," *Welding Journal*, vol. 77, no. 7, pp. 307-s–314-s, 1998.

[8] V. Gunaraj and N. Murugan, "Prediction and comparison of the area of the heat-affected zone for the bead-on-plate and bead-on-joint in submerged arc welding of pipes," *Journal of Materials Processing Technology*, vol. 95, no. 1–3, pp. 246–261, 1999.

[9] D. C. Montgomery, *Design and Analysis of Experiments*, John Wiley & Sons, 2003.

[10] T. Kannan and J. Yoganandh, "Effect of process parameters on clad bead geometry and its shape relationships of stainless steel claddings deposited by GMAW," *International Journal of Advanced Manufacturing Technology*, vol. 47, pp. 1083–1095, 2010.

[11] P. Dutta and D. K. Pratihar, "Modeling of TIG welding process using conventional regression analysis and neural network-based approaches," *Journal of Materials Processing Technology*, vol. 184, no. 1–3, pp. 56–68, 2007.

[12] H. Ates, "Prediction of gas metal arc welding parameters based on artificial neural networks," *Materials and Design*, vol. 28, no. 7, pp. 2015–2023, 2007.

[13] J. I. Lee and K. W. Um, "Prediction of welding process parameters by prediction of back-bead geometry," *Journal of Materials Processing Technology*, vol. 108, no. 1, pp. 106–113, 2000.

[14] A. Mukhopadhyay and A. Iqbal, "Prediction of mechanical properties of hot rolled, low-carbon steel strips using artificial neural network," *Materials and Manufacturing Processes*, vol. 20, no. 5, pp. 793–812, 2005.

[15] D. K. Panda and R. K. Bhoi, "Artificial neural network prediction of material removal rate in electro discharge machining," *Materials and Manufacturing Processes*, vol. 20, no. 4, pp. 645–672, 2005.

[16] D. S. Nagesh and G. L. Datta, "Prediction of weld bead geometry and penetration in shielded metal-arc welding using artificial neural networks," *Journal of Materials Processing Technology*, vol. 123, no. 2, pp. 303–312, 2002.

[17] K. Manikya Kanti and P. Srinivasa Rao, "Prediction of bead geometry in pulsed GMA welding using back propagation neural network," *Journal of Materials Processing Technology*, vol. 200, no. 1–3, pp. 300–305, 2008.

[18] A. M. Zain, H. Haron, and S. Sharif, "Optimization of process parameters in the abrasive waterjet machining using integrated SA-GA," *Applied Soft Computing Journal*, vol. 11, no. 8, pp. 5350–5359, 2011.

[19] F. Cus and J. Balic, "Optimization of cutting process by GA approach," *Robotics and Computer-Integrated Manufacturing*, vol. 19, no. 1-2, pp. 113–121, 2003.

[20] A. M. Zain, H. Haron, and S. Sharif, "Estimation of the minimum machining performance in the abrasive waterjet machining using integrated ANN-SA," *Expert Systems with Applications*, vol. 38, no. 7, pp. 8316–8326, 2011.

[21] P. Sahoo, "Optimization of turning parameters for surface roughness using RSM and GA," *Advances in Production Engineering and Management*, vol. 6, no. 3, pp. 197–208, 2011.

[22] F. Kolahan and M. Heidari, "A new approach for predicting and optimizing weld bead geometry in GMAW," *World Academy of Science, Engineering and Technology*, vol. 59, pp. 138–141, 2009.

[23] K. Siva, N. Murugan, and R. Logesh, "Optimization of weld bead geometry in plasma transferred arc hardfaced austenitic stainless steel plates using genetic algorithm," *International Journal of Advanced Manufacturing Technology*, vol. 41, no. 1-2, pp. 24–30, 2009.

20

Experimental Evaluation of ZigBee-Based Wireless Networks in Indoor Environments

Jin-Shyan Lee[1] and Yuan-Ming Wang[2]

[1] *Department of Electrical Engineering, National Taipei University of Technology, Taipei 10608, Taiwan*
[2] *Information and Communication Research Labs, Industrial Technology Research Institute, Hsinchu 31040, Taiwan*

Correspondence should be addressed to Jin-Shyan Lee; jslee@mail.ntut.edu.tw

Academic Editor: Christos Bouras

ZigBee is an emerging standard specifically designed for wireless personal area networks (WPANs) with a focus on enabling the wireless sensor networks (WSNs). It attempts to provide a low-data rate, low-power, and low-cost wireless networking on the device-level communication. In this paper, we have established a realistic indoor environment for the performance evaluation of a 51-node ZigBee wireless network. Several sets of practical experiments have been conducted to study its various features, including the (1) node connectivity, (2) packet loss rate, and (3) transmission throughput. The results show that our developed ZigBee platforms could work well under multihop transmission over an extended period of time.

1. Introduction

For the past few decades, in order to access networks and services without cables, wireless communication is a rapid growing technology to provide the flexibility and mobility [1]. Obviously, reducing the cable restriction is clearly one of the benefits of wireless with respect to cabled devices. Other benefits include the dynamic network formation, easy deployment, and low cost in some cases. In general, the wireless networking has followed a similar trend due to the increasing exchange of data in services such as the Internet, e-mail, and data file transfer. The capabilities needed to deliver such services are characterized by an increasing need for data throughput. However, other applications in fields such as industrial [2], vehicular, and home sensors have more relaxed throughput requirements. Moreover, these applications require lower power consumption and low complexity wireless links for a low cost (relative to the device cost). ZigBee over [3–5] IEEE 802.15.4 [6] is the one that addresses these types of requirements.

Based on the IEEE 802.15.4 standard, ZigBee is a global specification created by a multivendor consortium called the ZigBee Alliance. Whereas 802.15.4 defines the physical (PHY) and medium access control sublayer (MAC) layers of an application, ZigBee defines the network and application layers, application framework, application profile, and the security mechanism. ZigBee provides users in specific applications with a simple, low-cost global network that supports a large number of nodes with an extremely low power drain on the battery. The ZigBee stack architecture, as shown in Figure 1, is based on the standard open systems interconnection (OSI) seven-layer model but defines only those layers relevant to achieving functionality in the intended market space.

Wireless links under ZigBee can operate in three license free industrial scientific medical (ISM) frequency bands, including 868 MHz in Europe, 915 MHz in the USA and Australia, and 2.4 GHz in most jurisdictions worldwide. Data transmission rates vary from 20 to 250 kbps. A total of 27 channels are allocated in 802.15.4, including 1 channel in the 868 MHz, 10 channels in the 915 MHz, and 16 channels in the 2.4 GHz band. The ZigBee network layer supports star, tree, and mesh topologies. Each network must have one ZigBee coordinator (ZC) to create and control the network. In the tree and mesh topologies, the ZigBee routers (ZRs) are used to extend the communication range at the network level. ZigBee devices are of three types, besides the mentioned ZC and ZR. For sensing applications, the sensors are usually

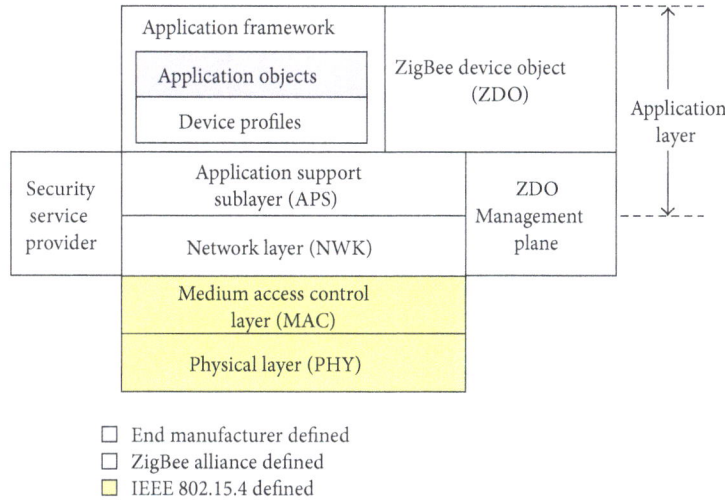

FIGURE 1: The ZigBee over IEEE 802.15.4 protocol stack.

FIGURE 2: Computing and networking module for ITRI SCAN-ZB32.

programmed as ZigBee end devices (ZED), which contain just enough functionality to communicate with their parent node (either a ZC or ZR) and cannot relay data from other devices.

Up to now, many performance analyses for wireless sensor networks (WSNs) have been proposed [7–10]. Kohvakka et al. [7] studied the performance of ZigBee-based sensor networks based on a cluster-tree topology. The study in [9] also analyzed the reliability of IEEE 802.15.4 cluster trees. For beacon-enabled IEEE 802.15.4 WPAN, Chen et al. [10] evaluated the performance for industrial monitoring applications using OMNeT++ tool. On the other hand, the evaluation of IEEE 802.15.4 for wireless medical applications was performed in [8] via systematic simulations. However, most of the previous work is based on simulation rather than practical experiments. Lee [11] established a five-node sensor network to experimentally evaluate the IEEE 802.15.4 performance with various features. In [12, 13], the performance of realistic IEEE 802.15.4 was evaluated and compared with existing simulation models in NS-2. They also analyzed the coexistence of IEEE 802.15.4 with IEEE 802.11 and Bluetooth which are operating in the same 2.4 GHz ISM band. Cano-García and Casilari [14] presented an empirical study of the effects of the channel occupation on the consumption of actual ZigBee/IEEE 802.15.4 motes. However, the WSN literature provides few experimental analyses for ZigBee-based sensor networks.

This paper evaluates the performance of a ZigBee-based wireless network under multihop transmission over an extended period of time. The ITRI ZBnode [15, 16], developed by the Industrial Technology Research Institute (ITRI), is applied and deployed in an indoor environment. Totally, 51

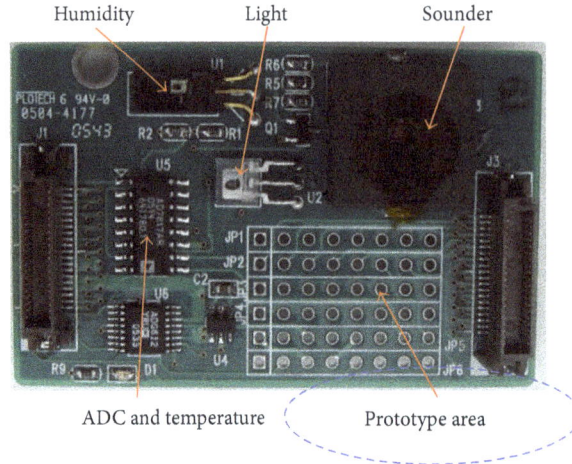

FIGURE 3: Sensing module for ITRI SCAN-ZB32. A prototype area is provided for other extended sensing functions.

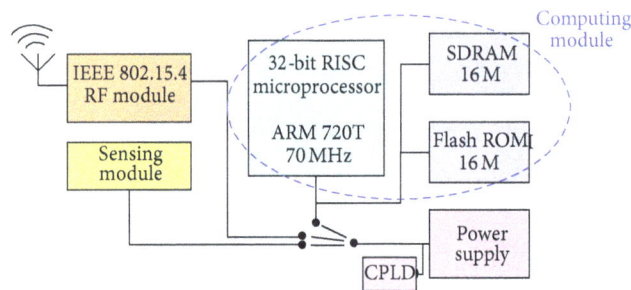

FIGURE 4: Power management mechanism for ITRI SCAN-ZB32.

sensor nodes are deployed in a hallway and a room. One of the nodes is a coordinator, and all the others are sensors with the routing capability. Each sensor regularly transmits packets to the coordinator during a long-term operation time. The practical performance of this wireless sensor network is evaluated in terms of (1) node connectivity, (2) packet loss rate, and (3) transmission throughput. This paper is an extension of our previous work [17] with newly added experiments on the node connectivity.

The rest of the paper is organized as follows: Section 2 introduces the developed ZigBee devices, and then experimental configurations are described in Section 3. The experimental results are shown in Section 4, and finally, Section 5 gives the conclusions.

2. Proposed Platform: ITRI ZBnode

Based on an IEEE 802.15.4 radio module and an ARM processor, the developed ZBnode [15, 16] is an autonomous, light weight wireless communication and computing platform. As ITRI performed a project for developing a small device with sensing, computing, and networking (SCAN) capabilities, the first version of ZBnode was designed and implemented. Since the first version used a powerful 32-bit processor, it was named SCAN-ZB32. Since individual sensor configurations are required depending on applications, the ZB32 device has

no integrated sensors for general-purpose applications. In the development stage, ZB32 can be used with various serial devices through predetermined sockets. The serial devices include sensors, actuators, power chargers, RFID readers, and even user interface components.

2.1. ZigBee Hardware Design. The ZBnode hardware adopts a 32-bit RISC processor which features an ARM 720T CPU core running up to 70 MHz, as shown in Figure 2. Several integrated peripherals including timers, counters, 10-bit AD converter, USB, UARTs, LCD controllers, infrared communications, controller area network (CAN) interfaces, pulse-width modulation (PWM), and JTAG for debugging are built around the processor. The external memory comprises an in-system programmable Flash ROM (16 MB) and an SDRAM (16 MB). To implement a visual user application interface, ZB32 provides four buttons and five LEDs. An IEEE 802.15.4 compliant RF transceiver, Chipcon CC2420, is connected to an on-board 2.4 GHz chip antenna and to one of the serial peripheral interfaces of the processor. Moreover, the RF module can be switched to an external antenna via a standard SMA (subminiature type A) connector for a better performance. The sensing module, where several sensors are supplied for detecting environment conditions (such as temperature, humidity, and light), is shown in Figure 3. For general-purpose development, a prototype area is provided

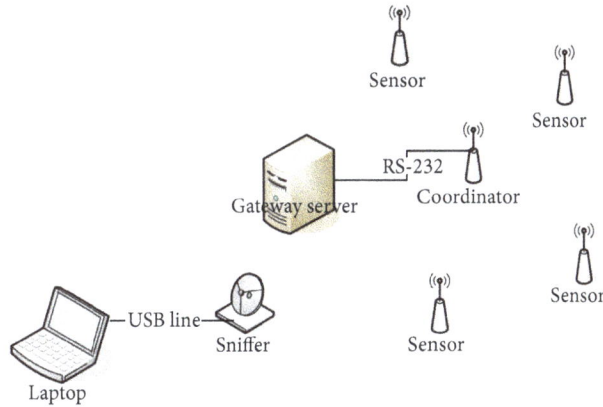

FIGURE 5: Experimental network structure.

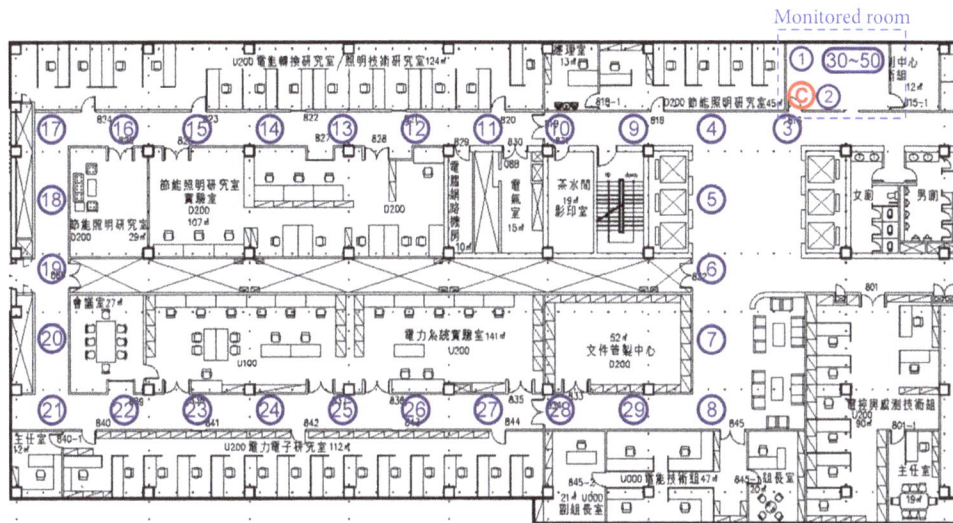

FIGURE 6: The deployment layout for sensor nodes.

for other extended sensing functions. In addition, a sounder which could be used as an alarm notification is also available. About the energy source, the ZBnode can be powered with an AC adapter or an Li-ion battery with a DC input range from 3.5 to 5.5 V.

2.2. Power Management Mechanism.

In in-door applications to environmental monitoring, for the most time only the sensing module is active while the computing and communication modules are power down. As shown in Figure 4, ZB32 uses a separate power-gating circuit to perform the power management. The power streams of the computing module (including ARM processor, ROM, and SDAM), communication module (IEEE 802.15.4 RF), and sensing module (temperature, humidity, and light sensors) when not in use can be completely disconnected from the energy source. A complex programmable logic device (CPLD) is adopted to control power stream via three power switches and also to handle power management requests in the developed ZBnode platform. The power management requests could be issued

from either the computing, communication, or the sensing module. For example, as soon as the sensing module detects an abnormal situation, it would signal the CPLD to wake up the computing module to process such events.

2.3. ZigBee Protocol and Software Stack.

In our design, the SCAN-ZB32 platform uses an ARM Linux kernel 2.4. In addition, the system software is a Linux-based framework written in ANSI C. For general-purpose development, an open ARM-Linux compiler suite is applied. Within this framework, applications are typically partitioned into (1) a ZigBee protocol layer for communicating with the IEEE 802.15.4 front end through commands/events and keeping track of point-to-point connections with individual state machines, (2) a command line terminal for debugging and control, and (3) a user-defined application object. The whole application is defined at compile time and then programmed into the flash memory using an Xmodem terminal via the RS-232 port if the Linux kernel is loaded in advance. In addition, an in-system programmer through the JTAG interface is

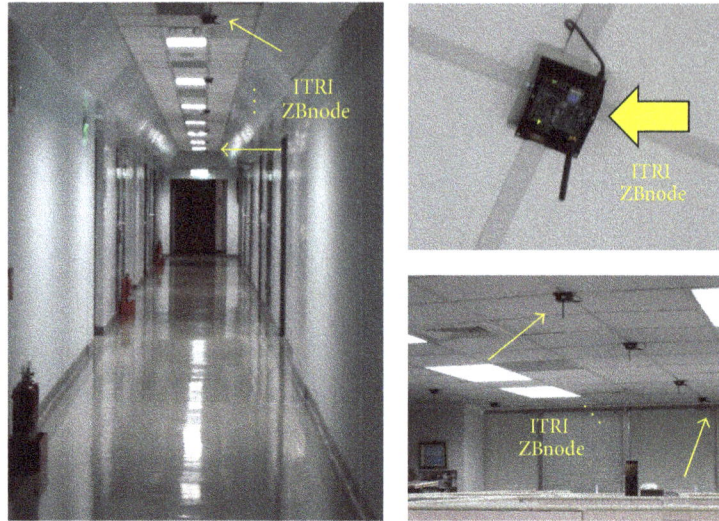

FIGURE 7: Sensor nodes attached to the ceiling of the hallway and room.

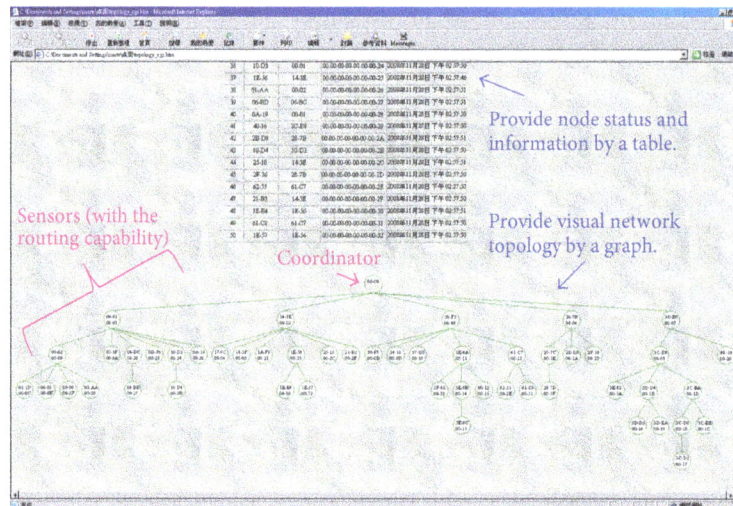

FIGURE 8: Node status and network topology monitored via internet.

provided to download the application program. Furthermore, the developed protocol stack is embedded into the driver with convenient APIs available for application developers.

3. Experimental Configurations

3.1. Performance Metrics. In this paper, the node connectivity, packet loss rate, and transmission throughput are used as the performance metrics.

(i) *Node Connectivity.* In our work, if a node can send out packets to the coordinator within a specified period of time, the node is defined as "connected." Otherwise, the node is "disconnected." Obviously, if more nodes in a sensor network are connected, the network is more stable.

(ii) *Packet Loss Rate.* The packet loss rate of a node is defined as the number of packets lost by the coordinator divided by the number of packets transmitted by the node. The less a packet loss rate is, the better a network performance is. Moreover, the ZigBee standard provides an optional acknowledged service in application support sublayer (APS) for reliable transmission.

(iii) *Transmission Throughput.* For simplicity, the transmission throughput is measured under a two-hop communication (from a sensor to the coordinator) in our experiments. In a specific period, the more packets received by the coordinator results in a higher transmission throughput.

3.2. Node Deployment. The experimental network structure is presented in Figure 5. The coordinator stores data received

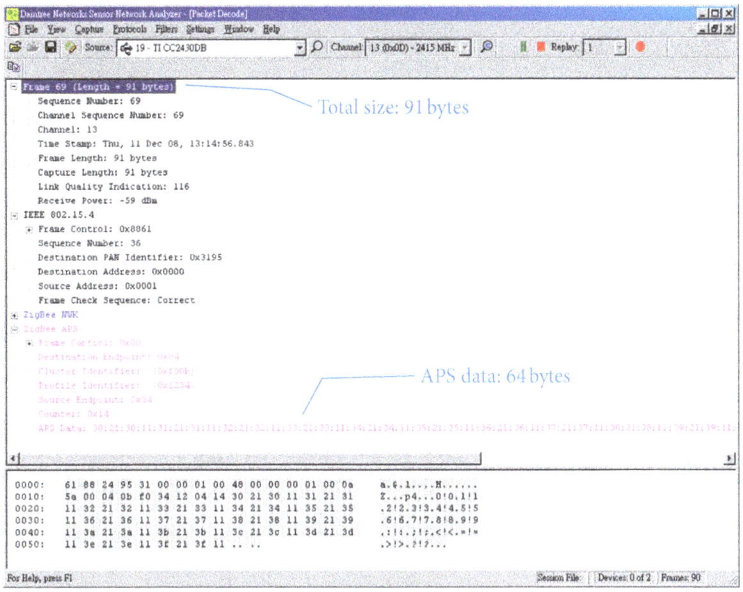

FIGURE 9: Sniffed packet: the packet size used for performance analysis is 91 bytes.

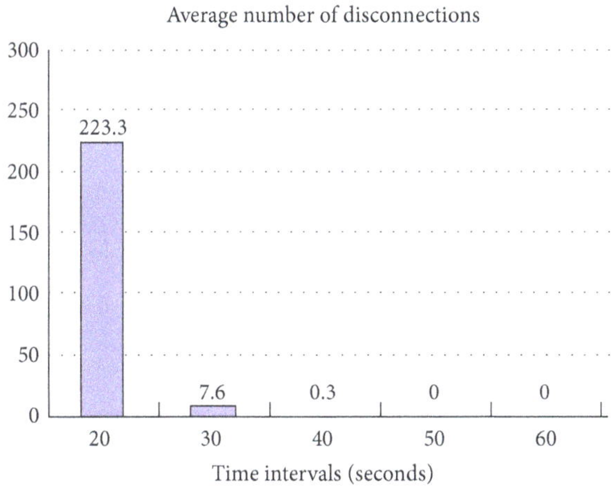

FIGURE 10: Average number of disconnections with varied time intervals.

from each node in an MySQL database of a gateway server via RS-232 port. Meanwhile, there is a sniffer near the coordinator (within one hop) to show and record the over-the-air data.

Figure 6 shows the deployment layout for sensor nodes. There are totally 51 sensor nodes in the network, of which one is the coordinator (node number C) and the others are sensors with the routing capability (node number 1–50). Nodes 3–29 are attached to the ceiling of the hallway. The coordinator and nodes 1, 2, and 30–50 are in the same room due to the security policy. In our experiments, each node is equipped with an AC power source for long-term operation time. Figure 7 shows several sensor nodes attached to the ceiling of the hallway and rooms. Figure 8 draws the node status and network topology monitored via internet using a web browser. Each circle respects a sensor, and the coordinator is the root node. Obviously, a cluster tree network with multihop transmission can be successfully formed.

3.3. Transmitted Packets. Figure 9 shows the content of sniffed data packets during transmission. The APS overhead in packets used for analyzing network performance is 12 bytes (8 bytes for destination address and 4 bytes for serial number), and the APS payload in a packet is 64 bytes. Moreover, with the overhead of other protocol layers (network, MAC, and physical layers), the total data size of a packet is 91 bytes.

4. Experimental Results

4.1. Node Connectivity. In this experiment, the transmission rate is 1 packet per 10 seconds for each node, and the operation time is seven days. As defined before, if a node can send out packets to the coordinator within a specified period of time, the node is defined as connected. Figure 10 shows the average number (50 nodes) of disconnections for varied specified period of time (20 seconds to 1 minute). Obviously, as the specified period of time is tight as 20 seconds, the number of

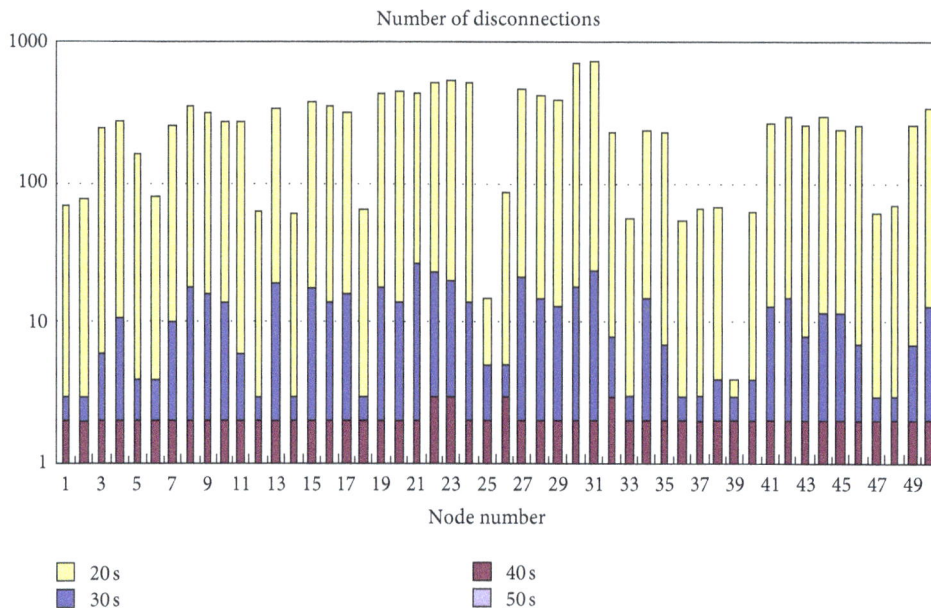

FIGURE 11: Number of disconnections for each node with varied time interval.

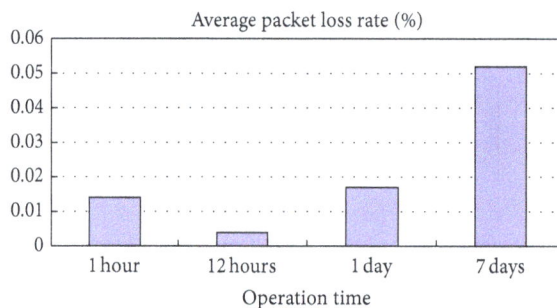

FIGURE 12: Average packet loss rate with varied operation time.

disconnections is the highest. Also, the fact that the average disconnection of 30 seconds is significantly less than that of 20 seconds shows that sensors with the routing functions are able to retransmit packets via APS ACK mechanism.

Figure 11 shows the number of disconnections for each node with varied time interval. According to the empirical results, when the time interval changes from 20 seconds to 30 seconds, the number of disconnections between the coordinator and each sensor node significantly decreases.

4.2. Packet Loss Rate. The average packet loss rate (50 nodes) with varied operation time has been recorded as shown in Figure 12. In this experiment, the transmission rate is 1 packet per 10 minutes for each node with APS ACK. When the operation duration is 7 days, the packet loss rate has the highest value. Overall, the shorter an operation time is, the lower packet a loss rate is. However, the reason why the packet loss rate of 1-hour operation time is higher than that of 12-hour one is the fact that only a small number of packets are

transmitted of 1-hour operation time so that losing one packet increases the packet loss rate a lot.

Figure 13 shows the average packet loss rate with and without APS ACK. The operation time for this experiment is 1 hour. The maximum loss rate is 0.147% and 3.86% with and without APS ACK, respectively. The result shows that packet loss rate could be significantly decreased to almost zero by using the APS ACK mechanism.

Figure 14 shows the packet loss rate of each node with APS ACK. The operation duration for this experiment is 7 days. The experimental results show that the more hops to the coordinator, the higher packet loss rate will be.

4.3. Transmission Throughput. In this experiment, the transmitted packets are with and without APS ACK. The empirical results are shown in Table 1. If the APS ACK is applied, the coordinator totally received 1300 packets in 27 seconds, indicating the transmission rate is 24.65 Kbits/sec. As the transmission is without APS ACK, the result shows the

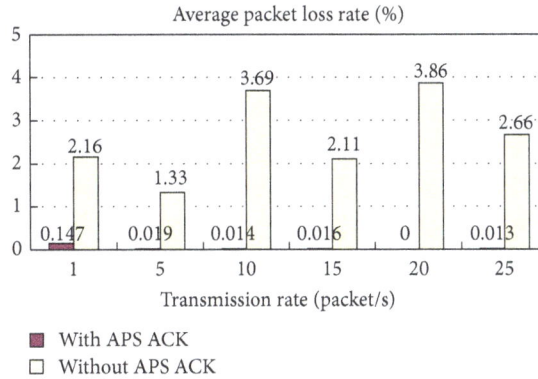

FIGURE 13: The effect of APS ACK to packet loss rates.

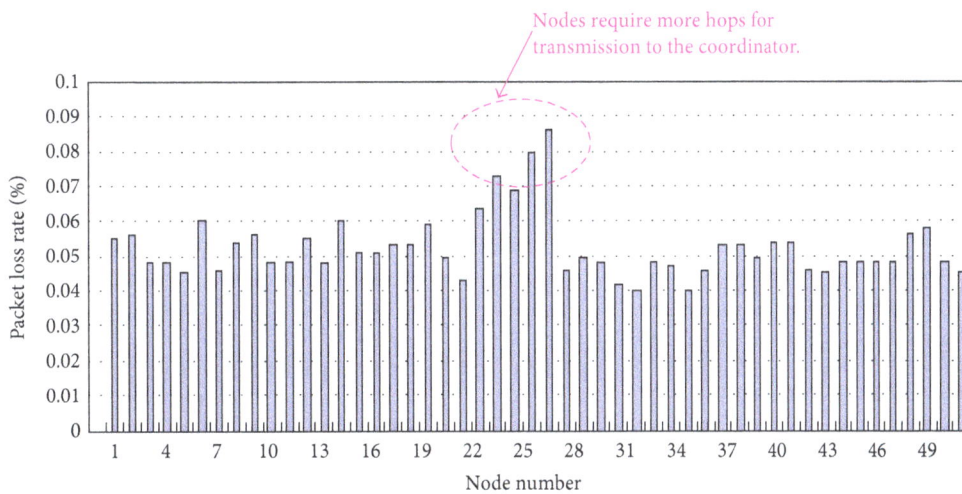

FIGURE 14: Packet loss rate of each node for seven-day transmission.

TABLE 1: Empirical results of transmission throughput.

	With APS ACK	Without APS ACK
Number of packets received by the Coordinator	1300	1097
Duration (second)	27	23
Transmission rate (Kbps)	**24.65**	**24.42**

coordinator receives fewer packets at the similar transmission rate.

5. Conclusion

This paper presents the practical performance of a ZigBee wireless network with multihop transmission in an indoor environment during a long-term operation time. Totally, 51 sensor nodes are deployed in a hallway and a room. Several sets of practical experiments are conducted to study its various features, including the (1) node connectivity, (2) packet loss rate, and (3) transmission throughput. The results show that our developed ZigBee platforms could work well under multi-hop transmission over an extended period of time.

The overall goal of this paper is to contribute and help through realistic measurements towards the dimensioning of the sensor networks for future applications using Zig-Bee/IEEE 802.15.4 technology. The developed ITRI ZBn-odes are employed for realistic experiments. During our experiments, we found that the achieved transmission rate is around 25 kbps. Note that this is substantially below the nominal value of 250 kbps, and reasons may be due to the transmission overhead (such as frame headers), the CSMA-CA random backoffs, the presence of interframe spacing, and the concurrent transmission of multiple nodes. Note that the CSMA-CA mechanism in 802.15.4 automatically backs off initially when a transmission is imminent; that is, each data and command frame transfer will at least have one backoff. This would also reduce the performance.

Future work includes the measurements on different deployments to evaluate the impact of the deployment on the ZB32 platform's performance, including the evaluation of network parameters under a mesh (data flow) topology [18] and the power consumption under a node mobility condition

[19]; Moreover, a kind of performance comparison between the presented ITRI ZBnodes and other platforms (or results from network simulators or mathematic models) should be further investigated.

Acknowledgments

This research was supported by the Ministry of Economic Affairs, Taiwan, under Grant 100-EC-17-A-02-01-0617, and in part by the National Science Council, Taiwan, under Grant NSC-100-2221-E-027-010. The authors declare that there is no conflict of interests.

References

[1] J. S. Lee, Y. W. Su, and C. C. Shen, "A comparative study of wireless protocols: bluetooth, UWB, ZigBee, and Wi-Fi," in *Proceedings of the 33rd Annual Conference of the IEEE Industrial Electronics Society (IECON'07)*, pp. 46–51, Taipei, Taiwan, November 2007.

[2] J. S. Lee, C. C. Chuang, and C. C. Shen, "Applications of short-range wireless technologies to industrial automation: a ZigBee approach," in *Proceedings of the 5th Advanced International Conference on Telecommunications (AICT'09)*, pp. 15–20, Venice, Italy, May 2009.

[3] C. H. Wu, H. S. Liu, Y. F. Lee, M. S. Wei, Y. J. Li, and J. S. Lee, "A gateway-based inter-PAN binding mechanism for ZigBee sensor networks," in *Proceedings of the 37th Annual Conference on IEEE Industrial Electronics Society (IECON'11)*, pp. 3808–3813, Melbourne, Australia, November 2011.

[4] ZigBee Alliance, *ZigBee Document 053520r25: Home Automation Profile Specification*, ZigBee Alliance, San Ramon, Calif, USA, 2007.

[5] ZigBee Alliance, *ZigBee Document 053474r17: ZigBee Specification*, ZigBee Alliance, San Ramon, Calif, USA, 2007.

[6] IEEE 802. 15. 4, Wireless Medium Access Control (MAC), and Physical Layer (PHY), *Specifications for Low-Rate Wireless Personal Area Networks (LR-WPANs)*, IEEE, New York, NY, USA, 2006.

[7] M. Kohvakka, M. Kuorilehto, M. Hännikäinen, and T. D. Hämäläinen, "Performance analysis of IEEE 802. 15. 4 and ZigBee for large-scale wireless sensor network applications," in *Proceedings of the 3rd ACM International Workshop on Performance Evaluation of Wireless ad hoc, Sensor and Ubiquitous Networks (PE-WASUN'06)*, pp. 48–57, Torremolinos, Spain, October 2006.

[8] H. López-Fernández, P. Macedo, J. A. Afonso, J. H. Correia, and R. Simões, "Performance evaluation of a ZigBee-based medical sensor network," in *Proceedings of the 3rd International Conference on Pervasive Computing Technologies for Healthcare (PervasiveHealth'09)*, pp. 1–4, London, UK, March 2009.

[9] G. Anastasi, M. Conti, M. Di Francesco, and V. Neri, "Reliability and energy efficiency in multi-hop IEEE 802.15.4/ZigBee wireless sensor networks," in *Proceedings of the 15th IEEE Symposium on Computers and Communications (ISCC'10)*, pp. 336–341, Riccione, Italy, June 2010.

[10] F. Chen, N. Wang, R. German, and F. Dressler, "Simulation study of IEEE 802.15.4 LR-WPAN for industrial applications," *Wireless Communications and Mobile Computing*, vol. 10, no. 5, pp. 609–621, 2010.

[11] J. S. Lee, "Performance evaluation of IEEE 802.15.4 for low-rate wireless personal area networks," *IEEE Transactions on Consumer Electronics*, vol. 52, no. 3, pp. 742–749, 2006.

[12] M. Petrova, J. Riihijärvi, P. Mähönen, and S. Labella, "Performance study of IEEE 802.15.4 using measurements and simulations," in *Proceedings of the IEEE Wireless Communications and Networking Conference (WCNC'06)*, pp. 487–492, Las Vegas, Nev, USA, April 2006.

[13] K. Shuaib, M. Alnuaimi, M. Boulmalf, I. Jawhar, F. Sallabi, and A. Lakas, "Performance evaluation of IEEE 802. 15. 4: experimental and simulation results," *Journal of Communications*, vol. 2, no. 4, pp. 29–37, 2007.

[14] J. M. Cano-Garcia and E. Casilari, "An empirical evaluation of the consumption of 802.15.4/ZigBee sensor motes in noisy environments," in *Proceedings of the International Conference on Networking, Sensing and Control (ICNSC'11)*, pp. 439–444, Delft, The Netherlands, April 2011.

[15] J. S. Lee and Y. C. Huang, "Design and implementation of ZigBee/IEEE 802.15.4 nodes for wireless sensor networks," *Measurement and Control*, vol. 39, no. 7, pp. 204–208, 2006.

[16] J. S. Lee and Y. C. Huang, "ITRI ZBnode: a ZigBee/IEEE 802.15.4 platform for wireless sensor networks," in *Proceedings of the IEEE International Conference on Systems, Man and Cybernetics*, pp. 1462–1467, Taipei, Taiwan, October 2006.

[17] J. S. Lee, Y. M. Wang, and C. C. Shen, "Performance evaluation of ZigBee-based sensor networks using empirical measurements," in *Proceedings of the IEEE International Conference on Cyber Technology in Automation, Control, and Intelligent Systems (CYBER'12)*, pp. 58–63, Bangkok, Thailand, May 2012.

[18] C. Antonopoulos and S. Koubias, "Congestion control framework for Ad-hoc wireless networks," *Wireless Personal Communications*, vol. 52, no. 4, pp. 753–775, 2010.

[19] T. Sun, N. C. Liang, L. J. Chen, P. C. Chen, and M. Gerla, "Evaluating mobility support in ZigBee networks embedded and ubiquitous computing," in *Proceedings of the International Conference on Embedded and Ubiquitous Computing (EUC'07)*, vol. 4808 of *Lecture Notes in Computer Science*, pp. 87–100, Springer, 2007.

21

One More Tool for Understanding Resonance and the Way for a New Definition

Emanuel Gluskin,[1] Doron Shmilovitz,[2] and Yoash Levron[2]

[1] *Electrical Engineering Department, Faculty of Engineering, Kinneret College on the Sea of Galilee, Jordan Valley 15132, Israel*
[2] *Electrical Engineering Department, Faculty of Engineering, Tel-Aviv University, Israel*

Correspondence should be addressed to Emanuel Gluskin; gluskin@ee.bgu.ac.il

Academic Editor: H. P. S. Abdul Khalil

We propose the application of graphical convolution to the analysis of the resonance phenomenon. This time-domain approach encompasses both the finally attained periodic oscillations and the initial transient period. It also provides interesting discussion concerning the analysis of nonsinusoidal waves, based not on frequency analysis but on direct consideration of waveforms, and thus presenting an introduction to Fourier series. Further developing the point of view of graphical convolution, we arrive at a new definition of resonance in terms of time domain.

1. Introduction

1.1. General. The following material fits well into an "Introduction to Linear Systems," or "Mechanics," and is relevant to a wide range of technical and physics courses, since the resonance phenomenon has long interested physicists, mathematicians, chemists, engineers, and, nowadays, also biologists.

The complete resonant response of an initially unexcited system has two different, distinguishable parts, and there are, respectively, two basic definitions of resonance, significantly distanced from each other.

In the widely adopted textbook [1] written for physicists, resonance is defined as a *linear increase of the amplitude of oscillations in a lossless oscillatory system*, obtained when the system is pumped with energy by a sinusoidal force at the correct frequency. Figure 1 schematically shows the "envelope" of the resonant oscillations being developed.

Thus, a lossless system under resonant excitation absorbs more and more energy, and a steady state is never reached. In other words, in the lossless system, the amplitude of the steady state and the "quality factor" Q (having a somewhat semantic meaning in such a system) are infinite at resonance.

However, the slope of the envelope is always finite; it depends on the amplitude of the input function, and not on

Q. Though the steady-state response will never be reached in an ideal lossless system, the linear increase in amplitude by itself has an important sense. When a realistic physical system absorbs energy resonantly, say in the form of photons of electromagnetic radiation, there indeed is some period (still we can ignore power losses, say, some back radiation) during which the system's energy increases linearly in time. The energy absorption is immediate upon appearance of the influence, and the rate of the absorption directly measures the intensity of the input.

One notes that the energy pumping into the system at the initial stage of the resonance process readily suggests that the sinusoidal waveform of the input function is not necessary for resonance; it is obvious (think, e.g., about swinging a swing by kicking it) that the energy pumping can occur for other input waveforms as well. This is a heuristically important point of the definition of [1].

The physical importance of the initial increase in oscillatory amplitude is associated not only with the energy pumping; the informational meaning is also important. Assume, for instance, that we speak about the *start* of oscillations of the spatial positions of the atoms of a medium, caused by an incoming electromagnetic wave. Since this start is associated with the *appearance* of the wave, it can be also associated with

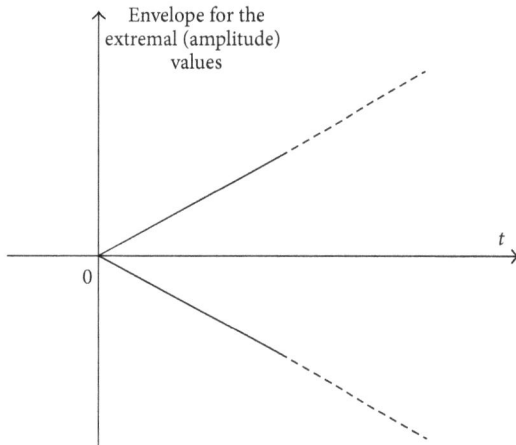

FIGURE 1: The definition of resonance [1] as *linear increase of the amplitude*. (The oscillations fill the angle of the envelope.) The *infinite* process of increase of the amplitude is obtained because of the assumption of *losslessness* of the system.

the registration of a signal. Later on, the *established* steady-state oscillations (that are associated, because of the radiation of the atoms, with the refraction factor of the medium) influence the velocity of the electromagnetic wave in the medium. As [2] stresses—even if this velocity is larger than the velocity of light (for refraction factor $n < 1$, i.e., when the frequency of the incoming wave is *slightly* higher than that of the atoms oscillators)—this does not contradict the theory of relativity, because there is already no signal. Registration of any signal and its group velocity is associated with a (forced) transient process.

A more pragmatic argument for the importance of analysis of the initial transients is that for any application of a steady-state response, especially in modern electronics, we have to know how much time is needed for it to be attained, and this relates, in particular, to the resonant processes. This is relevant to the frequency range in which the device has to be operated.

Contrary to [1], in textbooks on the theory of electrical circuits (e.g., [3–5]) and mechanical systems, resonance is defined as the *established* sinusoidal response with a relatively high amplitude proportional to Q. Only this definition, directly associated with *frequency domain analysis*, is widely accepted in the engineering sciences. According to this definition, the envelope of the resonant oscillations (Figure 2) looks even simpler than in Figure 1; it is given by two horizontal lines. This would be so for *any* steady-state oscillations, and the uniqueness is just by the fact that the oscillation amplitude is proportional to Q.

After being attained, the steady-state oscillations continue "forever," and the parameters of the "frequency response" can be thus relatively easily measured. Nevertheless, the simplicity of Figure 2 is a seeming one, because it is not known *when* the steady amplitude becomes established, and, certainly, the "frequency response" is *not* an immediate response to the input signal.

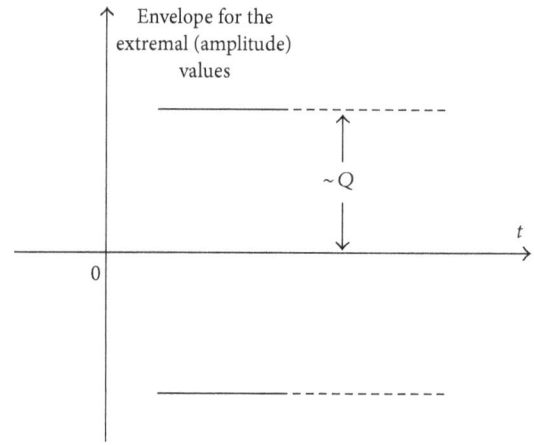

FIGURE 2: The envelope of resonant oscillations, according to the definition of resonance in [3, 4] and many other technical textbooks. When this steady state is attained?

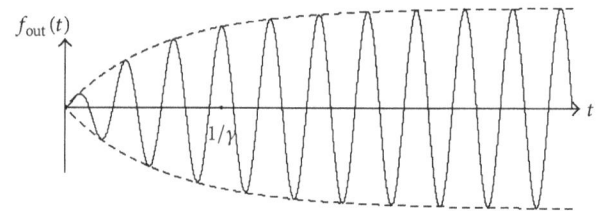

FIGURE 3: The illustration, for $Q = 10$, of the resonant response of a second-order circuit. Note that we show a case when the excitation is precisely at the resonant frequency, and the notion of "purely resonant oscillations" applies here to the *whole* process, and not only to the final steady-state part.

Thus, we do not know via the definition of [1] when the slope will finish, and we do not know via the definition of [3–5] when the steady state is obtained.

We shall call the definition of [1] "the "Q-t" definition," since the value of Q can be revealed via *duration* of the initial/transient process in a real system. The commonly used definition [3–5] of resonance in terms of the parameters of the sustained response will be called "the "Q-a" definition," where "a" is an abbreviation for "amplitude."

Figure 3 illustrates the actual development of resonance in a second-order circuit. The damping parameter γ will be defined in Section 3.

The Q-t and Q-a parts of the resonant oscillations are well seen. For such a not very high Q (i.e., $1/\gamma$ not much larger than the period of the oscillations) the period of fair initial linearity of the envelope includes only some half period of oscillations, but for a really high Q it can include many periods. The *whole curve* shown is the resonant response. This response can be obtained when the external frequency is closing the self-frequency of the system, from the beats of the oscillations (analytically explained by the formulae found in Section 3) shown in Figure 4.

Note that the usual interpretation is somewhat different. It just says that the linear increase of the envelope, shown in Figure 1, can be obtained from the first beat of the *periodic*

$f_{out}(t)$

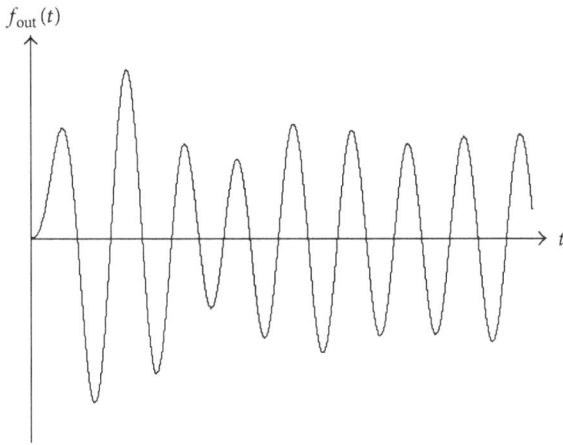

FIGURE 4: Possible establishing of the situation shown in Figure 3 through beats while adjustment of the frequency. We can interpret resonance as "filtration" of the beats when the resonant frequency is found.

beats observed in a *lossless* system. Contrary to that, we observe the beats in a system with losses, and after adjustment of the external frequency obtain the *whole* resonant response shown in Figure 3.

Our treatment of the topic of resonance for teaching purposes is composed of three main parts shown in Figure 5. The first part briefly recalls traditional "phasor" material relevant only to the *Q-a* part, which is necessary for introduction of the notations. The next part includes some simple, *though usually omitted*, arguments showing why the phasor analysis is insufficient. Finally, the third part includes the new tool, which is complementary to the classical approach of [1], and leads to a nontrivial generalization of the concept of resonance.

Our notations need minor comments. As is customary in electrical engineering, the notation for $\sqrt{-1}$ is j. The small italic Latin "v," v, is *voltage* in the *time domain* (i.e., a real value), $\widehat{V}$ means *phasor*, that is, a complex number in the frequency domain. λ is the *dummy variable* of integration in a definite integral of the convolution type. It is measured in seconds, and the difference $t - \lambda$, where t is time, often appears.

2. Some Advice to the Teacher

First, we deal here with a lot of *pedagogical* science—in principle the issues are not new but are often missed in the classroom; as far as we know, no such complete scheme of the necessary arguments for teaching resonance exists. Perhaps this is because some issues indeed require a serious revisiting and time is often limited due to overloaded teaching plans and schedules. That the results of this "economy" are not bright is seen first of all from the already mentioned fact that electrical engineering (EE) students often learn resonance only via phasors and are not concerned with the time needed for the very important steady state *to be established*. The resonance phenomenon is so physically important that it is taught to

technical students many times: in mechanics, in EE, in optics, and so forth. *However*, all this repeated teaching is actually equivalent to the use of phasors, that is, relates only to the established steady state.

Furthermore, the teachers (almost all of them) miss the very interesting possibility to exhibit the power of *the convolution-integral analysis* for studying the development of a resonant state. In our opinion, this demonstration makes the convolution integral a more interesting tool; this really is one of the best applications of the "graphical convolution," which should not be missed in any program. The convolution outlook well *unites* the view of resonance as a steady state by engineers and the view of resonance as energy pumping into a system, by physicists. The arguments of the graphical convolution also enable one to easily see (before knowing Fourier series) that a nonsinusoidal periodic input wave can cause resonance *just as* the sinusoidal one does. Thus, these arguments can be used also as an explanation of the physical meaning of the Fourier expansion. Our classroom experience shows that the average student can understand this material and finds it interesting.

Thus, regarding the use of the pedagogical material, we would advise the teacher of the EE students, *to return to the topic of resonance (previously taught via phasors), when the students start with convolution*.

Finally, the present work includes some new science, which can be also related to teaching, but perhaps at graduate level, depending on the level of the students or the university. We mean the generalization of the concept of resonance considered in Section 5. It is logical that if the convolution integral can show resonance (or resonant conditions) *directly*, not via Fourier analysis, then this "showing" exposes a general definition of resonance. Furthermore, since mathematically the convolution integral can be seen—with a proper writing of the impulse response in the integrand—as a *scalar product*, it is just natural to introduce into the consideration the outlook of Euclidean space.

The latter immediately suggests a geometric interpretation of resonance in functional terms, because it is clear what is the condition (here, the resonant one) for optimization of the scalar product of two normed vectors. As a whole, we simply replace the traditional requirement of *equality of some frequencies* to the condition of *correlation of two time functions*, which includes the classical sinusoidal (and the simplest oscillator) case as a particular one.

The geometrical consideration leads to a symmetry argument: since the impulse response $h(t)$ is the only given "vector," any optimal input "vector" has to be similarly oriented; there simply is no other selected direction. The associated writing $f_{inp} \sim h$, that is often used here just for brevity, precisely means the adjustment of the waveform of $f_{inp}(t)$ to that of $h(t)$ by the following two steps.

(1) Set $f_{inp}(t) \sim -h(T - t)$ in the interval $0 < t < T$.

(2) Continue this waveform periodically for $t > T$.

It is relevant here that for weak power losses typical for all resonant systems, the damping of $h(t)$ *in the first period* can be ignored, which should be a simplifying circumstance for

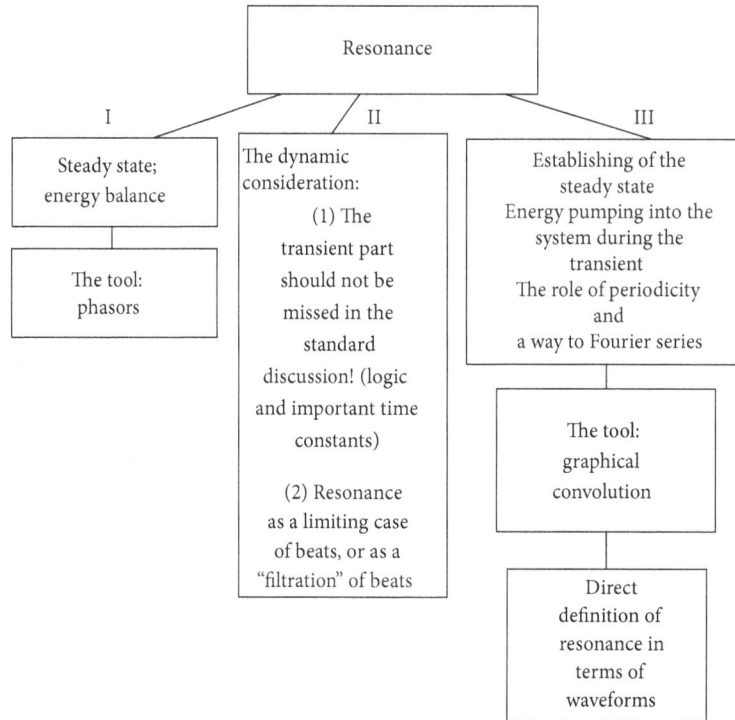

FIGURE 5: The methodological points regarding the study of resonance in the present work.

creation of the periodic $f_{\text{inp}}(t)$. The way of the adjustment of $f_{\text{inp}}(t)$ reflects the fact that the Euclidean space can relate to one period.

Both because of the somewhat higher level of the mathematical discussion and some connection with the theory of "matched filters," usually related to special courses (which could not be discussed here), it seems that this final material should be rather given for graduate students. However, we also believe that a teacher will find here some pedagogical motivation and will be able to convey more lucid treatment than we succeeded to doing. Thus, the question regarding the possibility of teaching the generalized resonance to undergraduate students remains open.

Some other nontrivial points, deserving pedagogical judgement or analytical treatment, appear already in the use of the convolution. This means the replacement of the weakly damping $h(t)$ of an oscillatory system by the *not damping but cut* function $h_S(t)$, shown in Figure 11, and the problem of definition of the damping parameter γ for the tending to zero $h(t)$ of a complicated oscillatory circuit. A possible way for the latter can be by observation (this is not yet worked out) of some averages, for example, how the integral of h^2, or of $|h|$, over the fixed-length interval $(t, t + \Delta)$ is decreased with increase in t.

3. Elementary Approaches

3.1. The Second-Order Equation. The background formulae for both the Q-t and Q-a parts of the resonant response can be given by the Kirchhoff *voltage equation* for the electrical current $i(t)$ in a series RLC (resistor-inductor-capacitor) circuit driven from a source of sinusoidal voltage with amplitude v_m:

$$L\frac{di}{dt} + Ri + \frac{1}{C}\int i(t)\,dt = v_m \sin\omega t. \tag{1}$$

Differentiating (1) and dividing by $L \neq 0$, we obtain

$$\frac{d^2i}{dt^2} + 2\gamma\frac{di}{dt} + \omega_o^2 i(t) = \frac{\omega v_m}{L}\cos\omega t, \tag{2}$$

with the damping factor $\gamma = R/2L$ and the resonant frequency $\omega_o = 1/\sqrt{LC}$.

For purely resonant excitation, the input sinusoidal function is at frequency $\omega = \omega_o$, or at a very close frequency ω_d, as defined below in (6).

3.2. The Time-Domain Argument. The full solution of (2) can be explicitly composed of *two* terms; the first, denoted as i_h, originates from the *homogeneous* (h) equation, and the second, denoted as i_{fs}, represents the *finally obtained* (fs) periodic oscillations, that is, is the *simplest* (but not the only possible!) partial solution of the *forced* equation:

$$i(t) = i_h(t) + i_{fs}(t). \tag{3}$$

It is important that the zero initial conditions cannot be fitted by the second term in (3), $i_{fs}(t)$, continued backward in time to $t = 0$. (Indeed, no sinusoidal function satisfies both the conditions $f(t) = 0$ and $df/dt = 0$, at any point.) Thus, it is obvious that a nonzero term $i_h(t)$ is needed in (3). This term is

$$i_h(t) = e^{-\gamma t}\left(K_1 \cos \omega_d t + K_2 \sin \omega_d t\right), \qquad (4)$$

where at least one of the constants K_1 and K_2 nonzero.

Furthermore, it is obvious from (4) that the time needed for $i_h(t)$ to decay is of the order of $1/\gamma \sim QT_o$ (compare to (9)). However, *according to the two-term structure of (3), the time needed for $i_{fs}(t)$ to be established, that is, for $i(t)$ to become $i_{fs}(t)$, is just the time needed for $i_h(t)$ to decay. Thus, the established "frequency response" is attained only after the significant time of order $QT_o \sim Q$.*

Unfortunately, this elementary logic argument following from (3) is missed in [3–5] and many other technical textbooks that ignore the Q-t part of the resonance and directly deal only with the Q-a part.

However form (3) is also not optimal here because it is not explicitly shown that for zero initial conditions not only $i_{fs}(t)$ but also the decaying $i_h(t)$ are directly proportional to the amplitude (or scaling factor) v_m of the input wave.

That is, from the general form (3) alone it is not obvious that, when choosing zero initial conditions, we make the response function as a whole (including the transient) to be proportional to v_m, appearing in (1), that is, to be a tool for studying the input function, at least in the scaling sense.

It would be better to have *one* expression/term from which this feature of the response is well seen. Such a formula appears in Section 4.

3.3. The Phasor Analysis of the Q-a Part. Let us now briefly recall the standard phasor (impedance) treatment of the final Q-a (steady-state) part of a system's response. We can focus here only on the results associated with the amplitude, the phase relations follow straightforwardly from the expression for the impedance [3, 4].

In order to characterize the Q-a part of the response, we use the common notations of [3, 4]: the damping factor of the response $\gamma \equiv R/2L$, the resonant frequency $\omega_o = 1/\sqrt{LC}$, the quality factor

$$Q = \frac{\omega_o}{2\gamma} = \frac{\omega_o L}{R} = \frac{\sqrt{L/C}}{R}, \qquad (5)$$

and the frequency at which the system self-oscillates:

$$\omega_d \equiv \sqrt{\omega_o^2 - \gamma^2} \approx \omega_o - \frac{\gamma^2}{2\omega_o} = \omega_o\left(1 - \frac{1}{4Q^2}\right) \approx \omega_o. \qquad (6)$$

Note that it is assumed that $4Q^2 \gg 1$, and thus ω_d and ω_o are practically indistinguishable. Thus, although we *never ignore* γ *per se*, the much smaller value $\gamma/Q \sim \gamma^2/\omega_o$ can be ignored. When speaking about "precise resonant excitation," we shall mean setting ω with *this* degree of precision, but when writing $\omega \neq \omega_o$, we shall mean that $\omega - \omega_o = O(\gamma)$, and not $O(\gamma/Q)$. Larger than $O(\gamma)$ deviations of ω from ω_o are irrelevant to resonance.

The impedance of the series circuit is $Z(j\omega) = R + j\omega L + (1/j\omega C)$, and the phasor approach simply gives the *amplitude* of the steady-state solution of (2) as:

$$i_m(\omega) = \left|\widehat{I}(j\omega)\right| = \left|\frac{\widehat{V}}{Z(j\omega)}\right|$$

$$= \frac{\left|\widehat{V}\right|}{\left|R + j\omega L + (1/j\omega C)\right|} = \frac{v_m}{\sqrt{R^2 + (\omega L - (1/\omega C))^2}}. \qquad (7)$$

For $\omega - \omega_o \ll \omega_o$, when $\omega^2 - \omega_o^2 \approx 2\omega_o(\omega - \omega_o) \approx 2\omega(\omega - \omega_o)$,

$$i_m(\omega) \approx \frac{v_m}{2L\sqrt{\gamma^2 + (\omega - \omega_o)^2}}. \qquad (8)$$

From (8), the frequencies at "half-power level," for which $i(\omega) = (1/\sqrt{2})(i_m)_{\max}$, are defined by the equality $(\omega - \omega_o)^2 = \gamma^2$, from which we obtain $\omega_1 = \omega_o - \gamma$ and $\omega_2 = \omega_o + \gamma$, that is, for the circuit's frequency "pass-band" $\Delta\omega \equiv \omega_2 - \omega_1$ we have, with the precision taken in the derivation of (8), that $\Delta\omega = 2\gamma$.

It is remarkable that *however small is γ, it is easy, while working with the steady state, to detect differences of order γ between ω and ω_o, using the resonant curve/response described by (8).*

Figure 6 illustrates the resonance curve. Though this figure is well known, it is usually not stressed that since each point of the curve corresponds to some steady state, a certain time is needed for the system to pass on from one point of the curve to another one, and the sharper the resonance is the more time is needed. The physical process is such that for a small γ the establishment of this response takes a (long) time of the order of

$$\frac{1}{\gamma} = \frac{\omega_o}{\gamma}\frac{1}{\omega_o} = 2Q\frac{T_o}{2\pi} = \frac{1}{\pi}QT_o \sim QT_o, \quad \left(T_o = \frac{2\pi}{\omega_o}\right), \quad (9)$$

which is not directly seen from the resonance curve.

The relation $1/\gamma \sim QT_o$ for the transient period should be remembered regarding *any* application of the resonance curve, in any technical device. The case of a mistake caused by assuming a quicker performance for measuring input frequency by means of passing on from one steady state to another is mentioned in [2]. This mistake is associated with using only the resonance curve, that is, thinking only in terms of the frequency response.

4. The Use of Graphical Convolution

We pass on to the constructive point, the convolution integral presenting the resonant response, and its graphical treatment. It is desirable for a good "system understanding" of the topic that the concepts of *zero input response* (ZIR) and *zero state response* (ZSR), especially the latter one, be known to the reader.

Briefly, ZSR is the *partial response* of the circuit, which satisfies the zero initial conditions. As $t \rightarrow \infty$ (and only

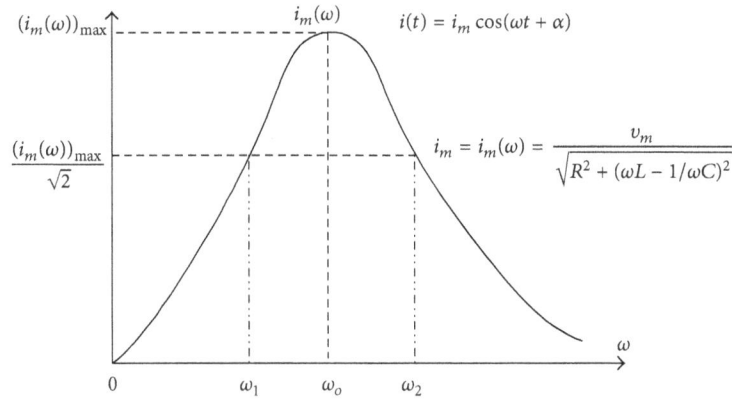

FIGURE 6: The resonance curve: $\Delta\omega \equiv \omega_2 - \omega_1 = 2\gamma$; $Q = \omega_o/\Delta\omega$.

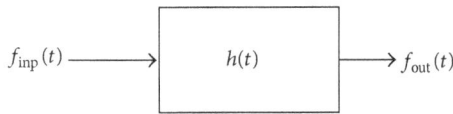

FIGURE 7: The input-output map $(f_{\text{inp}} \rightarrow f_{\text{out}}(t) = \text{ZSR}(t))$ given by "impulse response" $h(t)$.

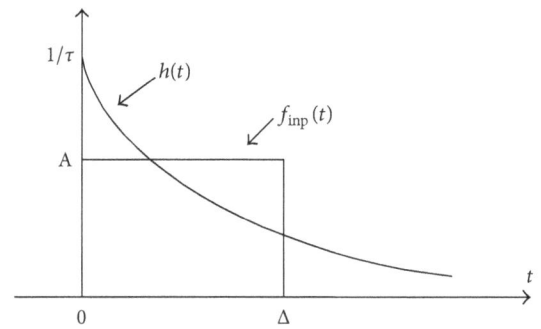

FIGURE 8: The functions for the simplest example of convolution. (A first-order circuit with an input block pulse.)

then), it becomes the final steady-steady response, that is, becomes the *simplest* partial response (whose waveform can be often guessed).

The appendix illustrates the concepts of ZIR and ZSR in detail, using a first-order system and stressing the distinction between the forms ZIR + ZSR and (3) of the response.

Our system-theory tools are now the *impulse (or shock) response* $h(t)$ (*or Green's function*) and the integral response to $f_{\text{inp}}(t)$ for zero initial conditions:

$$f_{\text{out}}(t) = \left(h * f_{\text{inp}}\right)(t) = \int_0^t h(\lambda) f_{\text{inp}}(t - \lambda) \, d\lambda$$

$$= \left(f_{\text{inp}} * h\right)(t).$$
(10)

The convolution integral (10) is an example of ZSR, and it is the most suitable tool for understanding the resonant excitation.

It is clear (contrary to (3)) that the total response (10) is directly proportional to the amplitude of the input function.

Figure 7 shows our schematic system.

Of course, the *system-theory outlook* does not relate only to electrical systems; this "block-diagram" can mean influence of a mechanical force on the position of a mass, or a pressure on a piston, or temperature at a point inside a gas, and so forth.

Note that if the initial conditions are zero, they are simply not mentioned. If the input-output map is defined solely by $h(t)$ (e.g., when one writes in the domain of Laplace variable $F_{\text{out}}(s) = H(s)F_{\text{inp}}(s)$), it is always ZSR.

In order to treat the convolution integral, it is useful to briefly recall the simple example [5] of the first-order circuit influenced by a single square pulse. The involved

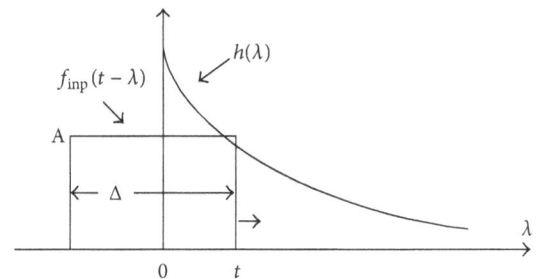

FIGURE 9: The functions appearing in the integrand of the convolution integral (10). The "block" $f_{\text{inp}}(t - \lambda)$ is riding (being moved) to the right on the λ-axes, as time passes. We multiply the present curves in the interval $0 < \lambda < t$ and, according to (10), take the area under the result, in this interval. When $t < \Delta$, only the interval $(0, t)$ is relevant to (10). When $t > \Delta$, only the interval $(t - \Delta, t)$ is actually relevant, and because of the decay of $h(t)$, $f_{\text{out}}(t)$ becomes decaying.

physical functions are shown in Figure 8, and the associated "*integrand* situation" of (10) is shown in Figure 9.

It is *graphically obvious* from Figure 9 that the maximal value of $f_{\text{out}}(t)$ is obtained for $t = \Delta$, when the rectangular pulse already fully overlaps with $h(\lambda)$, but still "catches" the initial (highest) part of $h(\lambda)$. This simple observation shows the strength of the graphical convolution for a qualitative analysis.

4.1. The (Resonant) Case of a Sinusoidal Input Function Acting on the Second-Order System.

For the second-order system with weak losses, we use for (10)

$$h(t) = \frac{\omega_o^2}{\omega_d} e^{-\gamma t} \sin \omega_d t \tag{11}$$

$$\sim e^{-\gamma t} \sin \omega_d t \approx e^{-\gamma t} \sin \omega_o t, \quad \gamma \ll \omega_o, \ (Q \gg 1).$$

As before, we apply

$$f_{\text{inp}}(t) = f_m \sin \omega_d t \approx f_m \sin \omega_o t. \tag{12}$$

Figure 10 builds the solution (10) step by step; first our $h(\lambda)$ and $f_{\text{inp}}(t-\lambda)$ (compare to Figure 9), then the product of these functions, and finally the integral, that is, $f_{\text{out}}(t) = S(t)$.

On the upper graph, the "train" $f_{\text{inp}}(t - \lambda)$ travels to the right, starting at $t = 0$, on the middle graph we have the integrand of (10). The area $f_{\text{inp}}(t) = S(t)$ under the integrand's curve appears as the final result on the third graph.

The extreme values of $S(t)$ are $S(k(\pi/\omega_d))$, obviously. For k odd these are positive maxima because the overlaps in the upper drawing are then "+" with "+" and "−" with "−." For k even these are negative minima because we multiply the opposite polarities in the overlap $f_{\text{inp}}(t - \lambda)h(\lambda)$ each time. Thus, $S(\pi/\omega_d) > 0$ and $S(2\pi/\omega_d) < 0$.

In view of the basic role of the overlapping of $f_{\text{inp}}(t - \lambda)$ with $h(\lambda)$, it is worthwhile to look forward a little and compare Figure 10 to Figures 14 and 15 that relate to the case of an input *square wave*. For the upper border of integration in (10) be $t = k(\pi/\omega_d)$ and for very weak damping of $h(\lambda)$ the situations being compared are very similar. The distinction is that, in order to obtain the extremes of $f_{\text{inp}}(t)$, we integrate in Figure 15 the absolute value of several *sinusoidal pieces (half-waves)*, while in Figure 10 we integrate the *squared sinusoidal pieces*. Since we integrate, in each case, k similar pieces (all positive, giving a maximum of $f_{\text{out}}(t)$, or all negative, giving a minimum), the result of each such integration is directly proportional to k.

Thus, if $\gamma = 0$, when $h(\lambda)$ is strictly periodic, from the *periodic nature* of also $f_{\text{inp}}(t)$, it follows that

$$f_{\text{out}}\left(k\frac{\pi}{\omega_d}\right) \sim (-1)^{k+1} k \sim k, \tag{13}$$

for any integer k, which is a linear increase in the envelope for the two very different input waves, in the spirit of Figure 1.

For a small but finite γ, $0 < \gamma \ll \omega_o$, the initial linear increase has high precision only for some first few k when $t \sim T_o \sim 1/\omega_o \ll 1/\gamma$, that is, $\gamma t \ll 1$, or $e^{-\gamma t} \approx 1$. (The damping of $h(t)$ may be ignored *for these k*.)

Observe that the finally obtained periodicity of $f_{\text{out}}(t)$ follows only from that of $f_{\text{inp}}(t)$, while the linear increase requires periodicity of both $f_{\text{inp}}(t)$ and $h(t)$.

The above discussion suggests the following simplification of the impulse response of the circuit, useful for analysis of the resonant systems. This simplification is a useful preparation for the rest of the analysis.

4.2. A Simplified h(t) and the Associated Envelope of the Oscillations.

Considering that the parameter $1/\gamma$ appears in the above (and in Figure 3) as some symbolic border for the linearity, let us take a constructive step by suggesting a geometrically clearer situation when this border is artificially made sharp by introducing an idealization/simplification of $h(t)$, which will be denoted as $h_S(t)$.

In this idealization—that seems to be no less reasonable and suitable in qualitative analysis than the usual use of the vague expression "*somewhere at "t" of order $1/\gamma$*", we replace $h(t)$ by a finite "piece" of nondamping oscillations of total length $1/\gamma$.

We thus consider that however weak the damping of $h(t)$ is, for sufficiently large t, when $t \gg 1/\gamma \sim QT_o$, we have $e^{-\gamma t} \ll 1$, that is, the oscillations become strongly damped with respect to the first oscillation. For $t > 1/\gamma$ the further "movement" of the function $f_{\text{inp}}(t - \lambda)$ to the right (see Figure 10 again) becomes less effective; the exponentially decreasing tail of the oscillating $h(t)$ influences (10), via the overlap, more and more weakly, and as $t \to \infty$, $f_{\text{out}}(t)$ ceases to increase and becomes periodic, obviously.

We simplify this qualitative vision of the process by assuming that up to $t = 1/\gamma$, there is no damping of $h(t)$, but, starting from $t = 1/\gamma$, $h(t)$ completely disappears. That is, we replace the function $e^{-\gamma t} \sin \omega_d t$ by the function $h_S(t) = [u(t) - u(t - 1/\gamma)] \sin \omega_o t$, where $u(t)$ is the unit step function. The factor $u(t) - u(t - 1/\gamma)$ here is a "cutting window" for $\sin \omega_o t$. This is the formal writing of the "piece" of the nondamping self-oscillations of the oscillator. See Figure 11.

For $h_S(t)$, it is obvious that when the "train" $f_{\text{inp}}(t - \lambda)$ crosses in Figure 10 the point $t = 1/\gamma$, the graphical construction of (10), that is, $f_{\text{out}}(t)$, becomes a periodic procedure. Figuratively speaking, we can compare $h_S(t)$ with a railway station near which the infinite train $f_{\text{inp}}(t - \lambda)$ passes; some wagons go away, but similar new ones enter and the total overlapping is repeated periodically.

The same is also analytically obvious, since when setting, for $t > 1/\gamma$, the upper limit of integration in (10) as $1/\gamma$, we have, because of the periodicity of $f_{\text{inp}}(\cdot)$, the integral:

$$f_{\text{out}}\left(t > \frac{1}{\gamma}\right) = \int_0^{1/\gamma} h(\lambda) f_{\text{inp}}(t - \lambda) \, d\lambda, \tag{14}$$

as a periodic function of t.

As is illustrated by Figure 12—which is an approximation to the envelope shown in Figure 3—the envelope of the output oscillations becomes completely saturated for $t > 1/\gamma$.

Figure 12 clearly shows that both the amplitude of the finally established steady-state oscillations and the time needed for establishing these oscillations are proportional to Q, while the initial slope is obviously independent of Q.

It is important that $h_S(t)$ can be also constructed for more complicated functions $h(t)$ (for which it may be, for instance, $h(t + T/2) \neq -h(t)$) and also then the graphical convolution is easier formulated in terms of $h_S(t)$. As an example relevant to the theoretical investigations—approximately presenting the

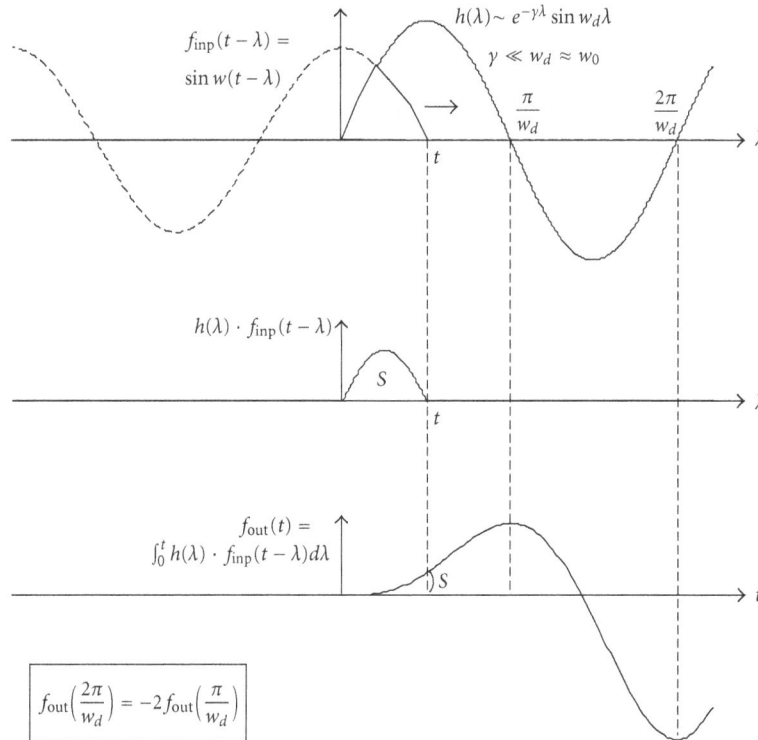

FIGURE 10: Graphically obtaining the resonant response for a second-order oscillatory system and a sinusoidal input, according to (10). The envelope (not shown) has to pass via the maxima and minima of $f_{out}(t)$ appearing in the last graph.

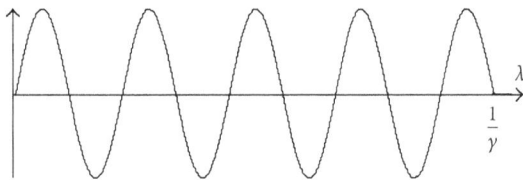

FIGURE 11: The simplified $h(t)$ (named $h_S(t)$): there is no damping at $0 < t < 1/\gamma$, but for $t > 1/\gamma$, it is identically zero, that is, we first ignore the damping of the real $h(t)$ and then cut it completely. This idealization expresses the undoubted fact that the interval $0 < t < 1/\gamma$ is dominant and makes the treatment simpler. A small change in $1/\gamma$ which makes the oscillatory part more pleasing by including in it just the (closest) integer number of the half waves, as shown here, may be allowed, and when using $h_S(t)$ in the following we shall assume for simplicity that the situation is such.

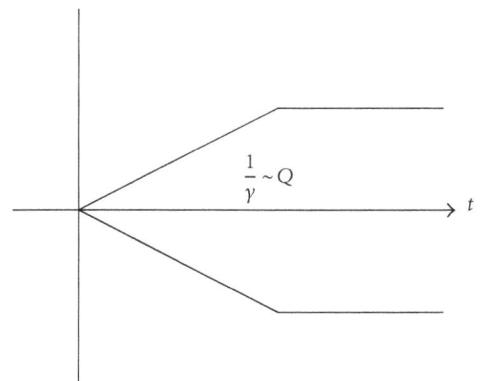

FIGURE 12: The envelope of $f_{out}(t)$ obtained for the simplified $h(t)$ shown in Figure 11.

maximal values of the established oscillations, obtained for $t_k \gg 1/\gamma$,

$$\left| f_{out}\left(t_k\right) \right| = \left| \int_0^\infty f_{inp}\left(t_k - \lambda\right) h\left(\lambda\right) d\lambda \right|, \quad t_k \gg 1/\gamma, \quad (15)$$

as

$$\left| f_{out}\left(\gamma^{-1}\right) \right| = \left| \int_0^{1/\gamma} f_{inp}\left(\gamma^{-1} - \lambda\right) h_S\left(\lambda\right) d\lambda \right|, \quad (16)$$

we can easily reduce, using periodicity of $f_{inp}(t)$ for any oscillatory $h(t)$ (and $h_S(t)$), the analysis of the interval $(0, 1/\gamma)$ to that of a small interval, as was for $(0, \pi/\omega_d)$ in Figure 10.

4.3. Nonsinusoidal Input Waves. The advantage of the graphical convolution is not so much in the calculation aspect. It is easy for imagination (insight) procedure, and it is a flexible tool in the qualitative analysis of the time processes. The graphical procedure makes it absolutely clear that the really basic point for a resonant response is not sinusoidality, but *periodicity* of the input function. Not being derived from the spectral (Fourier) approach, this observation heuristically completes this approach and may be used (see the following) in an introduction to Fourier analysis.

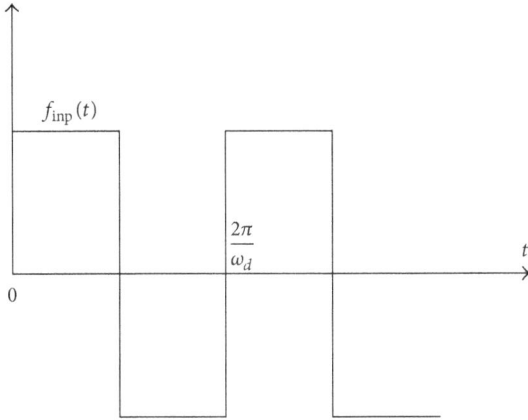

FIGURE 13: The rectangular wave at the input.

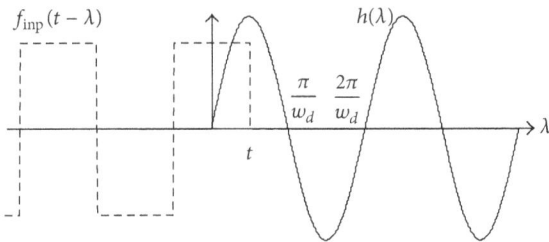

FIGURE 14: Convolution with a rectangular wave at the input. Compare to Figures 9 and 10.

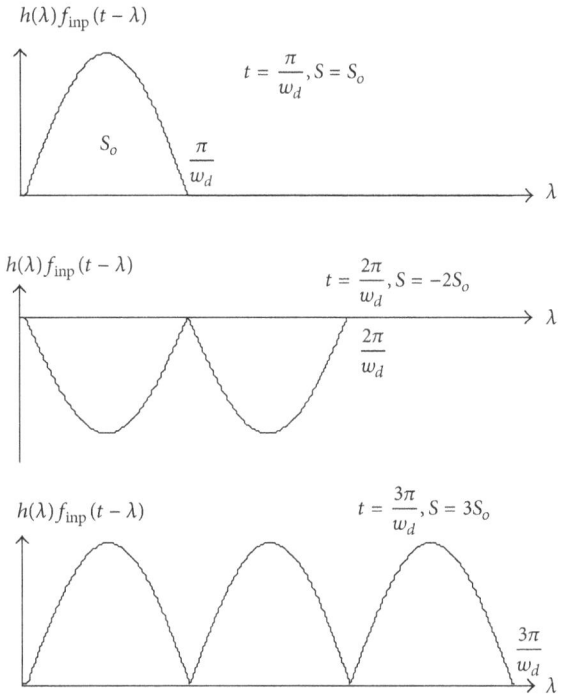

FIGURE 15: Continuation of the creation of the convolution value, after Figure 14. The function $h(\lambda)f_{\text{inp}}(t - \lambda)$ is shown at three intervals $0 < \lambda < t = k(\pi/\omega_d) \approx k(\pi/\omega_o)$, $k = 1, 2, 3$, for which the area under this function of t has local extremes, $S_o = S(2\pi/\omega_d)$, denotes the area under a half-wave of $h(\lambda)f_{\text{inp}}(t - \lambda)$. Damping of $h(t)$ is ignored, and we have here the cases of $S = S_o$, $S = -2S_o$, and $S = 3S_o$, which represent the output function at its extremes; see Figure 16.

Thus, let us now take $f_{\text{inp}}(t)$ as the rectangular wave shown in Figure 13 and follow the way of Figures 9 and 10, in the sequential Figures 14 and 15.

Here too, the envelope of the resonant oscillations can be well outlined by considering $f_{\text{out}}(t)$ at instances $t_k = k\pi/\omega_d$; first of all π/ω_d, $2\pi/\omega_d$, and $3\pi/\omega_d$, for which we, respectively, have the first maximum, the first minimum, and the second maximum of $f_{\text{out}}(t)$.

There are absolutely the same qualitative (geometric) reasons for resonance here, and Figure 15 explains that if the damping of $h(t)$ is weak, that is, some first sequential half-waves of $f_{\text{inp}}(t-\lambda)h(t)$ are similar, then the respective extreme values of $S(t) = f_{\text{out}}(t)$ form a linear increase in the envelope.

Figure 16 shows $f_{\text{out}}(t) = S(t)$ at these extreme points.

Though it is not easy to find the precise $f_{\text{out}}(t)$ everywhere, for the envelope of the oscillations, which passes through the extreme points, the resonant increase in the response amplitude is absolutely clear.

Figures 10, 14, 15, and 16 make it clear that many other waveforms with the correct period would likewise cause resonance in the circuit. Furthermore, for the overlapping to remain good, we can change not only $f_{\text{inp}}(t)$, but also $h(t)$. Making the form of the impulse response more complicated means making the system's structure more complicated, and thus graphical convolution is also a valuable starting point for studying resonance in complicated systems in terms of the waveforms. This point of view will be realized in Section 5 where we generalize the concept of resonance.

Thus, using the algorithm of the graphical convolution, we make two more methodological steps; a pedagogical one in Section 4.4 and the constructive one in Section 5.

4.4. Let Us Try to "Discover" the Fourier Series in Order to Understand It Better. The conclusion regarding the possibility of obtaining resonance using a nonsinusoidal input reasonably means that when pushing a swing with a child on it, it is unnecessary for the father to develop a sinusoidal force. Moreover, the nonsinusoidal input even has some obvious advantages. While the sinusoidal input wave leads to resonance only when its frequency has the correct value, exciting resonance by means of a nonsinusoidal wave can be done at very different frequencies (one need not to kick the swing at every oscillation), which is, of course, associated with the Fourier expansions of the force.

Let us see how, using graphical convolution, we can reveal harmonic structure of a function, *still not knowing anything about Fourier series*. For that, let us continue with the case of square wave input, but take now such a waveform with a period that is 3 times longer than the period of self-oscillations of the oscillator. Consider Figure 17.

This time, the more distant instances, $t = 3\pi/\omega_d$, $6\pi/\omega_d$, and $9\pi/\omega_d$, are obviously most suitable for understanding how the envelope of the oscillations looks.

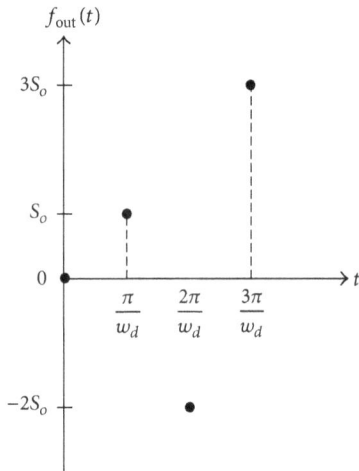

FIGURE 16: Linear increase of the envelope (ideal in the in the lossless situation) for the square wave input. Compare to Figures 1, 3, and 12.

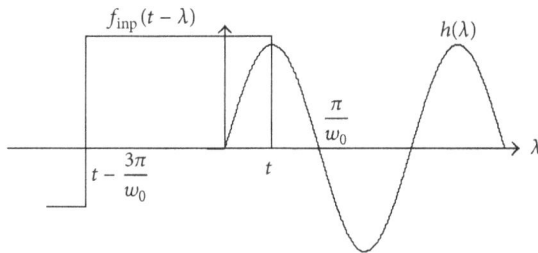

FIGURE 17: We "discover" the Fourier series using graphical convolution. The convolution of $h(t)$ with the square wave having $T = 3T_o$.

One sees that also for $T = 3T_o$, *the same* geometric "resonant mechanism" exists, but the transfer from $T = T_o$ to $T = 3T_o$ makes the excitation significantly less intensive. Indeed, see Figure 18 comparing the present extreme case of $t = 3\pi/\omega_d$ to the extreme case of $t = \pi/\omega_d$ of Figure 15.

We see that each extreme overlap is now only *one-third* as effective as was the respective maximum overlap in the previous case. That is, at $t = 3\pi/\omega_d$, we now have what we previously had at $t = \pi/\omega_d$, which means a much slower increase in the amplitude in time.

Since $f_{out}(t)$ is now increased at a much slower rate, but $1/\gamma$ is the same (i.e., the transient lasts the same time), the amplitude of the final periodic oscillations is respectively smaller, which means weaker resonance in terms of frequency response.

Let us compare the two cases of the square wave thus studied to the initial case of the sinusoidal function. The case of the "nonstretched" square wave corresponds to the input $\sin \omega_o t$, while according to the conclusions derived in Figure 18, the case of the "stretched" wave corresponds to the input $(1/3) \sin \omega_o t$. We thus simply (and roughly) reduce the change in period of the nonsinusoidal function to the equivalent change in amplitude of the sinusoidal function.

Let us now try—as a tribute to Joseph Fourier—to speak not about the same circuit influenced by different waves, but about the same wave influencing different circuits. Instead

of increasing T, we could decrease T_o, thus testing the ability of the same square wave to cause resonance in the different oscillatory circuits. For the new circuit, the graphical procedure remains the same, obviously, and the ratio 1/3 of the resonant amplitudes in the compared cases of $T/T_o = 3$ and $T/T_o = 1$ remains.

In fact, we are thus testing the square wave using *two* simple oscillatory circuits of different self-frequencies. Namely, connecting in parallel to the source of the square wave voltage two simple oscillatory circuits with self-frequencies ω_o and $3\omega_o$, we reveal for one of them the action of the square wave as that of $\sin \omega_o t$ and for the other as that of $(1/3) \sin 3\omega_o t$.

This associates the square wave of height A, with the series

$$f(t) \sim A \left(\sin \omega t + \frac{1}{3} \sin 3\omega t + \frac{1}{5} \sin 5\omega t \cdots \right) \quad (17)$$

(which precisely is $f = (4A/\pi)(\sin \omega t + \cdots)$).

Let us check this result by using the arguments in the inverse order. The first sinusoidal term of series (17) roughly corresponds to the square wave with $T = T_o$ (i.e., $\omega = \omega_o$), and in order to make the *second term* resonant, we have to change the self-frequency of the circuit to $\omega_o = 3\omega$, that is, make $\omega = (1/3)\omega_o$, or $T = 3T_o$, which is our second "experiment" in which the reduced to 1/3 intensity of the resonant oscillations is indeed obtained, in agreement with (17).

It is possible to similarly graphically analyze a triangular wave at the input, or a sequence of periodic pulses of an arbitrary form (more suitable for the father kicking the swing) with a period that is an integer of T_o.

One notes that such figures as Figure 18 are relevant to the standard integral form of Fourier coefficients. However on the way of graphical convolution, this similarity arises *only* for the extremes $(f_{out}(t))_{max} = |f_{out}(t_k)|$, and this way is independent and visually very clear.

5. A Generalization of the Definition of Resonance in Terms of Mutual Adjustment of $f_{inp}(t)$ and $h(t)$

After working out the examples of the graphical convolution, we are now in position to formulate a wider t-domain definition of resonance.

In terms of the graphical convolution, the analytical symmetry of (10):

$$\left(h * f_{inp} \right)(t) = \left(f_{inp} * h \right)(t), \quad (18)$$

means that besides observing the overlapping of $f_{inp}(t - \lambda)$ and $h(\lambda)$, we can observe overlapping of $h(t - \lambda)$ and $f_{inp}(\lambda)$. In the latter case, the graph of $h(-\lambda)$ starts to move to the right at $t = 0$, as was in the case with $f_{inp}(-\lambda)$.

Though equality (18) is a very simple mathematical fact, similar to the equalities $ab = ba$ and $(\vec{a}, \vec{b}) = (\vec{b}, \vec{a})$, in the context of graphical convolution, there is a nontriviality in the *motivation* given by (18), because the possibility to move $h(-\lambda)$ also suggests changing the *form* of $h(\cdot)$, that is, starting to deal with a *complicated system* (or *structure*) to be resonantly excited. We thus shall try to define resonance, that

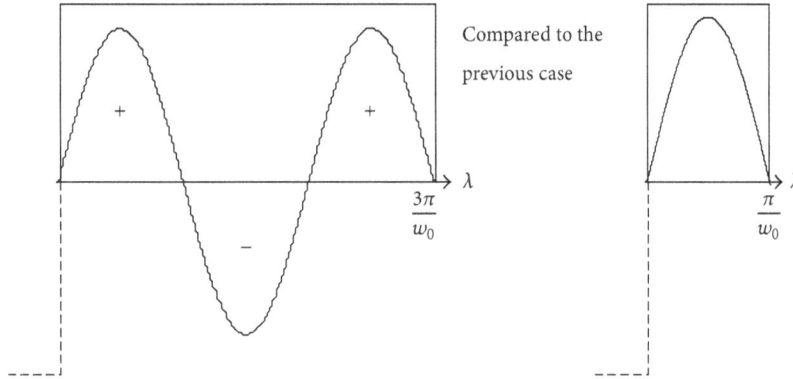

FIGURE 18: Because of the mutual compensation of two half-waves of $h(\lambda)$, only each third half-wave of $h(t)$ contributes to the extreme value of $f_{\text{out}}(t)$, and the maximum overlaps between $h(\lambda)$ and $f_{\text{inp}}(t)$ are now one-third as effective as before. The reader is asked (this will soon be needed!) to similarly consider the cases of $T = 5T_o$ and so forth.

is, the optimization of the peaks of $f_{\text{out}}(t)$ (or its *r.m.s.* value), in the terms of more arbitrary waveforms of $h(t)$, while the case of the sinusoidal $h(\cdot)$, that is, of simple oscillator, appears as a particular one.

5.1. The Optimization of the Overlapping of $f(\lambda) \equiv f_{inp}(t-\lambda)$ and $h(\lambda)$ in a Finite Interval and Creation of the Optimal Periodic $f_{\text{inp}}(t)$. Let us continue to assume that the losses in the system are small, that is, that $h(t)$ is decaying so slowly that we can speak about at least few oscillatory spikes (labeled by k) through which the envelope of the oscillations passes during its linear increase.

Using notation S_o of Figures 15 and 16, we speak about the extreme points of the graph of the resulting function $f_{\text{out}}(t)$, that is, about the points whose coordinates are $(t_k, f_{\text{out}}(t_k))$, or

$$\left(t_k, (-1)^{k+1} S_o k\right), \quad k = 1, 2, \ldots. \tag{19}$$

In view of the examples studied, the extreme points of $f_{\text{out}}(t)$ are obtained when t_k are the zero-crossings of $h(t)$, because only then the overlapping of $f_{\text{inp}}(t_k - \lambda)$ with $h(\lambda)$ can be made maximal.

Comment. Assuming that the parameters of the type γ/ω_0 of the different harmonic components of $h(t)$ are different, one sees that for a nonsinusoidal damping $h(t)$, the distribution of the zero-crossings of $h(t)$ can be changed with the decay of this function, and thus for a periodic $f_{\text{inp}}(t)$ the condition

$$\text{sign}\left[f_{\text{inp}}\left(t_k - t\right)\right] = \text{sign}\left[h(t)\right], \tag{20}$$

or

$$\text{sign}\left[f_{\text{inp}}\left(t_k - t\right)\right] = -\text{sign}\left[h(t)\right], \tag{21}$$

considered for $k \gg 1$ need not be satisfied in the whole interval of the integration $(0 < \lambda < t_k)$ related to the case of $t_k \gg T$. However since both the amplitude-type decays and the change in the intervals between the zeros are defined by the same very small damping parameters, the resulted effects

of imprecision are of the same smallness. Both problems are not faced when we use the "generating interval" and employ $h_S(t)$ instead of the precise $h(t)$. The fact that any use of $h_S(t)$ is anyway associated with error of order $Q^{-1} \sim \gamma T_o$ points to the expected good precision of the generalized definition of resonance.

Thus, $\{t_k\}$, measured with respect to the time origin, that is, with respect to the moment when $f_{\text{inp}}(t)$ and $h(t)$ arise, is assumed to be given by the known $h(t)$. Of course, we assume the system to be an oscillatory one, for the parameters t_k and S_o of our graphical constructions to be meaningful.

Having the linearly increasing sequence $|f_{\text{out}}(t_k)| = S_o k$ belonging to the envelope of the oscillations and wishing to increase the finally established oscillations, obviously we have to increase the factor S_o.

However since S_o and the whole intensity of $f_{\text{out}}(t)$ can be increased not only by the proper wave-form of $f_{\text{inp}}(\cdot)$, but also by an amplitude-type *scaling factor*, for the general discussion, some *norm* for $f_{\text{inp}}(\cdot)$, has to be introduced.

For the definitions of the norm and the scalar products of the functions appearing during adjustment of $f_{\text{inp}}(t)$ to $h(t)$, it is sufficient to consider a *certain* (for a fixed, not too large k) *interval* (t_k, t_{k+1})—the one in which we can calculate S_o. This interval can be simply $(0, t_1)$ or $(0, T)$.

The norm over the chosen interval is taken as

$$\|f\| = \sqrt{\int_{t_k}^{t_{k+1}} f^2(t)\, dt}. \tag{22}$$

For instance, $\|\sin \omega t\|$ calculated over interval $(0, T/2 = \pi/\omega)$ or $(\pi/\omega, 2\pi/\omega)$ is $\sqrt{\pi/2\omega}$, as is easy to find by using the equality $\sin^2 \alpha = (1/2)(1 - \cos 2\alpha)$.

Respectively, the scalar product of two functions is taken as

$$\left(f_1, f_2\right) = \int_{t_k}^{t_{k+1}} f_1(t)\, f_2(t)\, dt. \tag{23}$$

With these definitions, the set of functions defined for the purpose of the optimization in the interval (t_k, t_{k+1}) forms an (infinite-dimensional) Euclidean space.

For the quantities that interest us, we have from (23) for the absolute values

$$S_o = |(f, h)| = \left| \int_{t_k}^{t_{k+1}} f(t) h(t) \, dt \right|, \qquad (24)$$

where (see Figures 10, 14, and 15) it is set for simplicity of writing

$$f(t) \equiv f_{\text{inp}}(t_{k+1} - t), \quad t_k < t < t_{k+1}. \qquad (25)$$

Not ascribing to "$f(\cdot)$" index "k" is justified by the fact that the *particular* interval (t_k, t_{k+1}) to be actually used, is finally chosen very naturally.

The basic relation $|f_{\text{out}}(t_k)| = S_o k$ means that any local extremum of $f_{\text{out}}(t)$ is a sum of such scalar products as (23).

Observe that the physical dimensions of $\| \cdot \|$ and $(\cdot, \cdot)$ are

$$[\|\cdot\|] = [f] \left[t^{1/2} \right] = \frac{[f]}{[\omega^{1/2}]}, \qquad (26)$$

$$[(\cdot, \cdot)] = [f_1][f_2][t] = \frac{[f_1][f_2]}{[\omega]}.$$

Observe also from (22) and (23) that

$$(f, f) = \|f\|^2, \qquad (27)$$

and that if we take $f_2 \sim f_1$, that is,

$$f_2(t) = K f_1(t), \quad K \in \mathbb{R}, \qquad (28)$$

then (27) is generalized to

$$(f_1, f_2) = \text{sign}[K] \|f_1\| \cdot \|f_2\|. \qquad (29)$$

Indeed, using (27), and then the obvious equalities $K = |K| \text{sign}[K]$ and $|K| \cdot \|f\| = \|Kf\|$, we obtain

$$(f_1, f_2) = (f_1, K f_1) = K(f_1, f_1) = K\|f_1\|^2$$
$$= \text{sign}[K] |K| \cdot \|f_1\| \cdot \|f_1\| = \text{sign}[K] \|f_1\| \cdot \|K f_1\|$$
$$= \text{sign}[K] \|f_1\| \cdot \|f_2\|. \qquad (30)$$

The factor $\text{sign}[K]$ means, in particular, that excitation of an oscillatory circuit can be equivalently done by either an $f_{\text{inp}}(t)$ or $-f_{\text{inp}}(t)$. (Consider the concept of "overlapping" in this view.)

It follows from (30) that if (28) is provided, then

$$|(f_1, f_2)| = \|f_1\| \cdot \|f_2\| \quad \text{(for (28) provided)}. \qquad (31)$$

Furthermore, we use that the following general inequality

$$|(f_1, f_2)| \leq \|f_1\| \cdot \|f_2\| \qquad (32)$$

takes place. In view of (22) and (23), (32) is just the known Cauchy-Bunyakovsky integral inequality.

Comparing (32) with (31), we see that condition (28) provides optimization of $|(f_1, f_2)|$. Applied to f and h, that is, to $S_o = |(f, h)|$, this conclusion regarding optimization says that the condition $f \sim h$ optimizes S_o. Thus, $f \sim h$ optimizes the extremes $f_{\text{out}}(t_k) \sim S_o k$ of the system's response.

Thus, we finally have the following two points.

(a) We find the proper interval (t_k, t_{k+1}) for creating the optimal periodic $f_{\text{inp}}(t)$.

(b) The proportionality $f \sim h$ in this interval is the optimal case of the influence of an oscillatory circuit by $f_{\text{inp}}(t)$.

Items (a) and (b) are our definition of the generalized resonance. The case of sinusoidal $h(t)$ is obviously included since the proportionality to $h(t)$ requires $f(t)$ to also be sinusoidal of the same period.

This mathematical situation is the constructive point, but the discussion of Sections 5.3 and 5.5 of the optimization of S_o from a more physical point of view is useful, leading us to very compact formulation of the extended resonance condition. However, let us first of all use the simple oscillator checking how essential is the direct proportionality of f to h, that is, what may be the quantitative miss when the waveform of $f(t)$ differs from that of $h(t)$ in the chosen interval (t_k, t_{k+1}).

5.2. An Example for a Simple Oscillator. Let us compare the cases of the *square* (Figures 13, 14, and 15) and *sinusoidal* (Figure 10) input waves of the same period for S_o defined in the interval $(0, T/2 = \pi/\omega)$. Of course, the norms of the input functions have to be equal for the comparison of the respective responses. (Note that in the consideration of the above figures, equality of the norms was *not* provided, and thus the following result cannot be derived from the previous discussions.)

Let the height of the square wave be 1. Then, $f_{\text{inp}}^2(\pi - \lambda) = 1$ everywhere, and according to (22) the norm is obtained as $\sqrt{\pi/\omega}$. For obtaining the same norm for a sinusoidal input, we write it as $K \sin \omega t$ and find $K > 0$ so that

$$\|K \sin \omega t\| = \sqrt{\frac{\pi}{\omega}}, \qquad (33)$$

that is,

$$K = \frac{\sqrt{\pi}}{\sqrt{\omega} \|\sin \omega t\|}. \qquad (34)$$

Because of the symmetry of the sinusoidal and square-wave inputs, in both cases $f_{\text{inp}}(\lambda) = f_{\text{inp}}(t - \lambda) \equiv f(\lambda)$ in the interval $(0, \pi/\omega)$. For either of the input waveforms the norm of $f_{\text{inp}}(t)$ now equals $\sqrt{\pi/\omega}$, and for $h(t) = \sin \omega t$ of the simple oscillator (the damping in this interval is ignored), we have, according to (24) and (32),

$$S_o = (f, h) \leq \left(\sqrt{\frac{\pi}{\omega}} \right) \|\sin \omega t\| = \sqrt{\frac{\pi}{\omega}} \sqrt{\frac{\pi}{2\omega}} = \frac{\pi}{\omega\sqrt{2}} \approx \frac{2.221}{\omega}, \qquad (35)$$

as the upper bund.

Thus, while for the response to the square wave we have

$$S_o = \int_0^{\pi/\omega} 1 \cdot \sin \omega t \, dt = \frac{1}{\omega} \int_0^{\pi} 1 \cdot \sin x \, dx = \frac{2}{\omega} \qquad (36)$$

only, for the response to the input $K \sin \omega t$ we have, for the K found,

$$
\begin{aligned}
S_o &= \int_0^{\pi/\omega} \left(\frac{\sqrt{\pi}}{\sqrt{\omega} \, \|\sin \omega t\|} \sin \omega t \right) \sin \omega t \, dt \\
&= \frac{\sqrt{\pi}}{\sqrt{\omega} \, \|\sin \omega t\|} \|\sin \omega t\|^2 = \sqrt{\frac{\pi}{\omega}} \, \|\sin \omega t\|,
\end{aligned}
\tag{37}
$$

as (35).

The "relative *missing* the optimality", in the sence of $f \sim h$, in the case of the square wave, which we wanted to find, is

$$
\left| \frac{2 - 2.221}{2.221} \times 100\% \right| = |-9.95\%| \approx 10\%.
\tag{38}
$$

5.3. Analogy with the Usual Vectors.

In the mathematical sense, the set of functions that can be used for the optimization of S_o is analogous to the set of usual vectors.

For the scalar product $(\vec{a}, \vec{b})$ of two usual vectors $\vec{a}$ and $\vec{b}$, we have (compare to (32))

$$
\left| (\vec{a}, \vec{b}) \right| \le |\vec{a}| \, |\vec{b}|
\tag{39}
$$

(meaning "$\cos \theta \le 1$"), where the *equality* is obtained only when the vectors are mutually proportional ("θ" = 0), that is, similarly (may be opposite) directed:

$$
\vec{a} \sim \vec{b}.
\tag{40}
$$

The latter relation is *obvious*, in particular because it is obvious that while rotation of the usual vector (say a pencil) when directing it in parallel to another vector (another pencil), the length of this vector is unchanged. This point is much more delicate regarding the norm of a function being adjusted to $h(t)$, which is the "rotation" of the "vector" in the function space. Since the waveform of the function is being changed, its norm can be also changed.

Thus, the usual physical space *very simply* gives the extreme value of $|(\vec{a}, \vec{b})|$ as $|\vec{a}||\vec{b}|$:

$$
\max \left| (\vec{a}, \vec{b}) \right| = |\vec{a}| \, |\vec{b}|.
\tag{41}
$$

Since our "vectors" are the time functions and the functional analog of (40) is (for simplicity, we sometimes write t instead of $T - t$)

$$
f(t) \sim h(t),
\tag{42}
$$

we very simply obtain, *by the mathematical equivalence of the function and the vector spaces*, condition (31); that is, only an $f(t)$ that is directly proportional to $h(t)$ can give an extreme value for S_o.

For the vectors of the same length (e.g., for unit vectors) $|\vec{b}| = |\vec{a}|$, and the condition of optimality, $\vec{a} \sim \vec{b}$, becomes $\vec{a} = \pm \vec{b}$. In the functional space, the latter means that if $\|f(t)\| = \|h(t)\|$, then in order to have S_o maximal we should take $f(t) = \pm h(t)$.

5.4. Comments.

One can consider $f \sim h$ to be both a generalization and a direct *analogy* to the condition $\omega = \omega_o$ of the standard definitions of [1, 3–5]. Then, both of the equalities, $\omega = \omega_o$ and $f = Kh$, appear in the associated theories as sufficient conditions for obtaining resonance in a linear oscillatory system. The norms become important at the next step, namely, regarding the theoretical conditions of system's linearity, which always include some limitations on intensity of the function/process, in any application. For applications, the real properties of the physical source of $f_{\mathrm{inp}}(t)$ (e.g., a voltage source) whose *power* will here be proportional to K^2 obviously require K to be limited.

The requirement of preserving the norm $\|f(t)\|$ during realization of $f(t) \sim h(t)$ also necessarily originates from the practically useful formulation of the resonance problem as the *optimization problem* that requires *calculation* of the optimized peaks (or rms value) of $f_{\mathrm{inp}}(t)$.

If $h(t + T/2) \ne -h(t)$, then the interval in which the scalar products (i.e., the Euclidean functional space) are defined has to be taken over the whole period of $h(t)$, that is, as $S_o = \int_0^T f_{\mathrm{inp}}(T - \lambda) h(\lambda) d\lambda$. ($T = T_o$ is a necessary condition.)

The interval in which we define S_o can be named the "generating" interval.

We can finally write the optimal $f_{\mathrm{out}}(t)$ that resulted from the optimal $f_{\mathrm{inp}}(t)$ as

$$
\begin{aligned}
f_{\mathrm{out}}(t) &= \int_0^t h(\lambda) f_{\mathrm{inp}}(t - \lambda) \, d\lambda \\
&= \int_0^t h(\lambda) h_{(0,T)\mathrm{periodic}}(\lambda) \, d\lambda,
\end{aligned}
\tag{43}
$$

where the function $h_{(0,T)\mathrm{periodic}}(t)$ is $h(t)$ in the generating interval, periodically continued for $t > T$.

We turn now to an informal "physical abstraction," suggested by the comparison of the two Euclidean spaces. This abstraction leads us to a very compact formulation of the generalized definition of resonance.

5.5. A Symmetry Argument for Formulation of the Generalized Definition of Resonance.

For the usual vector space, we have well-developed vectorial analysis, in which *symmetry arguments* are widely employed. The mathematical equivalence of the two spaces under consideration suggests that such arguments—as far as they are related to the scalar products—are legitimized also in the functional space.

Recall the simple field problem in which the scalar field (e.g., electrical potential)

$$
\varphi(\vec{r}) = (\vec{a}, \vec{r})
\tag{44}
$$

is given by means of a constant vector $\vec{a}$, and it is asked in what direction to go in order to have the steepest change of $\varphi(\vec{r})$.

As the methodological point, one need *not* know how to calculate gradient. It is just *obvious* that only $\vec{a}$, or a *proportional vector*, can show the direction of the gradient, since there is only *one fixed vector given*, and it is simply impossible to "construct" from the given data any other constant vector, defining another direction for the gradient.

We thus consider the *axial symmetry* introduced by $\vec{a}$ in the physical space that can be seen *ad hoc* as the "space of the radius-vectors" and conclude that while catching the steepest increases of $(\vec{a}, \vec{r})$, we must go with some $\vec{r} \sim \vec{a}$.

Let us compare this very lucid situation with that of the functional space. In the problem of making the envelope of the *convolution* $f_{\text{out}}(t)$ (for the whole interval $0 < \lambda < t$) to increase as steep as possible, we have, in view of the relation $|f_{\text{out}}(t_k)| = S_o k$, to optimize the scalar product $S_o = (f, h)$. This is quite similar to (44), because here $h(t)$ is the only fixed "vector" involved, that is, no other "directions" in the functional space are given.

Thus, by the direct analogy to the fact that the gradient must be proportional to $\vec{a}$, the optimal $f(t)$ *must be* proportional to $h(t)$.

We thus can say that *in terms of ZSR, that is, in terms of the convolution integral response, resonance is a use of (or "obeying") the axial symmetry introduced by $h(t)$ in the space of the input functions convolving with $h(t)$, or $h(T - t)$.*

This argument makes the generalized definition compact and easy to remember. One just should not forget that we optimize the factor S_o in a certain interval, say the first period of $h(t)$.

6. Discussion

The traditional teaching of resonance in technical textbooks in terms of a purely steady-state, that is, frequency response and phasors, not deepening into the time process, that is, into the *establishment* of the steady-state, is seen to be unsatisfactory.

The general tendency of engineering teachers to work only in the frequency domain is explained but is not justified by the importance of the fields of communication and signal processing. A good understanding of the time processes is needed in physics, chemistry, biology, and also power electronics. We hope that the use of convolution integral suggested here can, to some extent, close any such logical gap when it appears and can make the topic of resonance more interesting to a student. The described graphical application of convolution is also important for understanding the convolution integral per se. Last but not least, we hope that our generalized definition of resonance in terms of optimization of a scalar product in an interval will be useful.

On the way to the generalized definition, our hero was the father swinging a swing and not the definitions of [1, 3, 4]. Everything relevant (even the Fourier series) can be directly understood from the *freedom* that the father has when enhancing the swing's oscillations.

In the historical plane, the simplicity of the mathematical treatment of the sinusoidal case once defined the general point of view on resonance and the standard classroom treatment, but we see that the convolution integral has become a sufficiently simple and common tool to make this definition wider.

The present criticism of the usual teaching resonance well correlates with the "old" pedagogical advice by Guillemin

[6] not to hurry with the frequency-domain analysis and to let the physical reality first be well understood in the time domain.

Direct study of waveforms (not necessarily using the graphical convolution) also reveals some specific resonant effects that are *not obtained at all* for a sinusoidal input [7, 8]. Thus, for some rectangular-wave periodic input waves, a *resonant suppression* of the response oscillations of a simple oscillator can occur *at certain, periodically repeated time intervals*, and only a direct analysis of the *waveforms* reveals this suppression [7–9]. It appears that the singularity of the waveform and its symmetry [7–9], and *not* Fourier (spectral) representation, reveal these "pauses" in the oscillatory function. Remarkably, since singularity and symmetry aspects are applied also to a nonlinear oscillatory circuit, these "pauses" in the oscillations can be similarly simply explained [7–9] for such a nonlinear circuit.

The topic of resonance is an important scientific and pedagogical point from which different mathematical and physical interpretations can be developed, and it should be revisited by a teacher.

Appendix

A. The Representation of the Circuit Response as ZIR(t) + ZSR(t) (Some Basic System-Theory Terminology for Physicists)

Besides the standard mathematical representation (3) of the solution of a linear equation, system theory commonly uses another representation in which the output function is composed of a *Zero Input Response* (ZIR) and a *Zero State Response* (ZSR).

The ZSR is influenced by the generator inputs and satisfies *zero initial conditions* (this is the meaning of the words "zero state"), and the ZIR is defined *only* by nonzero initial conditions, that is, is *not* influenced by the generator's inputs, which is the meaning of the words "zero-input response".

Since both the generator-type input functions and the initial conditions can be defined freely, they are both legitimized inputs and altogether form a *generalized input*.

A.1. The Superposition with Respect to the Generalized Input. The concept of *generalized input* (Figure 19) fully explains the construction of ZIR and ZSR via the superposition. Indeed, in the classical way of (3), $i_h(t)$ is found from the homogeneous equation which is *not* the given one but is artificially introduced. That is, the determination of $i_h(t)$ is an *auxiliary problem* in which the generator's inputs (that define the right-hand side of the given equation) are zero. The concept of generalized input requires doing *the same* also for the initial conditions, that is, to additionally use the given equation with the artificially introduced initial zero conditions. Thus, according to the two different groups of the inputs we have two parts of the whole solution, obtained from the following auxiliary *independently solvable* problems.

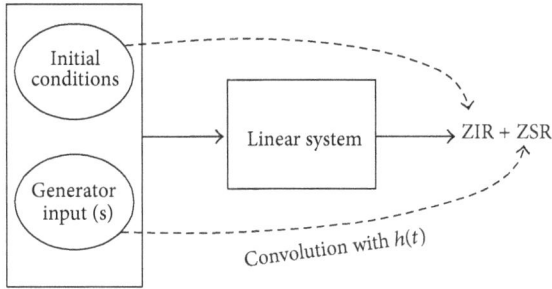

FIGURE 19: The *generalized input* includes both generator inputs and initial conditions. (This becomes trivial when the Laplace transform is used to transform a system *scheme*.) Considering the superposition of the output function of a linear system *for such an input*, we obtain the associated structure of the solution of the linear system in a form that is different from (3). Respectively, the output function is written as ZIR(t) + ZSR(t).

For ZIR. Homogeneous equation (zero generator inputs) plus the needed initial conditions.

For ZSR. Given equation with zero initial conditions.

Figure 19 schematically illustrates this presentation of the linear response.

Figure 7 reduces Figure 19 to what we actually need for the processes with zero initial conditions. The logical advantage of the presentation ZIR + ZSR over (3) becomes clear in the terms of the superposition.

The ZSR includes both, the decaying transient needed to satisfy zero initial conditions and the final steady state given in its general form by the following integral (A.8). The oscillations shown in Figure 3 are examples of ZSR.

The separation of the solution function into ZIR and ZSR is advantageous, for example, when the circuit is used to analyze the input signal, that is, when we wish to work only with the ZSR, when nonzero initial conditions are just redundant inputs.

The convolution integral (10) is ZSR. When speaking about system with constant parameters having one input and one output, the Laplace transform of ZSR(t) equals $H(s)F_{\text{inp}}(s)$ where $H(s)$ is the "transfer function" of the system, that is, the Laplace transform of $h(t)$. *Each time when we speak about transfer function, we speak about ZSR, that is, zero initial conditions.*

It is easy to write $F_{\text{out}}(s)$ for our problem. Using the known formula for Laplace transform of periodic function and setting the optimal $f_{\text{inp}}(T - \lambda) \equiv -h(\lambda)$, $\lambda \in [0, T]$, that is, $f_{\text{inp}}(t) \equiv -h(T - t)$, $t \in [0, T]$, where T is the period (in the sense of the generating interval) of $h(t)$, we have, for the periodically continued $f_{\text{inp}}(t)$, the Laplace transform of our $f_{\text{out}}(t)$ as (see (43))

$$F_{\text{out}}(s) = H(s) F_{\text{inp}}(s) = H(s) \frac{\int_0^T f_{\text{inp}}(t) e^{-st} dt}{1 - e^{-sT}}$$

$$= H(s) \frac{-\int_0^T h(T - t) e^{-st} dt}{1 - e^{-sT}} = H(s) \frac{\int_0^T h(t) e^{st} dt}{1 - e^{sT}} \tag{A.1}$$

(the integration only over *the first* period and, finally, "+s" everywhere), which is relevant to different oscillatory $h(t)$.

A.2. Example. Consider for $t > 0$ the following simplest example of the first-order system/equation

$$\frac{dy}{dt} + ay(t) = A, \quad y(0) \neq 0 \text{ is given,} \tag{A.2}$$

where a and A are constants. Here, the solution of type (8), $i_h(t) + i_{fs}(t)$, is first $K \exp(-at) + A/a$, and when involving the initial condition, finally,

$$y(t) = \left(y(0) - \frac{A}{a} \right) e^{-at} + \frac{A}{a}, \quad t > 0, \tag{A.3}$$

with the initial conditions and the generator function "mixed" in the first term.

The ZIR + ZSR representation is obtained by rewriting this expression as

$$y(t) = y(0) e^{-at} + \frac{A}{a} \left(1 - e^{-at} \right), \quad t > 0. \tag{A.4}$$

The first term depends on the initial condition, that is, is ZIR, and the second term depends on the generator input $Au(t)$ ($u(t)$ is the unit-step function), that is, is ZSR.

It is easy to check that ZIR can be *independently* found from the equation $y' + ay = 0$ and the given initial condition, and ZSR can be *independently* found from the given equation $y' + ay = A$ and the zero initial condition.

For $y(0) = 0$, $y(t) = \text{ZSR}(t) = (A/a)(1 - e^{-at})$, which can be also written as

$$y(t) = \int_0^t e^{-a(t-\lambda)} A \, d\lambda, \tag{A.5}$$

that is (as (10)), as

$$y(t) = \int_0^t h(t - \lambda) f_{\text{inp}}(\lambda) \, d\lambda, \tag{A.6}$$

where $h(t) = e^{-at} u(t)$ is the impulse response of the first-order circuit.

Considering (A.4), one sees that the ZSR includes (as $t \to \infty$) decaying components of the same *type* as the ZIR and that the asymptotic response A/a originates from the ZSR as $t \to \infty$ and *not at all* from the ZIR.

A.3. $f_{out}(t)$ as $t \to \infty$. If (as in the above example) $f_{\text{out}}(\infty)$ exists, then it is obtained as

$$f_{\text{out}}(\infty) = \lim_{t \to \infty} \int_0^t h(t - \lambda) f_{\text{in}}(\lambda) \, d\lambda, \tag{A.7}$$

but if $f_{\text{out}}(\infty)$ does not exist, then as $t \to \infty$ the time function of the final state is given by making the upper limit of the integration infinity:

$$f_{\text{out}}(t) \underset{t \to \infty}{\sim} \int_0^\infty h(t - \lambda) f_{\text{in}}(\lambda) \, d\lambda \tag{A.8}$$

(i.e., the roles of the argument "t" in (A.6) are different for the different places in which it appears).

The integral in (A.8) can be rewritten as

$$\int_{-\infty}^{t} h(\lambda) f_{\text{in}}(t - \lambda) d\lambda. \qquad (A.9)$$

Dealing with the asymptotic solution (A.9) is typical for *stochastic problems*, where, contrary to our statement of the resonance problem, the initial conditions are not important.

When speaking about convolution only in the form

$$\int_{-\infty}^{\infty} h(\lambda) f_{\text{in}}(t - \lambda) d\lambda, \qquad (A.10)$$

one misses the effects of the initial conditions which are important for our analysis, and it is inevitable that only the spectral approach appears be relevant.

A.4. A Case When ZIR + ZSR Is Directly Obtained. When a differential equation can be directly solved by integration, the solution is directly obtained in the form of ZIR + ZSR. Thus, for Newton's equation written in the usual notations,

$$m\frac{d\vec{v}}{dt} = \vec{F}(t), \qquad (A.11)$$

we have

$$\vec{v}(t) = \vec{v}(0) + \frac{1}{m}\int_{0}^{t} \vec{F}(\lambda) d\lambda, \qquad (A.12)$$

which obviously is ZIR + ZSR. Superposition *with respect to the force $\vec{F}(t)$* is realized only by the ZSR.

Consider also

$$m\frac{d^2\vec{r}}{dt^2} = \vec{F}(t) \qquad (A.13)$$

for $\vec{r}(0)$ and $(d\vec{r}/dt)(0)$ given.

The presentation ZIR + ZSR is generally relevant to *linear time-variant* (LTV, "parametric") equations that include the equations with constant parameters as a special case. For instance, if the mass in (A.11) depends on time, the integrand in ZSR in (A.12) would be $\vec{F}(\lambda)/m(\lambda)$. Generally, LTV equations are very difficult, but for *any linear homogenous equation* (*e.g., equation of parametric resonance*), for which ZSR need not be found, *it follows from the linearity* that the solution (which then is just ZIR) has the form

$$\begin{aligned} y(t) = \text{ZIR}(t) &= F\left(y(0), y'(0), \ldots, t\right) \\ &= y(0) f_1(t) + y'(0) f_2(t) + \cdots \end{aligned} \qquad (A.14)$$

with all the functions known. Since $y(0), y'(0), \ldots$ are legitimized inputs (Figure 19), this is the usual linear superposition.

References

[1] L. D. Landau and E. M. Lifschitz, *Mechanics*, Pergamon, New York, Ny, USA, 1974.

[2] L. I. Mandelstam, *Lectures on the Theory of Oscillations*, Nauka, Moscow, Russia, 1972.

[3] W. H. Hayt and J. E. Kemmerly, *Engineering Circuit Analysis*, McGraw-Hill, New York, NY, USA, 1993.

[4] J. D. Irwin, *Basic Engineering Circuit Analysis*, Wiley, New York, NY, USA, 1998.

[5] C. A. Desoer and E. S. Kuh, *Basic Circuit Theory*, McGraw Hill, New York, NY, USA, 1969.

[6] E. A. Guillemin, "Teaching of system theory and its impact on other disciplines," *Proceedings of the IRE*, pp. 872–878, 1961.

[7] E. Gluskin, "The internal resonance relations in the pause states of a nonlinear LCR circuit," *Physics Letters A*, vol. 175, no. 2, pp. 121–132, 1993.

[8] E. Gluskin, "The asymptotic superposition of steady-state electrical current responses of a nonlinear oscillatory circuit to certain input voltage waves," *Physics Letters A*, vol. 159, no. 1-2, pp. 38–46, 1991.

[9] E. Gluskin, "The symmetry argument in the analysis of oscillatory processes," *Physics Letters A*, vol. 144, no. 4-5, pp. 206–210, 1990.

Permissions

The contributors of this book come from diverse backgrounds, making this book a truly international effort. This book will bring forth new frontiers with its revolutionizing research information and detailed analysis of the nascent developments around the world.

We would like to thank all the contributing authors for lending their expertise to make the book truly unique. They have played a crucial role in the development of this book. Without their invaluable contributions this book wouldn't have been possible. They have made vital efforts to compile up to date information on the varied aspects of this subject to make this book a valuable addition to the collection of many professionals and students.

This book was conceptualized with the vision of imparting up-to-date information and advanced data in this field. To ensure the same, a matchless editorial board was set up. Every individual on the board went through rigorous rounds of assessment to prove their worth. After which they invested a large part of their time researching and compiling the most relevant data for our readers. Conferences and sessions were held from time to time between the editorial board and the contributing authors to present the data in the most comprehensible form. The editorial team has worked tirelessly to provide valuable and valid information to help people across the globe.

Every chapter published in this book has been scrutinized by our experts. Their significance has been extensively debated. The topics covered herein carry significant findings which will fuel the growth of the discipline. They may even be implemented as practical applications or may be referred to as a beginning point for another development. Chapters in this book were first published by Hindawi Publishing Corporation; hereby published with permission under the Creative Commons Attribution License or equivalent.

The editorial board has been involved in producing this book since its inception. They have spent rigorous hours researching and exploring the diverse topics which have resulted in the successful publishing of this book. They have passed on their knowledge of decades through this book. To expedite this challenging task, the publisher supported the team at every step. A small team of assistant editors was also appointed to further simplify the editing procedure and attain best results for the readers.

Our editorial team has been hand-picked from every corner of the world. Their multi-ethnicity adds dynamic inputs to the discussions which result in innovative outcomes. These outcomes are then further discussed with the researchers and contributors who give their valuable feedback and opinion regarding the same. The feedback is then collaborated with the researches and they are edited in a comprehensive manner to aid the understanding of the subject.

Apart from the editorial board, the designing team has also invested a significant amount of their time in understanding the subject and creating the most relevant covers. They scrutinized every image to scout for the most suitable representation of the subject and create an appropriate cover for the book.

The publishing team has been involved in this book since its early stages. They were actively engaged in every process, be it collecting the data, connecting with the contributors or procuring relevant information. The team has been an ardent support to the editorial, designing and production team. Their endless efforts to recruit the best for this project, has resulted in the accomplishment of this book. They are a veteran in the field of academics and their pool of knowledge is as vast as their experience in printing. Their expertise and guidance has proved useful at every step. Their uncompromising quality standards have made this book an exceptional effort. Their encouragement from time to time has been an inspiration for everyone.

The publisher and the editorial board hope that this book will prove to be a valuable piece of knowledge for researchers, students, practitioners and scholars across the globe.

List of Contributors

Dzeti Farhah Mohshim, Hilmi bin Mukhtar, Zakaria Man and Rizwan Nasir
Chemical Engineering Department, Universiti Teknologi Petronas, Bandar Seri Iskandar, Perak Darul Ridzuan, 31750 Tronoh, Malaysia

Marcos Arias-Acuña, Antonio García-Pino and Oscar Rubiños-López
Departamento de Teoría de la Senal y Comunicaciones, Universidade de Vigo, 36310 Vigo, Spain

A. S. Haynes, J. A. Cordes and J. Krug
US Army RDECOM ARDEC, Picatinny Arsenal Building 94, Morris County, NJ 07806, USA

Venkateswara Rao Surisetty and Janusz Kozinski
Faculty of Science & Engineering, York University, 4700 Keele Street, Toronto, ON, Canada M3J 1P3

L. Rao Nageswara
Department of Chemical Engineering, R.V.R & J.C College of Engineering, Guntur 522019, India

Agostino Poggi and Michele Tomaiuolo
Dipartimento di Ingegneria dell'Informazione, Università degli Studi di Parma, Viale U. P. Usberti 181/A, 43100 Parma, Italy

Davide Forcellini
Dipartimento Economia e Tecnologia, Universita degli Studi della Repubblica di San Marino, Via Salita alla Rocca, 44. San Marino Rep., San Marino

Angelo Marcello Tarantino
Dipartimento Ingegneria Meccanica e Civile DIMeC, Universita degli Studi di Modena e Reggio Emilia, Via Vignolese 905, Modena, Italy

John Di Cicco and Ayodeji Demuren
Department of Mechanical and Aerospace Engineering, Old Dominion University, Norfolk, VA 23529, USA

Sanjoy Das Neogi
Department of Civil Engineering, Meghnad Saha Institute of Technology, Kolkata 700150, India

Amit Karmakar
Department of Mechanical Engineering, Jadavpur University, Kolkata 700032, India

Dipankar Chakravorty
Department of Civil Engineering, Jadavpur University, Kolkata 700032, India

Sanjeev Jain
Department of Electronics & Communication Engineering, Motilal Nehru National Institute of Technology, Allahabad, India

Vijay Shanker Tripathi and Sudarshan Tiwari
Department of Electronics & Communication Engineering, NIT, Raipur, India

Ashutosh Kumar Singh and Rajiv Saxena
Department of Electronics and Communication Engineering, Jaypee University of Engineering and Technology, Guna, Raghogarh, Madahya Pradesh 473226, India

Sang-Wook Ui, In-Seok Choi and Sung-Churl Choi
Division of Materials Science and Engineering, College of Engineering, Hanyang University, 17 Haengdang-dong, Seongdong-gu, Seoul 133-791, Republic of Korea

Selvaraj Raja and Vytla Ramachandra Murty
Department of Biotechnology, Manipal Institute of Technology, Manipal, Karnataka 576104, India

Sarmila Sahoo
Department of Civil Engineering, Heritage Institute of Technology, Kolkata 700107, India

Olivier Pantalé and Babacar Gueye
Universite de Toulouse, INP/ENIT, Laboratoire Genie de Production, 47 Avenue d'Azereix, 65016 Tarbes, France

Haluk Çeçen
Division of Construction Management, Department of Civil Engineering, Yildiz Technical University, Esenler, 34220 Istanbul, Turkey

Begüm Sertyeşilışık
Liverpool John Moores University, Liverpool L3 2AJ, UK
Department of Architecture, Istanbul Technical University, Taskısla Taksim, 34437 Istanbul, Turkey

Mohamad M. Awad
National Council for Scientific Research, National Center for Remote Sensing, P.O. Box 11-8281, Beirut 11072260, Lebanon

Maryam Mardfekri and Jose M. Roesset
Zachry Department of Civil Engineering, Texas A&M University, College Station, TX 77843-3136, USA

Paolo Gardoni
Department of Civil and Environmental Engineering, University of Illinois at Urbana-Champaign, IL 61801, USA

Abbas Mahmoudabadi
Department of Industrial Engineering, Payame Noor University (PNU), Shahnaz Alley, Nourian Street, North Dibagi Avenue, Tehran, Iran

Arezoo Abolghasem
Road Maintenance and Transportation Organization, Number 12 Dameshq Street, Vali-e-Asr Avenue, Tehran, Iran

P. Sreeraj
Department of Mechanical Engineering, Valia Koonambaikulathamma College of Engineering and Technology, Trivandrum, Kerala 692574, India

T. Kannan
SVS College of Engineering, Coimbatore, Tamil Nadu 642109, India

Subhashis Maji
Department of Mechanical Engineering, IGNOU, New Delhi 110068, India

Jin-Shyan Lee
Department of Electrical Engineering, National Taipei University of Technology, Taipei 10608, Taiwan

Yuan-Ming Wang
Information and Communication Research Labs, Industrial Technology Research Institute, Hsinchu 31040, Taiwan

Emanuel Gluskin
Electrical Engineering Department, Faculty of Engineering, Kinneret College on the Sea of Galilee, Jordan Valley 15132, Israel

Doron Shmilovitz and Yoash Levron
Electrical Engineering Department, Faculty of Engineering, Tel-Aviv University, Israel